AGING, REPRODUCTION, AND THE CLIMACTERIC

AGING, REPRODUCTION, AND THE CLIMACTERIC

Edited by
Luigi Mastroianni, Jr.
University of Pennsylvania School of Medicine
Philadelphia, Pennsylvania

and

C. Alvin Paulsen
University of Washington School of Medicine
Seattle, Washington

PLENUM PRESS • NEW YORK AND LONDON

Library of Congress Cataloging in Publication Data

Main entry under title:

Aging, reproduction, and the climacteric.

Proceedings of a conference held at the National Institutes of Health in June 1984, sponsored by the American Fertility Society, the National Institute on Aging, and the National Institute of Child Health and Human Development.
Includes bibliographies and index.
1. Human reproduction—Age factors—Congresses. 2. Climacteric—Congresses. 3. Aging—Congresses. 4. Generative organs—Aging—Congresses. I. Mastroianni, Luigi. II. Paulsen, C. Alvin (Charles Alvin), 1924-   . III. American Fertility Society. IV. National Institute on Aging. V. National Institute of Child Health and Human Development (U.S.) [DNLM: 1. Aging—congresses. 2. Climacteric—congresses. 3. Reproduction—congresses. WP 580 A267 1985]

QP251.A34 1986	612.6	85-28299

ISBN-13: 978-1-4684-5049-1   e-ISBN-13: 978-1-4684-5047-7
DOI: 10.1007/978-1-4684-5047-7

© 1986 Plenum Press, New York
A Division of Plenum Publishing Corporation
233 Spring Street, New York, N.Y. 10013

Softcover reprint of the hardcover 1st edition 1986

CONTRIBUTORS

WILLIAM J. BREMNER   Endocrinology Section, Veterans Administration Medical Center, Seattle, Washington 98108

DONALD K. CLIFTON   Departments of Obstetrics and Gynecology, and Physiology and Biophysics, University of Washington School of Medicine, Seattle, Washington 98195

DANIEL M. DORSA   Gerontology Research, Education, and Clinical Center, Veterans Administration Medical Center, Seattle, Washington 98108

CALEB E. FINCH   Department of Biological Sciences, Andrus Gerontology Center, University of Southern California, Los Angeles, California 90089

DIANE S. FORDNEY   Health Sciences Center, College of Medicine, University of Arizona, Tucson, Arizona 85724

ROGER G. GOSDEN   Department of Physiology, University Medical School, Edinburgh EH8 9AG, United Kingdom

ERNEST B. HOOK   Bureau of Maternal and Child Health, New York State Department of Health, Albany, New York 12237; and Department of Pediatrics, Albany Medical College, Albany, New York 12208

HOWARD L. JUDD   Department of Obstetrics and Gynecology, School of Medicine, University of California, Los Angeles, Los Angeles, California 90024

NATHAN KASE   Department of Obstetrics, Gynecology, and Reproductive Sciences, Mount Sinai Medical Center, New York, New York 10029

ULLA LARSEN   Office of Population Research, Princeton University, Princeton, New Jersey 08544

NANCY C. LEE   Division of Reproductive Health, Center for Health Promotion and Education, Centers for Disease Control, Atlanta, Georgia 30333

MORTIMER B. LIPSETT   National Institute of Child Health and Human Development, National Institutes of Health, Bethesda, Maryland 20205

LUIGI MASTROIANNI, Jr.   Department of Obstetrics and Gynecology, Hospital of the University of Pennsylvania, Philadelphia, Pennsylvania 19104

ALVIN M. MATSUMOTO   Endocrinology Section, Veterans Administration Medical Center, Seattle, Washington 98108

JOHN B. McKINLAY   Cambridge Research Center, American Institutes for Research, Cambridge, Massachusetts 02138 and Boston University, Boston, Massachusetts 02215

SONJA M. McKINLAY   Cambridge Research Center, American Institutes for Research, Cambridge, Massachusetts 02138 and Boston University, Boston, Massachusetts 02215

JANE MENKEN   Office of Population Research, Princeton University, Princeton, New Jersey 08544

ERIK MICHEL   Max Planck Clinical Research Unit for Reproductive Medicine and Department of Experimental Endocrinology, University Women's Hospital, D-4400 Münster, Federal Republic of Germany

KENNETH L. MINAKER   Division on Aging, Harvard Medical School; Gerontology Division, Joint Department of Medicine, Beth Israel and Brigham and Women's Hospitals, Boston, Massachusetts 02215; Geriatric Research Education Clinical Center, West Roxbury/Brockton Veterans Administration Medical Center, West Roxbury, Massachusetts 02132

SANTO V. NICOSIA   Department of Obstetrics and Gynecology and Department of Pathology, University of Pennsylvania, School of Medicine, Philadelphia, Pennsylvania 19104. Present address: Department of Pathology, University of South Florida, College of Medicine, Tampa, Florida 33612

EBERHARD NIESCHLAG   Max Planck Clinical Research Unit for Reproductive Medicine and Department of Experimental Endocrinology, University Women's Hospital, D-4400 Münster, Federal Republic of Germany

C. ALVIN PAULSEN   University of Washington, Seattle, Washington

HERBERT B. PETERSON   Division of Reproductive Health, Center for Health Promotion and Education, Centers for Disease Control, Atlanta, Georgia 30333

ROBERT RESNIK   Department of Reproductive Medicine, School of Medicine, University of California, San Diego, San Diego, California 92103

GEORGE S. ROTH   Molecular Physiology and Genetics Section, Gerontology Research Center, National Institute on Aging, National Institutes of Health, Baltimore City Hospitals, Baltimore, Maryland 21224

JOHN W. ROWE   Division on Aging, Harvard Medical School; Gerontology Division, Joint Department of Medicine, Beth Israel and Brigham and Women's Hospitals, Boston, Massachusetts 02215; Geriatric Research Education Clinical Center, West Roxbury/Brockton Veterans Administration Medical Center, West Roxbury, Massachusetts 02132

GEORGE L. RUBIN   Division of Reproductive Health, Center for Health Promotion and Education, Centers for Disease Control, Atlanta, Georgia 30333

RICHARD W. SATTIN   Division of Reproductive Health, Center for Health Promotion and Education, Centers for Disease Control, Atlanta, Georgia 30333

PENTTI K. SIITERI   Reproductive Endocrinology Center, Department of Obstetrics, Gynecology, and Reproductive Sciences, University of California, San Francisco, San Francisco, California 94143

ROBERT A. STEINER   Departments of Obstetrics and Gynecology, and Physiology and Biophysics, University of Washington School of Medicine, Seattle, Washington 98195

PHYLLIS A. WINGO   Division of Reproductive Health, Center for Health Promotion and Education, Centers for Disease Control, Atlanta, Georgia 30333

PHYLLIS M. WISE   Department of Physiology, School of Medicine, University of Maryland, Baltimore, Maryland 21201

PREFACE

Interest in sexuality and reproductive function does not cease when people begin to age. Instead, a new set of questions arises. Women want to know if it is safe to have babies in their late thirties and early forties. They want to know more about hot flashes and other symptoms of menopause—which ones are dangerous and which are merely uncomfortable. They are eager to learn about the relative risks and benefits of estrogen replacement therapy. Men, too, are concerned about age-related changes in their sexual function.

Experts in reproductive physiology, gerontology, and genetics met at the National Institutes of Health in June of 1984 to discuss these and other concerns about aging and the reproductive system. The conference on Aging, Reproduction, and the Climacteric was sponsored by the American Fertility Society, The National Institute on Aging, and the National Institute of Child Health and Human Development. This volume is based on the proceedings of that conference.

In his introductory remarks, T. Franklin Williams, M.D., Director of the National Institute on Aging, emphasized the vital interest of the institute in the subjects addressed at the meeting. Of particular concern are ways to identify age-related changes and to make the distinction between what is normal aging and what are disease conditions in older people. He made the point that, as we have learned more about normal aging, it has become increasingly clear that in most organ systems there is continuation of virtually full function into very late years—the eighties and nineties. The reproductive system, particularly in women, seems clearly to be an exception, with the climacteric marking very specific changes, which in turn have implications for other organ systems.

Dr. Williams further addressed the need for the NIA to be concerned with reproduction. He stated,

There is another reason to focus on these subjects [he continued]. Letters and calls to the NIA as well as surveys on health concerns have shown that people are very concerned about issues related to health and continued sexuality. People have questions and it is not easy for them to find the answers they seek ... This conference addressed very important physical, social and behavioral issues. All of this occurs against the backdrop of a rapidly expanding older population. Every doctor and health care provider will increasingly be serving older patients. Some say that every doctor becomes a geriatrician as he or she reaches 65. I would extend that and say that we must all become researchers in aging and take an active role in accelerating the accumulation of basic knowledge about the total life cycle. Conferences such as this not only serve to introduce us to each other but allow a degree of "consciousness raising" about aging. If older women have felt "written off," if the diseases common to the later years have gone unstudied, if the basic process of aging is misunderstood, we have all suffered. This conference offers an excellent forum for redefining aging ... that is, life, and I know we will make good use of this opportunity.

Dr. Charles Hammond, the 1984 President of the American Fertility Society, in his introductory comments pointed to the new emphasis and new focus that have been evident at NIH, as represented by this conference. Speaking as a representative of some 8000 scientists and physicians who are interested in reproduction, he reminded us that shortly after the turn of the next century 20% or more of the population will be older than 65. They represent a population at risk for a variety of medical problems and changes in their bodies, and this conference was designed to address their needs.

The discussions that followed each of the presentations were lively and provocative. Of particular note was the emphasis on the aging germ cell. The importance of evaluating age-related factors in both genders was highlighted, with special consideration of possible genetic implications in the male.

Impetus for the conference came initially from the Public Affairs Committee of the American Fertility Society. The possibility of a jointly sponsored program emerged from discussions with Dr. William Sadler, Chief of the Reproductive Sciences Branch, Center for Population Research, NICHD, with the enthusiastic support of its Director, Dr. Mortimer Lipsett. Efforts of the National Institute on Aging were spearheaded by Dr. Norman Braveman. Both Dr. Sadler and Dr. Braveman served on the organizing committee and worked with us to identify salient subjects and appropriate speakers. Broad topics selected for review included methodologic considerations in research on the aging reproductive system, age-related changes in regulatory systems, aging in the reproductive years, the climacteric, and management of menopausal women. The published work provides a compendium of information on the subjects of aging and reproduction as they relate to one another and on the climacteric. It will prove useful to those investigators and clinicians who have focused their efforts on these areas. It is hoped that it will also spark the interest of those now concentrating either on reproduction or on aging and will encourage them to address emerging needs in both of these areas, each of which is of such vital importance in modern society.

Luigi Mastroianni, Jr., M.D.
C. Alvin Paulsen, M.D.

Philadelphia and Seattle

CONTENTS

III. AGING AND THE REPRODUCTIVE YEARS

IV. THE CLIMACTERIC

V. MANAGEMENT OF THE MENOPAUSE

VI. CONCLUSION

I

METHODOLOGICAL CONSIDERATIONS IN RESEARCH ON THE AGING REPRODUCTIVE SYSTEM

1

ANIMAL MODELS FOR THE HUMAN MENOPAUSE

CALEB E. FINCH and ROGER G. GOSDEN

Introduction

This review emphasizes phenomena associated with reproductive aging changes in the human female that also occur in laboratory models. The decrease and loss of fertility during midlife is a universal characteristic of aging in the human female, as well as in many shorter-lived mammalian species (Finch, 1976; Harman and Talbert, 1985). Although it is often thought that reproductive aging in female laboratory rodents (Meites, 1982) might be a fundamentally different neuroendocrine process than reproductive aging in women (Guyton, 1981), recent studies suggest that many mammals share similar phenomena of reproductive senescence, in which ovarian aging changes play a key role. We refer to more comprehensive reviews where possible: see Gosden (1985), Krohn (1962), Finch *et al.* (1980, 1984), Meites (1982), Nelson and Felicio (1985), and Wise (1983). For general information on animal models of aging, see Gibson *et al.* (1979) and Anonymous (1981).

A controversial issue in comparing aging phenomena of humans and higher primates with laboratory rodents concerns species differences in the relative contributions of the hypothalamus, pituitary, and ovary in initiating the preovulatory surge of gonadotropins and in governing fertility cycle length. A prevalent view holds that rodents differ from higher mammalian spontaneous ovulators in the greater importance for an increased hypothalamic output of gonadotrophin-releasing hormone (GnRH) during the preovulatory gonadotro-

CALEB E. FINCH • Department of Biological Sciences, Andrus Gerontology Center, University of Southern California, Los Angeles, California 90089. **ROGER G. GOSDEN** • Department of Physiology, University Medical School, Edinburgh EH8 9AG, United Kingdom.

pin surge of rodents (e.g., Fink, 1979; Leadem and Kalra, 1984; Ramirez *et al.*, 1984). Concurrently, the rodent pituitary becomes increasingly responsive to GnRH just before the preovulatory surge (e.g., Fink, 1979; Crowder and Nett, 1984). In primates, the balance of control for the preovulatory surge appears to reside more at the ovarian and pituitary level, where elevations of E_2 (estradiol) greatly increase the sensitivity of the gonadotropes to GnRH; only a permissive role of the hypothalamus for continued pulsatile release of GnRH is hypothesized (Knobil, 1980; Ramirez *et al.*, 1984). The frequency of GnRH and gonadotropin pulses, however, has profound influences on the recruitment and development of ovarian follicles and or their secretion (Pohl *et al.*, 1983).

Age Changes in Estrous Cycles

Cycle Irregularity

Women. The monumental longitudinal studies of Alan Treloar and associates, which now span two generations (Treloar *et al.*, 1967; Treloar, 1974, 1981), suggest that menstrual histories can be divided into three phases according to cycle length and variability (Figure 1B). During *phase I* (first 5–10 years after menarche) the variability of cycle length gradually declines. Regular cycles prevail during *phase II*, and cycle variability decreases further during these 15–20 years. In *phase III*, the variability of cycles increases sharply during the approach to menopause; phase III includes the *perimenopause* of another analysis (Metcalf *et al.*, 1981a). Both abnormally short and long cycles occur; yet, normal ovulation can occur within 6 months of menopause (Metcalf *et al.*, 1981a). Because menarche and menopause can occur over a wide range of ages, the demarcation of phases can not be assigned by age alone and requires longitudinal analysis. These and other studies (Metcalf *et al.*, 1981b) document the great extent of individual variations in menstrual cycle length and stability throughout life, and the variable ages of menarche and menopause. Athough no analysis has predicted the detailed characteristics of later phases from earlier phases, long menopausal transitions are associated with slower increases of variability (Treloar, 1981).

Rodents. Phases of reproductive aging and variations similar to those observed in women were found in longitudinal records of estrous cycles in C57BL/6J mice, as determined from daily vaginal smear patterns in this laboratory (Nelson *et al.*, 1982, Felicio *et al.*, 1984) (Figures 1A and 2; Table I). The extent of variability in this highly inbred strain is largely unexplained, but may be due in part to intrauterine influences, such as effects from the sex of the fetal neighbors *in utero* (e.g., vom Saal *et al.*, 1981; Meisel and Ward, 1981). These effects involve a puzzling hormonal interaction between fetuses of different genders. There is no evidence for exchange of placental blood between neighboring fetuses, each of which normally has an autonomous circulatory system;

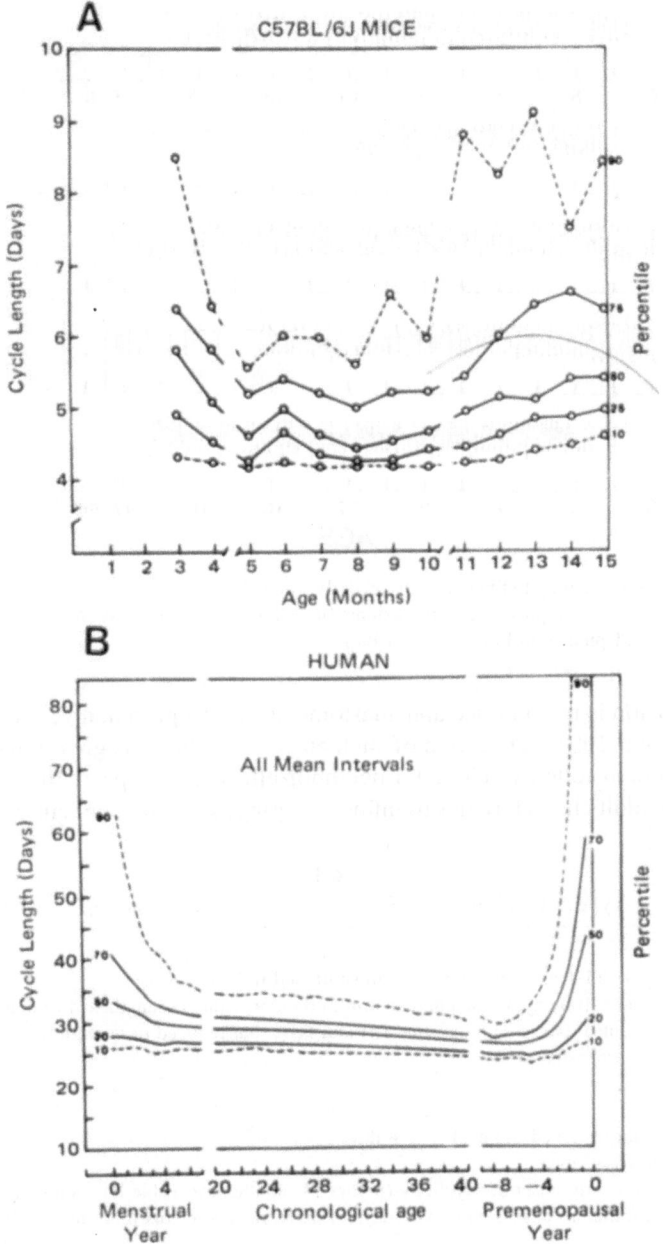

Figure 1. Comparison of age changes in menstrual cycles of C57BL/6J mice (A) and estrous cycles (vaginal smears) of human (B) from longitudinal studies. Redrawn from Nelson *et al.* (1982) (top) and Treloar *et al.* (1967) (bottom).

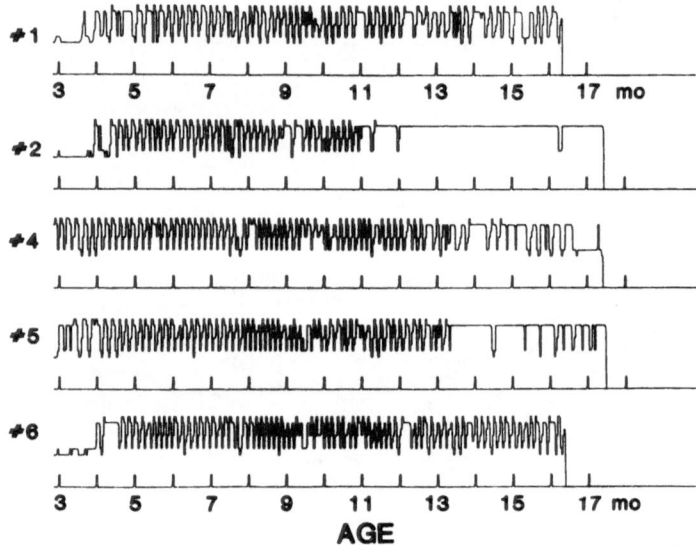

Figure 2. Individual longitudinal records of daily vaginal smears from C57BL/6J mice (same cohort). The elevations represent relative extents of leukocytes or cornified epithelial cells in the vaginal smear. Adapted from Finch *et al.* (1980).

placental fusion is rare in mice and anastomosis of feto-placental vessles is rarer still (vom Saal, 1983). Detection of such epigenetic effects is greatly facilitated by use of inbred rodents. The fetal neighbor effect now appears to extend beyond young adult characteristics to influence aging, since female fetuses flanked

Table I

Types of Variations in Ovulatory Cycles during Aging

Humans
  I. Shortening: (20–25 days) Commonly due to shortened follicular phase
 II. Lengthening: (35–100 days) Often anovulatory; commonly due to lengthened or defective
      luteal phase with elevated LH and FSH, but reduced or negligible
      progesterone

Rats and mice
  I. Shortening: Not observed; no cycles < 4 days
 II. Lengthening: (6–12 days)
      A. Strains with early onset of acyclicity (before 14 months; see Table II) commonly have an
         extended follicular phase, with extra days of proestrus and estrus (cornified vaginal
         cytology)
      B. Strains with later onset of acyclicity (after 10 months; see Table II) commonly have extra
         days of diestrus (leukocytic vaginal cytology, but distinct from pseudopregnancy)

---

in utero by males have an earlier loss of fertility than female-flanked fetuses (vom Saal and Moyer, 1985).

As observed in women, the cycle length variations observed during the approach to acyclicity in mice include interspersed normal short (4 days) and longer cycles. However, unlike humans, very short ovulatory cycles are not reported in rats or mice, during aging or in the young; 4 days is the observed lower limit of fertility cycle length in Myomorphs, and this is the most frequent cycle length in the young under optimum conditions (Hoffmann, 1973). During aging in C57BL/6J mice, repetitive pseudopregnancy is uncommon (Nelson *et al.*, 1982; Mobbs *et al.*, 1984a); however, repetitive pseudopregnancy is quite common in aging rats, as discussed below.

Mouse strains differ in their age-related patterns of cyclicity (Table II) (Thung *et al.*, 1956; Jones and Krohn, 1961a; Finch, 1978; Nelson *et al.*, 1982); the CBA family of strains are noteworthy for a relatively early loss of fertility and cyclicity (Jones and Krohn, 1961a), which is attributed to precocious exhaustion of ovarian oocytes (see below). Although studies on reproductive aging in widely used rat models (Fischer 344; Sprague–Dawley; Long–Evans) do not suggest major genotypic differences, comparisons from the literature are easily confounded by environmental variables; detection of genotypic influences requires concurrent study in the same environment.

Hamsters. In contrast to rats and mice, the estrous cycles of the golden hamster, *Mesocricetus auratus*, are remarkably regular at most ages under optimal photoperiods. Researchers often comment that most aging hamsters do not have the lengthening of cycles usually seen in mice and rats and may not become acyclic until just before death, when pathologic lesions are very common. In one study, most 19- to 22-month-old hamsters continued to have 4-day cycles and only 10–20% had irregular cycles (Chen, 1981; Blaha, 1967); only exceptional ($< 10\%$) mice or rats when 20 months old would continue to have 4-day cycles. Such species differences in effects of aging are not surprising in view of the major species differences between myomorphs in sex dimorphisms of hypothalamic neuroanatomy (Bleier *et al.*, 1982) and in the absence of E_2-induced daily luteinizing hormone (LH) surges in ovariectomized, young female mice (Gee *et al.*, 1984), whereas daily LH surges are readily induced by E_2 in ovariectomized rats and hamsters.

Gonadotropins and Steriods

Women. Both longitudinal and cross-sectional studies of women approaching menopause document a complicated range of endocrine states that parallel the variations of cycle lengthening. Decreasing cycle length can occur

Table II

Prolonged Cycles and Acyclicity in Different Strains of Mouse and Rat Ranked by Age: Incidence and Vaginal Cytological Status[a]

Species	Strain	Age at onset of acyclicity (month)	Prolonged cycles	Vaginal cytology of initial postcyclic state		Source
				Leukocytic	Cornified	
Rat	DA	6–7	Yes		X	Everett (1939)
Rat	Sprague–Dawley	6–8	Yes		X	Everett (1980, personal communication)
Rat	R×UF1	8–10	No		X	Van der Schoot (1976)
Rat	Long Evans	10–11	No		X	Copper et al. (1980)
Rat	Sprague–Dawley	12	Yes		X	R. Gosden (unpublished)
Rat	Sprague–Dawley		Yes			Gray and Wexler (1980)
Rat	Holtzman S/D		Yes		X	Butcher and Page (1981)
Rat	Wistar	12–15	Yes		X	Aschheim (1976)
Mouse	C57BL/6J	13–16	Yes	X	X	Nelson et al. (1981); Felicio et al. (1984)
Rat	Long Evans	14	Yes		X	Wilkes et al. (1979)
Rat	Long Evans	14–15	Yes		X	Lu et al. (1979)
Mouse	(DBA×CE)F1	12–19	Yes		X	Dickie et al. (1957)
Mouse	DBAf	15	Yes		X	Thung et al. (1956)
Rat	Fischer 344	16–18	Yes	X	X	Nelson, et al. (unpublished)
Mouse	C57BL/6NNia	16–20	Yes	X		Parkening et al. (1980)
Mouse	C3HfC57b/Se	>17	Yes	X		Caschera (1959)
Rat	Wistar	18	Yes	X		J. Clemens (personal communication)
Mouse	RIII/Dm/Se	<20	Yes	X		Caschera (1959)
Mouse	C57BL	20	Yes	X		Thung et al. (1956)
Mouse	020	20	Yes	X		Thung et al. (1956)
Mouse	(020×DBAf)F1	24	Yes	X		Thung et al. (1956)

[a] From Nelson et al. (1981) and Felicio et al. (1984).

because of a curtailed follicular phase, in association with decreased plasma E_2 and increased follicle-stimulating hormone (FSH). Nonetheless, blood levels of LH and progesterone (P) and the duration of the luteal phase in shortened cycles may be normal and ovulation may still occur (Sherman and Korenman, 1975; Sherman et al., 1976; Metcalf et al., 1981b).

There is a general trend for elevated FSH long before cycles cease; FSH rises to the postmenopausal range in the last menstrual cycles of many middle-aged women (Metcalf et al., 1981a); this is consistent with the hypothesis that circulating levels of the putative hormone folliculostatin decrease as the pool of growing follicles shrinks (Sherman et al., 1976). Prolonged cycles, lasting 50 days or more, are often anovulatory and associated with high plasma LH and FSH, but with low E_2 and P (Sherman and Korenman, 1975; Sherman et al., 1976; Metcalf et al., 1981b). Follicular maturation appears to continue even with elevated gonadotropins (Metcalf et al., 1981b). The next cycle may be of normal length, with apparent reinstatement of follicular growth (Sherman and Korenman, 1975). The LH surge in climacteric women has not been analyzed in detail, but sufficient LH is clearly secreted for ovulation (Sherman and Korenman, 1975). Even after menopause, normal surges of LH and FSH can be induced by E_2 and P (Odell and Swerdloff, 1968). In contrast, the induced LH surges of aging rodents can be markedly impaired.

Primates. Endocrine changes in aging chimpanizees (*Pan troglodytes*, Pongidae) and rhesus monkeys (*Macaca mulatta*, Circopithidae) demonstrate clear similarities to menopause, including increased cycle length and variability, decreased circulating E_2, and increased FSH and LH (Gould et al., 1981; Dierschke et al., 1983). Ovulation in rhesus monkeys ceases by about 31 years (Collins et al., 1983) and probably does so after 40 years in chimpanizees (Gould et al., 1981).

Rodents. Aging mice and rats can show a truncated LH surge on proestrus of a 4-day cycle (Gray et al., 1980; Miller and Riegle, 1980; Flurkey et al., 1982; Cooper et al., 1983; see Nass et al., 1984, for an exception). Reduced proestrous surges of LH and P also occur during lengthened cycles and may indicate a reduced sensitivity of the hypothalamus–pituitary unit to E_2 (Wise, 1983; Mobbs et al., 1984b).

Effects of aging on plasma E_2 suggest two different phenomena: aging rats with prolonged cycles tend to have additional days of elevated (proestrouslike) plasma E_2 before ovulation (e.g., Page and Butcher, 1982; Nass et al., 1984). However, in C57BL/6J mice the lengthened cycles are associated with a slower rise of plasma E_2 (Figure 3). The slower rise of E_2 with lengthening cycles in C57BL/6J mice might be a model for the lengthened follicular phase in some cycles of perimenopausal women (Sherman and Koreman, 1975).

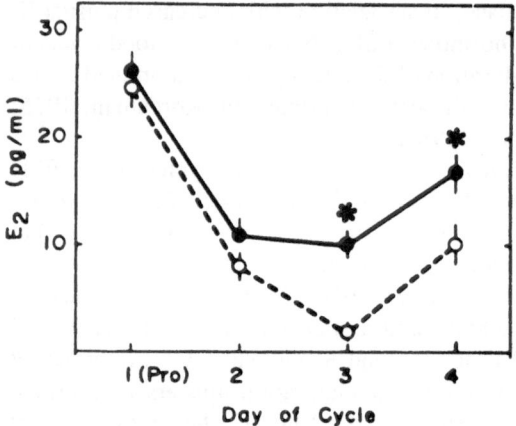

Figure 3. Plasma B2 in young and middle-aged C57BL/6J mice as a function of days from proestrus (day 1). Cross-sectional data from Nelson *et al.* (1981).

Pregnancy and Aging

Fertility

Women. Although decreased fertility of women with increasing age is widely acknowledged, confounding results from reduced coital frequency or male infertility are difficult to resolve. However, a French study clearly documented reduced fertility with age (live births) in a population of 2193 women, whose husbands were totally sterile and who received artificial insemination with donor semen (Federation CECOS, 1982). Although the causes of decreased fertility are complex, the frequency of ovulation probably does not change enough before 45 years to account for the loss of fertility, since ovulatory cycles continue to occur until just before menopause (Sherman and Korenman, 1975). Thus, the decreased fertility with age can be hypothesized to result in large part from defective oocytes or from uterine deficiencies, as found in aging rodents.

Primates. Menopause in laboratory primates is probably also preceded by declining fertility, as indicated by limited data on rhesus monkeys (van Wagenen, 1972) and chimpanzees (Flint, 1976).

Rodents. The decline of fertility with age is documented in great detail for laboratory rodents (Harman and Talbert, 1985). The initial stages of procreation seem relatively intact, including the normal incedence of insemination even in age groups with a big loss of fertility (Holinka and Finch, 1981). As is the case of humans, ovulatory cycles yielding normal or close-to-normal clutches of eggs continue until just before cycle cessation, despite the imminent exhaustion of the ovary at the onset of cyclicity in several mouse strains (e.g., Jones and Krohn, 1961a; Gosden *et al.*, 1983a). However, rats may have larger follicular reserves at acyclicity (Sopelak and Butcher, 1982a; Mandl and Zuckerman, 1951). Many studies show a sharp decline in viable embryos, shortly after fertilization. Depending upon maternal age, there can be a major sevenfold decrease in the number of implantation sites just after implantation, as well as increases of resorption in established implantation sites in 13- to 16-month-old hamsters (Connors *et al.*, 1972). Alternatively, in 11- to 12-month-old C57BL/6J mice the number of implantation sites is only slightly reduced from normal, whereas the fetal resorption increases by twofold (Holinka *et al.*, 1979a). Defective ova are generally considered to be the major factor in declining fertility of rodents and may be a consequence of delayed ovulation in association with lengthened cycles (Page and Butcher, 1982; Fugo and Butcher, 1971). There is also a large literature documenting uterine age-related impairments in the experimentally-induced decidual response of virgin and multiparous mice (Holinka and Finch, 1977; Finch and Holinka, 1982). Although uterine growth during implantation in old mice is not obviously impaired (Finch and Holinka, 1982), uterine scars from implantation can accumulate during successive pregnancies and may limit the number of implantations that the uterus can sustain (Gosden, 1979). A major candidate for decreased embryo viability with maternal age in mice seems to be defective ova and fetal aneuploidy (see below).

Another source of reduced fertility in aging rodents is the prolongation of gestation, or delay of parturition, as observed in 12-month-old C57BL/6J mice, which have average delays of 2 days (range 0–4 days) (Holinka *et al.*, 1978). This delayed parturition is associated with a slower preparturitional decrease of circulatory progesterone (Holinka *et al.*, 1978, 1979a), slower elevations of circulating estradiol (Holinka *et al.*, 1979b), and a great increase of stillbirths, which are rare in younger mothers (Holinka *et al.*, 1978). The prolonged gestation was not a consequence of delayed implantation in the older mice or of retarded growth of the fetus (Holinka *et al.*, 1979a). However, about 50% of fetuses in 12-month-old mice lagged by one day before optic cups were visible in the gross (Holinka *et al.*, 1979a). Detailed studies are needed to characterize the factors in the fetal mortality as a function of developmental stage.

In contrast to rodents, the incidence of prolonged pregnancy in humans (> 45 weeks) decreases with maternal age (Beischer *et al.*, 1969a). Nonetheless, the aging rodent may serve as a useful model for the fetal distress syndrome observed in labor after prolonged gestation (Beischer *et al.*, 1969b) and the greater incidence of postterm neurologic problems in the neonate (Field *et al.*, 1977).

Birth Defects

Increases of birth defects with maternal age in humans result predominantly from chromosomal aneuploidy, such as Down's syndrome (trisomy of chromosome 21). However, nonchromosomal developmental anomalies also occur. The frequency of clinically significant chromosomal abnormalities among new-born infants in the U.S. population rises from 0.2% for mothers younger than 30 years old to 5.4% at 45 years (Hook, 1981; see also Chapter 8 by Hook in this volume). The age distribution of several types of aneuploidies is J-shaped, with a steeply ascending limb above age 30; similar patterns occur in other human populations (Ferguson-Smith, 1983).

The incidence of aneuploidy at birth is a small proportion of those conceived. Among survivors the trisomies are numerically dominant. Monosomies are far less frequent at full term than trisomies. Studies of mice show that deletion of an autosome leads to embryonic death at an earlier stage in gestation than when the same chromosome is overrepresented in trisomic karyotypes (Gropp, 1976). The viability of conceptuses with aneuploidy of the sex chromosomes is greater than for the autosomes, although aneuploidy always confers some phenotypic disturbances. In contrast to 47-XXX and 47-XXY conceptuses, the proportion of 45-XO fetuses surviving to term is diminished and does not show the expected rise with maternal age (Court Brown *et al.*, 1969; Hook, 1981). In the future, human oocytes obtained from *in vitro* fertilization programs may help research on the etiology of aneuploidy, which has hitherto been confined to epidemiological studies in man. Nevertheless, animal models will continue to have a major role in the study of these phenomena, despite the lower incidence of fetal aneuploidy at most maternal ages (Bond and Chandley, 1983).

Maternal age-dependent aneuploidy is better documented in mice than in any other species (Table III). The variable estimates of age-specific frequency partly reflect technical difficulties and small sample sizes. The genotype also influences the susceptibility to aneuploidy (Fabricant and Schneider, 1978; Kram and Schneider, 1978). The higher incidence of aneuploidy among older mice (Fabricant and Schneider, 1978; Maudlin and Fraser, 1978; Tease, 1982) contradicts the hypothesis proposed for our own species that selection against defective fetuses becomes less stringent during middle age because of a compromising uterine environment. Indeed, competition by growing fetuses may even increase under these conditions, leading to differential elimination of aneuploidies (Parsons, 1964); this opposing hypothesis can now be tested by constructing mouse embryos with aneuploid karyotypes (Gropp *et al.*, 1974) with transfer to hosts of different ages.

A number of hypotheses for the age-related increase of aneuploid embryos are summarized (Table IV). Trisomy was once widely thought to arise from an accumulation of accidental damage to the cellular apparatus responsible for the meiotic spindle during the prolonged diplotene stage. However, there is no par-

Table III

Variation in the Incidence of Aneuploid Embryos in Mice According to
Maternal Age and Strain[a]

Strain	Age of embryo (days postcoitum)	Percent aneuploidy (maternal age, months)	Reference
CBA/Ca	4	5.3% (2–6); 12.1% (8–9); 19.6% (10–12)	Brook et al. (1984)
CBA	10–14	0% (2–5); 13.6% (7–10); 0% (>10)	Fabricant and Schneider (1978)
NzB/J	10–14	0% (2–10)	Fabricant and Schneider (1978)
C57BL/6J	10–14	0% (2–5); 9.0% (7–10); 5.5% (<10)	Fabricant and Schneider (1978)
A/J	10–14	0% (2–5); 5.6% (7–10); 0% (<10)	
C3H/H3J	10–14	0% (2–10); 8.0% (<10)	
CBA/H-T6	4	0% (2–7); 21.0% (8–12)	Gosden (1973)
TO	1 (fertilized in vitro)	3.3% (2–4); 7.5% (8–10)	Maudlin and Fraser (1978)
CBA	9.5–12.5	0% (2–6); 7.4% (>10)	Max (1977)
Q	9–10	0.7% (1.5–2); 1.2% (9–12)	Speed and Chandley (1981)
3H1	1	0.4% (2–4); 3.1% (11–13)	Tease (1982)
CF	10.5	1.3% (3–5); 4.3% (11–13); 12.8% (14–16)	Yamamoto et al. (1973)

[a] In cases where reliable estimates of monosomy could not be obtained, the incidence of aneuploidy was estimated by doubling the values for trisomy. This procedure assumes that univalent or nondisjoined chromosomes migrate randomly to either pole of the meiotic spindle.

allel increase in the vulnerability of ovarian oocytes to atresia (see next section). Moreover, aging oocytes are not hypersensitive to mutagens, as predicted by this hypothesis (Speed and Chandley, 1981; Tease, 1982; Golbus, 1983). Although environmental factors may account for temporal and geographical variations in trisomy (Collman and Stoller, 1962; Evans et al., 1978), many suspect that physiological (hormonal) changes underlie the age effect.

One hypothesis proposes that a proportion of oocytes are doomed to chromosomal nondisjunction as a result of prenatal factors during oogenesis. This is based on the inverse relationship between chiasma frequency and age in mouse oocytes at metaphase I and the correspondingly higher incidence of univalent

Table IV

Factors Proposed to Increase Risk of Aneuploidy among Older Mothers

1. Oocytes shed late in life, with defective chiasmata relationships determined during oogenesis
2. Persistent nucleoli associated with acrocentric chromosomes
3. Overripeness of oocytes; delayed ovulation
4. Physiological changes of the aging maternal environment
5. Relaxed selection in utero against aneuploid fetuses
6. Environmental agents (ionizing radiation, mutagenic chemicals, viruses)
7. Autoimmune changes

pairs that might segregate randomly at the ensuing anaphase (Henderson and Edwards, 1968; Luthardt *et al.*, 1973; Polani and Jagiello, 1976; Speed, 1977). Henderson and Edwards (1968) postulated that, since chiasmata are formed prenatally, a "production line" might exist in which the first oocytes to be formed are the first ovulated. However, a gradient of anomalies is not obvious in cytological studies of oogenesis (Speed and Chandley, 1983), nor is there a correlation between chromosome groups responsible for univalents and those involved in nondisjunction at the following division (Sugawara and Mikamo, 1983).

Alternatively, aneuploidy may not be predetermined during oogenesis, but may arise from physiological age changes in follicular oocytes involving changes in the timing of hormone-dependent steps (Butcher, 1975; Crowley *et al.*, 1979). Such hypotheses are consistent with data in humans (Jacobs and Hassold, 1980; Mikkelsen *et al.*, 1980) and mice (De Boer and van der Hoeven, 1980) showing that aneuploidy arises chiefly, but not exclusively, during the first meiotic division of the egg, i.e., before ovulation. It has been suggested that lengthening of the follicular phase of the human menstrual cycle is a risk factor for pregnancy loss and birth defects (Hertig, 1967). While major practical obstacles prevent rigorous testing for hypothetical effects of follicular overripening in humans, rodent oocytes can be obtained readily from Graafian follicles after delaying ovulation by blocking the gonadotropin surge with barbiturates (Butcher, 1975) or with an antiserum to GnRH (Laing *et al.*, 1984). Such treatments significantly increase the risk of triploid or mosaic karyotypes, but effects on the incidence of *meiotic* nondisjunction are still unresolved. Delayed fertilization has also been proposed as a factor in maternal age-related birth defects (German, 1968), but appear to be ruled out by evidence mentioned above, showing that nondisjunction mainly occurs during meiosis I. The risk of aneuploidy may also be affected by physiological age changes of the ovarian-hypophyseal axis. Removal of one ovary early in adulthood precociously increases the incidence of morphologically abnormal and aneuploid embryos, in association with an earlier onset of irregular cycles and acyclicity (Sopelak and Butcher, 1982b; Brook *et al.*, 1984). The explanation of these findings is still unclear, because high levels of gonadotropins may not account for aneuploidy (Hansmann and El-Nahass, 1979). However, such results may explain why mouse strains that lose oocytes precociously (e.g., CBA) also have a high incidence of trisomy and why the incidence of trisomic offspring is much higher than expected in Turner's syndrome patients, who sometimes are fecund (but rarely fertile) before their unusually early menopause (Reyes *et al.*, 1976; King *et al.*, 1978).

Two other hypotheses concerning autoimmune changes and persistent nucleoli (Table IV) are not well supported. Persistent nucleoli associated with chromosomal bivalents could increase the risk of nondisjunction, but this would not explain why most (perhaps all) chromosomes can be responsible for aneu-ploidy. Claims that autoimmune processes heighten the risk of aneuploidy were not supported by animal experiments (reviewed in Kram and Schneider, 1978).

Ovarian Aging: Rodents, Primates, and Humans

Follicular Attrition

Since oogenesis occurs exclusively at pre- or perinatal stages in most mam-mals, ovarian follicles are lost irreversibly. The follicular store in each newborn infant must serve the needs of reproduction throughout life, and numbers about 10,000 in mice (Jones and Krohn, 1961a) and 2 million in women (Block, 1953; Baker, 1963). Most oocytes are wasted; less than 0.1% are ovulated in humans. Age changes in the follicular store during the adult lifespan of women and mice are very similar when scaled to the lifespan (Figure 4). There is relatively little direct data on the follicle store at menopause (Nelson and Felicio, 1985). How-

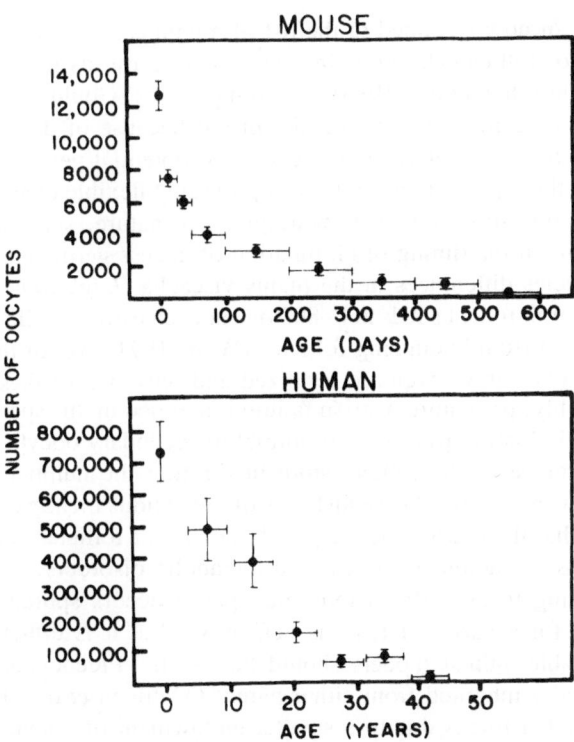

Figure 4. Oocyte loss as a function of age in A strain mice (top) and humans (bottom). Redrawn from Jones and Krohn (1961a) (top) and Block (1952) (bottom).

ever, some studies suggest that sporadic follicular growth and formation of corpora lutea continue after menopause (Novak, 1970; Costoff and Mahesh, 1975).

The combined processes of recruitment into the growing population and death (''atresia'') account for the steady loss of the primordial follicle store, which asymptotically approaches exhaustion in late midlife in most mammals. The rate of this decline therefore imposes an upper limit on the functional lifespan of the ovary that can be quantitatively modeled (Nelson and Felicio, 1985). In immature ovaries, more follicles die each day than are recruited, but this situation reverses after puberty in most mouse strains (Faddy et al., 1983). CBA mice become sterile by about one year because primordial follicles continue to die in relatively large numbers at adult ages. In CBA mice, as in C57BL/6J, few follicles remain after the last ovarian cycle (Gosden et al., 1983a), but oocyte depletion occurs earlier in CBA mice compared to most strains (Papadaki et al., 1979; Faddy et al., 1983). Thus several mouse strains provide models for the condition found in human and some subhuman primate ovaries after menopause (Costoff and Mahesh, 1975; Graham et al., 1979; Gould et al., 1981).

Primates can now be added to the list of mammals whose ovary is essentially depleted of follicles during aging, as shown in rhesus monkeys (Collins et al., 1983; Dierschke et al., 1983) and chimpanzees (Gould et al., 1981).

Although some follicles may be eliminated because of defects that arise during meiosis or from mutations, the excess of potential germ cells at young ages may serve the reproductive strategy of providing flexible ovarian responses (see below). While this strategy is widespread in nature, some species have different patterns in the timing of elimination of excess germ cells. A striking example of species differences in the plains viscacha (*Lagostomus maximus*). This hystricomorph rodent sheds 200–800 ova at each estrus, while many others fail to be released from luteinizing follicles (Weir, 1971); yet of this relatively huge clutch of ova, only seven are fertilized and only two of these survive to full term. Probably, premature ovarian failure is avoided in this species by shifting germ cell elimination from the primordial to secondary oocyte stage.

The importance of the follicle store in limiting the numbers of ovarian cycles is evident from clinical case histories of precocious menopause, in which biopsies show that the ovaries are frequently devoid of follicles. Early ovarian exhaustion leads to significant sexual and metabolic disorders, which vary in severity according to age. Precocious menopause occurs sporadically in the population and for a varity of reasons (Table V), but it is difficult to study. Therefore, suitable animal models should be sought. Precocious menopause may be genetically inherited from either parent (Mattison et al., 1984), but it is not clear whether this is due to a smaller endowment of oocytes, as occurs in some mutant mice (Mintz and Russell, 1957; McCoshen, 1982), or a higher rate of postnatal follicular death, as in CBA mice (Faddy et al., 1983).

Table V
Etiologies of Precocious Ovarian Failure in Women (Precocious Menopause)

1. Genetic predisposition with diploid or aneuploid karyotypes
2. Autoimmune destruction of ovarian tissues
3. Other disease effects, e.g., tuberculosis, mumps, oophoritis, metabolic disorders (17-α-hydroxylase deficiency)
4. Cytotoxic drugs
5. Ionizing radiation
6. Partial oophorectomy

Aneuploid karyotypes are a common cause of germ cell hypoplasia and streak gonads in humans, e.g., in Turner's syndrome (McDonough *et al.*, 1977). Further understanding of germ cell hypoplasia might be gained from XO mice; although these animals are more fertile than XO women, they have fewer oocytes and earlier loss of fertility (Burgoyne and Baker, 1981). Premature ovarian failure often has an immunological basis and may be associated with hypoplasia of the thymus gland, as in ataxia telangiectasia (Miller and Chatten, 1967), or with autoimmune disorders, particularly Addison's disease (Tulandi and Kinch, 1981). The importance of thymic-ovarian interactions during ovarian development in neonates is clearly demonstrated in mice (Michael *et al.*, 1981). Circulating antibodies against gonadotropin receptors occasionally are found but are not common (Tang and Faimen, 1983). Ovaries, like many other glands, can be damaged by mumps virus, but mumps oophoritis is considered rare (Morrison *et al.*, 1975). Nonetheless, a recent epidemiologic survey suggests a slightly earlier menopause (not significant) with a strong inverse correlation of age at menarche and menopause in those with childhood mumps (Cramer *et al.*, 1983). Besides these effects of disease, the ovaries can be damaged by a wide range of toxic environmental agents, including mutagens present in industrial pollution and tobacco smoke (Mattison and Thorgeirsson, 1978) and ionizing radiations (Baker, 1971). It is potentially important in this regard that menopause may occur earlier in cigarette smokers (Jick *et al.*, 1977). These agents can usefully be tested on rodent ovaries, providing that attention is paid to species differences in metabolism and timing of developmental stages of ooccytes.

Follicular Kinetics

Ovarian follicular growth continues at most ages or physiological conditions throughout most of postnatal life, even after the loss of cycles in mice with persistent vaginal cornification (Gosden *et al.*, 1983b). Primordial follicles enter the growing pool at a rate that is approximately constant at all ages; subsequent follicular development is interrupted only by atresia or ovulation. The rate of outflow from the primordial store is independent of the size of the Graafian population and does not depend on ovarian or pituitary hormones, though hypophysectomy retards it (Jones and Krohn, 1961b; Faddy *et al.*, 1977). Thus, small follicles are not conserved by repeated pregnancy and lactation (Shelton, 1959; Jones and Krohn, 1961a). These general characteristics of mammalian ovaries have far-reaching implications for ovarian function in middle age, because they imply that the numbers of follicles available for ovulation reflect the numbers of follicles beginning to grow a few weeks earlier and, hence are limited by the remaining follicular store. The number of potential follicles recruited substantially exceeds the ovulation quota needed in the young and provides the possibility of rapidly increasing the ovulation rate if the contralateral ovary fails or is removed, or in response to supplementary gonadotropins. However, there is a critical size of the follicular pool at which all available follicles are being recruited; from this time onward superovulation is not possible (Peppler, 1971; Gosden, 1985). Further attrition of follicles is responsible for a lowering of the numbers of ova that are shed at estrus just before acyclicity; thus, irregular cycles at this time may reflect the stochastic nature of follicular growth initiation. Some mice become acyclical with several hundred follicles remaining, whereas other continue to produce progressively smaller clutches at increasingly irregular intervals until monovular cycles are produced: this suggests that several different factors, including neuroendocrine, can cause acyclicity in inbred mice (Gosden *et al.*, 1983a; Nelson and Felicio, 1985). Nevertheless, the functional lifespan of the ovary is clearly determined by the follicular stores rather than by chronologic age *per se*, whether or not there is a prolonged period of anovulatory cycles with vaginal cornification.

The growth kinetics of preantral and small antral follicles appear to follow similar patterns in rodents and women, though the transit time between successive stages are much longer in the latter (Pedersen, 1972; Gougeon, 1982). In these species, the ratio of large/small follicles rises during aging (Block, 1952; Gosden *et al.*, 1983a). Therefore, since the follicular growth and transit times are similar at all adult ages (Pedersen, 1972; Gosden *et al.*, 1983b), the survivorship of follicles must be increased to maintain ovulatory constancy as the follicular store dwindles. This hypothesis accounts for the smaller proportion of atretic follicles in aged ovaries by rescue through FSH. However, the ovulation rate is only approximately constant throughout life and can vary considerably. In man, the frequency of dizygotic twinning (and presumably double ovulations) rises with parity and age (Bulmer, 1970); rodents have a corresponding rise with parity (Kennedy and Kennedy, 1972). Since these changes are opposed by the dwindling numbers of follicles available for recruitment, extrinsic factors must be responsible. These phenomena could involve subtle changes in the output of GnRH by the hypothalamus and/or changes in gonadotrope responsiveness. Further investigation is required.

Postcycling States

Humans

The predominant characteristics of menopause are major decreases of blood E_2 and P with elevations of LH and FSH to castrate levels (Sherman *et al.*, 1976). Similar endocrine states eventually occur at advanced ages in some laboratory rodent strains described below.

Rodents

Persistent Vaginal Cornification. The most common initial acyclic state in aging laboratory rats and mice is associated with the anovulatory, polyfollicular ovary, which secretes moderate amounts of E_2, at about the average blood levels in cycling young rodents (Lu *et al.*, 1979; Felicio *et al.*, 1980). The vaginal cytology characteristically has cornified epithelia (Table II), with thick vaginal smears, and is variously known as *persistant vaginal cornification, persistent estrus, or constant estrus*. Blood LH, progesterone, and 20-α-hydroxyprogesterone are usually low, whereas FSH may be moderately elevated (Huang *et al.*, 1976; Lu *et al.*, 1979; Flurkey *et al.*, 1982; Nelson *et al.*, 1982). Blood testosterone and androstenedione are usually low (Lu *et al.*, 1979). A characteristic of this state of some strains is the sustained (but modest) elevations of circulating E and low P. Unopposed estrogenic stimulation often causes a stromal and cystic glandular hyperplasia of the uterus (Christy *et al.*, 1951; Dickie *et al.*, 1957; Malinin and Malinin, 1972; Cosgrove *et al.*, 1978). This glandular hyperplasia may be similar in some regards to the endometrial hyperplasia that is common in premenopausal women, which is also a feature of the *dysfunctional uterine bleeding* (DUB) syndrome (Schroder, 1954; Fraser and Baird, 1974; Van Look *et al.*, 1977). The DUB syndrome in humans may be associated with elevated levels of FSH and LH, but its major characteristic is the sustained stimulation of estrogen target cells by moderately elevated E_2 and low P, as in the rodent syndrome. The rodent syndromes of persistent vaginal cornification should be distinguished from the polycystic ovarian syndrome (the Stein–Levinthal syndrome) of humans, since, in the latter, large pulses of LH are released without much FSH; also, in particular contrast to the aging rodent, androgens are commonly elevated in the human disorder (Goldzieher, 1981). During persistent vaginal cornification in rats, testosterone and androstenedione are at average values for young cycling rats; however, androgens can become elevated in very old rats with enlarged and hemorrhagic pituitaries (Lu *et al.*, 1979).

Repetitive Pseudopregnancy. Particularly in laboratory rats, persistent vaginal cornification is followed by strings of pseudopregnancies with marked elevations of circulating prolactin and P (Lu *et al.*, 1979; Huang *et al.*, 1976; Aschheim, 1976; Everett, 1980). Pseudopregnancy is much rarer in normal aging C57BL/6J mice, but it appears if middle aged mice are given young ovary transplants (Mobbs *et al.*, 1984b). A major cause is the prevalence of pituitary tumors (lactotrophe adenomas) in aging rats (reviewed in Duchen and Schurr, 1976; Clayton *et al.*, 1984), which can cause hyperprolactinemia (Huang *et al.*, 1976) and which may stimulate the corpora lutea of the ovary and the mammary glands, leading to galactorrhea and tumors in hyperprolactinemic rats. Suppression of prolactin by bromocriptine in repetitively pseudopregnant rats can return them to persistent vaginal cornification (Everett, 1980). There is no clear analogue of these phenomena during menopause in most women, since prolactin levels generally remain low (Yamaji *et al.*, 1976; Govoni *et al.*, 1983). However, slight elevations were observed in normal 70-year-old women in one population (Govoni *et al.*, 1983) and could imply disturbed neuroendocrine control of prolactin during aging, for which there is substantial evidence in aging rats (Gudelsky *et al.*, 1981; Reymond and Porter, 1981). It is of interest that hypothalamic defects are implicated in some prolactin secreting tumors of adult humans (Van Loon, 1978; Tucker *et al.*, 1980).

Anestrus. Eventually, secretion of E_2 and P by the ovary fall to castrate values at the end of persistent vaginal cornification or repetitive pseudopregnancy in both rats and mice; then, persistent anestrus ensues (Lu *et al.*, 1979; Gee *et al.*, 1983). The ovary of old persistent anestrus rodents is probably devoid of follicles (an inference only, for which data remain to be obtained) and vaginal cytology becomes atrophic as at menopause (see below); these changes are associated with thin leukocytic vaginal smears. However, the postcastration elevations of LH and FSH expected from reduced ovarian E_2 secretion are only observed under special circumstances in aging rodents. In the absence of lactotrophic pituitary adenomas and other gross pathologic lesions, the healthy anestrus subpopulations of 2-year-old C57BL/6J and C57BL/6NNia* mice have dramatic elevations of LH, close to postovariectomy levels (Figure 5) (Gee *et al.*, 1983; Parkening *et al.*, 1980; Collins *et al.*, 1979). The C57BL/6 family of mouse strains may be particularly favorable models for some aspects of menopause, because the incidence of prolactinemia (Flurkey *et al.*, 1982; Parkening *et al.*, 1980) is so much lower than in most aging rats, even if tumors are present (Nelson *et al.*, 1980). The failure to observe elevations of LH and FSH in old

*Two strains widely used in current aging studies are the C57BL/6J and C57BL/6NNia. C57BL/6J mice are, by definition, derived exclusively and obtained directly from the Jackson Laboratory (Bar Harbor, ME) where they have been inbred since 1937 (Staats, 1976). The C57BL/6N strain was derived from C57BL/6J mice at the National Institutes of Health, Laboratory Animal Genetic Center (Bethesda MD). The resulting C57BL/6N mice were sent in 1975 to Charles River Laboratories (Wilmington, MA) to establish a breeding stock for an aging colony initially contracted by the National Institute of Aging (Bethesda, MD), and are designated C57BL/6NNia. The C57BL/6NNia may not be strictly identical with C57BL/6J or C57BL/6N genotypes.

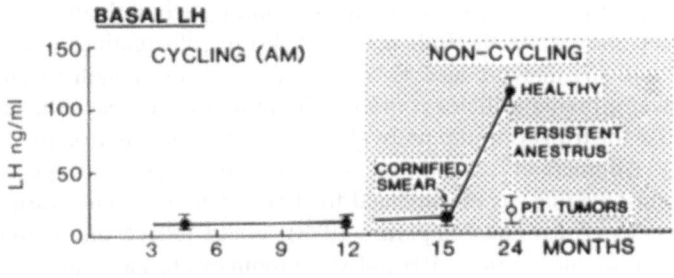

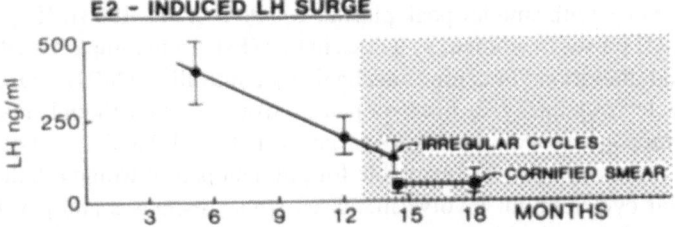

Figure 5. Effects of age on basal LH levels of intact C57BL/6J mice (top) and on the E$_2$-induced LH surge in ovariectomized C57BL/6J mice (bottom). Data redrawn from Gee *et al.* (1983) (top) and Mobbs *et al.* (1984a) (bottom).

anestrus rats (Huang *et al.*, 1976) is most likely a consequence of prolactinemia and pituitary tumors since prolactin elevations inhibit LH secretion (Cheung, 1983). CBA mice also provide a model for relatively early depletion of oocytes, acyclicity, and genital tract atrophy (see below).

Potential Neuroendocrine Variables in Human Menopause and Rodent Neuroendocrine Aging

Most strains of laboratory rats and mice manifest progressive neuroendocrine impairments during the approach to acyclicity at 12–16 months, which include diminished spontaneous LH surges at proestrus, diminished E$_2$-induced LH surges (Figure 5) and pulsatile LH in ovariectomized rodents, and altered responses of hypothalamic catecholamine metabolism to steroids (Finch *et al.*, 1980, 1984; Estes and Simpkins, 1982; Meites, 1982; Wise, 1983; Mobbs *et al.*, 1984b). Despite this extensive evidence, the causal relationship of altered gonadotropin output to the lengthening and loss of cycles is far from proven, e.g., the amount of LH released at proestrus greatly exceeds that required for ovulation (Turgeon, 1979).

Intriguing questions arise about the sources of variability in cycle lengthening, especially because effects of age on distribution of cycle frequency is so similar in humans and mice (Figure 1), species which appear to have different pacemakers governing the ovulatory surge of gonadotrophins (see above). A stochastic hypothesis is suggested by the dwindling size of the pool of growing follicles (see above): As the follicular pool shrinks, fluctuations in numbers of the large follicles that produce E_2 could influence the length of subsequent cycles; fewer growing follicles could require more time to reach the circulating E_2 threshold for triggering the preovulatory surge of gonadotropins (Nelson *et al.*, 1981). Alternatively, a reduced frequency of the pulsatile release of LH and FSH during aging, as documented for LH in rats that were ovariectomized for two weeks (Estes and Simpkins, 1982), could reduce the recruitment of follicles. The frequency of GnRH pulses in monkeys has a major influence on follicular development: slight reductions in GnRH pulse frequency cause anovulatory cycles with smaller peak plasma E_2 (Pohl *et al.*, 1983). If aging influences GnRH pulses in women, as it does in rats (Estes and Simpkins, 1982), then age-related alterations in central mechanisms controlling GnRH could contribute to cycle irregularity in premenopausal women, even though the primary cause of menopause would remain ovarian. It is hoped that detailed data on LH and FSH pulses will become available for premenopausal women. Other factors in irregular cycles might involve increased stress responses and psychological depressions, whose incidence increases with age, since cortisol can inhibit the pituitary response to GnRH (Padmanabhan *et al.*, 1983). Modest plasma corticosterone elevations are common in aging rats (Landfield *et al.*, 1978; Sapolsky *et al.*, 1983).

The most certain neural component of menopause is the hot flash, which is discussed below.

Pathophysiology of Menopause

Hot Flashes

Hot flashes are well characterized as an "organic," neurological disturbance of low E_2 after menopause and involve a coordinate thermogenic output in synchrony with episodes of LH secretion (Judd, 1983; see also Chapter 13 by Judd in this volume). Similar hot flashes also occur in recently ovariectomized adult rhesus monkeys (Dierschke, 1982). Laboratory monkeys should thus be very useful in studying ovariprival aspects of the menopause.

Uterine and Vaginal Atrophy

Uterine atrophy is a common consequence of decreased E_2 after menopause. Similar atrophic cellular changes in the uterus, cervix, and vagina were observed in 11-month-old CBA mice (Papadaki *et al.*, 1979), a strain with early ovarian exhaustion (see above), and can also be simulated in young rodents by ovariectomy.

Striking individual differences are reported in vaginal atrophy after menopause. Some women do not experience vaginal atrophy, even in the absence of steroid therapy (Masukawa, 1960; Novak *et al.*, 1965; Lin and So-Bosita, 1972, 1973; Lieblum *et al.*, 1983). Diet may contribute to individual differences since postmenopausal vegetarians had lower urinary estrogens than their carnivorous controls (Armstrong *et al.*, 1981). Sexual activity in women 6 years after menopause was correlated with reduced vaginal atrophy and higher levels of both androgens and LH; however circulating E_2 and estrone were not correlated with vaginal atrophy (Lieblum *et al.*, 1983). Genotype may also influence response to decreased E_2, since mouse strains show important differences in their vaginal cell sensitivity to E_2 (Trentin, 1950; Claringbold and Biggers, 1955; Drasher, 1955).

Receptors for E_2 are being characterized in the postmenopausal uterus. There were no differences in the E_2 cytosol-binding characteristics of uteri from pre- and postmenopausal women (not steroid treated) if endogenous proteases were inhibited (Pellika *et al.*, 1983). Moreover, *in vitro* autoradiography detected abundant E_2-binding sites in the postmenopausal vagina (Gould *et al.*, 1983). The maintenance of E_2 receptors is obviously crucial for the response of these target cells in estrogen-replacement therapy. In rodents, studies show significant loss of E_2 binding sites in the aging uterus (Papadaki *et al.*, 1979; Roth and Hess, 1982), but it is unclear whether the loss is due to a decrease of E_2-binding sites per cell, or to a loss of E_2-binding cells.

Osteoporosis

The view prevailing since the classic study of Albright, *et al.* (1941) holds that deficiences in estrogens are a major cause of postmenopausal osteoporosis. Many studies support this view, including the recent findings that postmenopausal women with hip fractures had lower free plasma E_2 and testosterone than nonfractured (not E_2 treated) controls (Davidson *et al.*, 1982) and that the favorable effects of E_2 therapy on postmenopausal bone loss is dose related, e.g., in metacarpal cortex (Horsman *et al.*, 1983). However, it must be stressed that loss of bone mineral content in adults of *both* sexes is a *universal* feature of aging (Garn *et al.*, 1967; Avioli, 1982). Decreases of bone mineral are progressive from the fourth decade in normal women (Riggs *et al.*, 1981; Marcus *et al.*, 1983; Parfitt *et al.*, 1983), which is 5–10 years *before* significant decreases in circulating E_2. Thus, osteoporosis in women is not precipitated only by sex steroid loss at menopause. Many factors may be involved in the "normal" osteoporosis of aging, including alterations of mineral intake, physical activity, adrenal steroids, parathyroid hormone, calcitonin, and growth hormone (see reviews by Urist, 1962; Avioli, 1982; Raisz, 1982; Zerwekh *et al.*, 1983; Armbrecht, 1984). The potential, multifactorial etiology of osteoporosis poses major difficulties in determining why osteoporosis leads to vertebral collapse and crush fractures in some older women and men, but not in others. One source of difficulty in identifying possible common denominators of age-related osteoporosis is the limited genetic uniformity in most of the noninbred animal models studied.

The deficiencies of plasma P that often precede menopause may also contribute to osteoporosis. In ovariectomized young rats (Aitken *et al.*, 1972) and in postmenopausal women (Nordin *et al.*, 1981; Mandel *et al.*, 1982), synthetic progestins were at least as effective as estrogens in ameliorating bone changes, as judged by mineral content or mineral metabolism respectively.

Osteoporosis occurs with age in C57BL/6J female mice, as indicated by decreased cortical thickness of the femur by 26 months and decreased Ca^{45} retention (Sha *et al.*, 1967; Silberberg and Silberberg, 1962). Decreased thickness of the femur and spinal vertebrae was detected in (C57BL/6J × DBA/2J) Fl hybrids, at 6 vs. 18 months (Krishna Rao and Draper, 1969). Osteoporosis also occurs during aging in other mouse strains (Silberberg and Silberberg, 1941; Krishna Rao and Draper, 1969). The premature onset of tone loss by castration of female rats at 3 months (Aitken *et al.*, 1972; Wronski *et al.*, 1985) or male rats at 12 months (Wink and Felts, 1980) suggests that osteoporosis in old female mice, at least in part, is related to changes in ovarian activity. Moreover, treatment of 11-month-old CBA mice with E_2 for 3 months tended to increase the amount of cortical bone in the femur (Papadaki *et al.*, 1979).

Investigation of the timing of osteoporosis in relation to the sequential decreases of plasma, E_2, and progesterone during reproductive senescence as C57BL/6J mice pass from persistent vaginal cornification to anestrus (Nelson *el al.*, 1982; Gee *et al.*, 1983) could provide a valuable animal model for this major debilitating condition. Plasma testosterone should also be characterized during aging because of some association of hip fractures in postmenopausal women with low plasma testesterone (Davidson *et al.*, 1982).

Summary

This review was intended to emphasize major features of menopause for which animal models are plausible. The extent of information on homologous changes of aging varies widely; it is much less for osteoporosis than for oocyte depletion or fertility cycle irregularities. Nonetheless, we conclude that laboratory animal models are now available to study most aspects of the human menopause. The types of individual variations of human aging can probably be found in the different rodent genotypes and primate species, for which detailed information should become available in the future.

ACKNOWLEDGMENTS

These studies were supported by research grants to C.E.F. from the National Institute on Aging (Bethesda) and to R.G.G. from the Medical Research Council (London).

References

Aitken, J. M., Armstrong, E., and Anderson, J. R., 1972, Osteroporosis after oophorectomy in the mature female rat and the effect of oestrogen and/or progestogen replacement therapy in its prevention, *J. Endocrinol.* **55**:79–87.

Albright, F., Smith, P. H., and Richardson, A. M., 1941, Postmenopausal osteoporosis, its clinical features, *J. Am. Med. Assoc.* **116**:2465–2474.

Anonymous, 1981, *Mannalian Models for Research on Aging*, National Academy Press, Washington, D.C.

Armbrecht, H. J., 1984, Changes in calcium and vitamin D metabolism with age, in: *Nutritional Intervention in the Aging Process* (H. J. Armbrecht, J. M. Prendergast, and R. M. Coe, eds.), Springer-Verlag, New York, pp. 69–83.

Armstrong, B. K., Brown, J. B., Clarke, H. T., Crooke, D. K., Hahnel, R., Masarei, J. R., and Ratajczak, T., 1981, Diet and Reproductive hormones: A study of vegetarian and nonvegetarian postmenopausal women, *J. Nat. Cancer Inst.* **67**:761–767.

Ascheim, P., 1976, Aging in the hypothalamic-hypophyseal-ovarian axis in the rat, in *Hypothalamus, Pituitary, and Aging* (A. V. Everitt and J. A. Burgess, eds.), Charles C. Thomas, Springfield, Illinois, pp. 376–418.

Avioli, L. V., 1982, Aging, bone, osteoporosis, in: *Endocrine Aspects of Aging* (S. G. Korenman, ed.), Elsevier, London, pp. 199–230.

Baker, T. G., 1963, A quantitative and cytological study of germ cells in human ovaries, *Proc. R. Soc. Lond. B.* **158**:417–433.

Baker, T. G., 1971, Radiosensitivity of mammalian oocytes with particular reference to the human female, *Am. J. Obstet. Gynecol.* **110**:746–761.

Beischer, N. A., Evans, J. H., and Townsend, L., 1969a, Studies in prolonged pregnancy. I. The incidence of prolonged pregnancy, *Am. J. Obst. Gynecol.* **103**:476–482.

Beischer, N. A., Brown, J. B., Smith, M. A., and Townsend, L., 1969b, Studies in prolonged pregnancy. II. Clinical results and urinary estriol excretion in prolonged pregnancy, *Am. J. Obstet. Gynecol.* **103**:483–495.

Blaha, G. C., 1967, Effects of age, treatment, and method of induction on deciduonata in the golden hamster, *Fertil. Steril.* **18**:477–485.

Bleier, R., Byne, W., and Siggelkow, 1982, Cytoarchetectonic sexual dimorphisms of the medial preoptic and anterior hypothalamic areas in guinea pig, rat, hamster, and mouse, *J. Comp. Neurol.* **212**:118–131.

Block, E., 1952, Quantitative morphological investigations of the follicular system in women. Variations at different ages, *Acta. Anta.* **14**:108–123.

Block, E., 1953, A quantitative morphological investigation of the follicular system in newborn infants, *Acta. Anat.* **17**:201–206.

Bond, D. J., and Chandley, A. C., 1983, *Aneuploidy*, Oxford University Press.

Brook, J. D., Gosden, R. G., and Chandley, A. C., 1984, Maternal aging and aneuploid embryos—evidence form the mouse that biological and not chronological age is the important influence, *Hum. Genet.* **66**:41–45.

Bulmer, M. G., 1970, *The Biology of Twinning in Man.* Clarendon Press, Oxford.

Burgoyne, P. S., and Baker, T. G., 1981, Oocyte depletion in XO mice and their XX sibs from 12 to 200 days *post partum*, *J. Reprod. Fertil.* **61**:207–212.

Butcher, R. L., 1975, The role of intrauterine environment and intrafollicular aging of the oocyte on implantation rates and development, in: *Aging Gametes: Their Biology and Pathology* (R. J. Blandau, ed.), S. Karger, Basel, pp. 201–218.

Butcher, R. L., and Page, R. D., 1981, Role of aging ovary in cessation of reproduction, in: *Dynamics of Ovarian Function* (N. B. Schwartz and M. Hunzicker-Dunn, eds.) Raven Press, New York, pp. 253–271.

Caschera, F., 1959, "La menopause" nei topi femmine vergini (RIII/Dm/Se, C3H6/Se, A/He/Se substrains) *Lav. Anat. Patol.* Perugia, **19**:13–30.

Chen, H. J., 1981, Effects of aging on luteinizing hormone release in different physiological states of the female golden hamster, *Neurobiol. Aging* **2**:215–220.

Cheung, C. Y., Prolactin suppresses luteinizing hormone secretion and pituitary responsiveness to luteinizing hormone-releasing hormone by a direct action at the anterior pituitary, *Endocrinology* **113**:632–638.

Christiansen, C., Mazess, R. B., Transbol, I., and Jensen, G. F., 1981, Factors in response to treatment of early postmenoupausal bone loss, *Calc. Tissue Int.* **33**:575–581.

Christy, N. P., Dickie, M. M., Atkinson, W. B., and Woolley, G. W., 1951, The pathogenesis of uterine lesions in virgin mice and in gonadectomized mice bearing adrenal cortical and pituitary tumors, *Cancer Res.* **11**:413–425.

Claringbold, P. J., and Biggers, J. D., 1955, The response of inbred mice to estrogens, *J. Endocrinol.* **12**:9–14.

Clayton, C. J., Schechter, J., and Finch, C. E., 1984, The development of mammotroph adenomas in pituitaries of aging female C57BL/6J mice, *Exp. Gerontol.* **19**:313–320.

Collins, K., Uno, H., and Dierschke, D. J., 1983, Ovarian follice populations in peri- and postmenopausal rhesus monkeys, *Fed. Am. Soc. Exp. Biol. Abstracts*, p. 316.

Collins, T. J., Parkening, T. A., and Smith, E. R., 1979, Plasma and pituitary concentrations of gonadotropins and prolactin in ovariectomized C57BL/5 mice, *Gerontologist* **19**:(Part II) 58(abstract).

Collman, R. D., and Stroller, A., 1962, A survey of mongoloid births in Victoria, Australia, 1942-1957, *Am. J. Public Health* **52**:813–829.

Connors, T. J., Thorpe, L. W., and Soderwall, A. L., 1972, An analysis of preimplantation embryonic death in senescent golden hamsters, *Biol. Reprod.* **6**:131–135.

Cooper, R. L., Conn, P. M., and Walker, R. F., 1980, Characterization of the LH surge in middle-aged female rats, *Biol. Reprod.* **23**:611–615.

Cosgrove, C. B., Satterfield, L. C., Bowles, N. D., and Klima, W. C., 1978, Diseases of untreated virgin female REM and BALB/c mice, *J. Gerontol.* **33**:178–183.

Costoff, A., and Mahesh, V. B., 1975, Primordial follicles with normal oocytes in the ovaries of postmenopausal women, *J. Am. Geriatr. Soc.* **23**:193–196.

Court Brown, W. M., Law, P., and Smith, P. G., 1969, Sex chromosome aneuploidy and parental age, *Ann. Hum. Gent.* **33**:1–14.

Cramer, D. W., Welch, W. R., Cassells, S., and Scully, R. B., 1983, Mumps, menarche, menopause, and ovarian cancer, *Am. J. Obstet. Gynecol.* **147**:1–6.

Crowder, M. E., and Nett, T. M., 1984, Pituitary content of gonadotropins and receptors for GnRH during the periovulatrory period of the ewe, *Endocrinology* **114**:234–239.

Crowley, P. H., Gulati, D. K., Hayden, T. L., Lopez, P., and Dyer, R., 1979, A chiasma-hormonal hypothesis relating Down's syndrome and maternal age, *Nature* **280**:417–419.

Davidson, J., Ross, R. K., Paganini-Hill, A., Hammond, C. D., Siiteri, P. K., and Judd, H. L., 1982, Total and free estrogen and androgens in postmenopausal women with hip fractures, *J. Clin. Endocrinol. Metab.* **54**:115–120.

De Boer, P., and van der Hoeven, F. A., 1980, The use of translocation-derived "marker-bivalents" for studying the origin of meiotic instability in female mice, *Cytogenet. Cell Genet.* **26**:49–58.

Dickie, M. M., Atkinson, W. B., and Fekete, B., 1957, The ovary, estrous cycle and fecundity of DBA × CE and reciprocal hybrid mice in relation to age and the hyperovarian syndrome, *Anat. Rec.* **127**:187–199.

Dierschke, D. J., 1982, Hot flushes in rhesus monkeys, *Gerontologist* **22**:99 (abstract).

Dierschke, D. J., Koenig, J., Krueger, G., and Robinson, J. A., 1983, Reproductive and hormonal patterns in perimonpausal rhesus monkeys, *65th Annual Meeting, Endocrine Society*, p. 250.

Drasher, M. L., 1955, Strain differences in the response of the mouse uterus to estrogens, *J. Hered.* **46:**190–192.

Duchen, L. W., and Schurr, P. H., 1976, Pathology of the pituitary glands in old age, in: *Hypothalamus, Pituitary, and Aging* (A. V. Everitt and J. A. Burgess, eds.), Charles C. Thomas, Springfield, Illinios, pp. 137–155.

Estes, K. S., and Simpkins, J. S., 1982, Resumption of pulsatile luteinizing hormone release after alpha-adrenergic stimulation in aging constant estrous rats, *Endocrinology* **111:**1778–1784.

Evans, J. A., Hunter, A. C. W., and Hamerton, J. L., 1978, Down syndrome and recent demographic trends in Manitoba, *J. Med. Gent.* **15:**43–47.

Everett, J. W., 1939, Spontaneous persistent estrus in a strain of albino rats, *Endocrinology* **25:**123–127.

Everett, J. W., 1980, Reinstatement of estrous cycles in middle-age persistent estrous rats: Importance of circulating prolactin and the resulting facilitative action of progesterone, *Endocrinology* **106:**1691–1696.

Fabricant, J. D., and Schneider, E. L., 1978, Studies of the genetic and immunologic components of the maternal age effect, *Dev. Biol.* **66:**337–343.

Faddy, M. J., Jones, E. C., and Edwards, R. G., 1977, Analytical model for ovarian follicular dynamics, *J. Exp. Zool.* **197:**173–186.

Faddy, M. J., Gosden, R. G., and Edwards, R. C., 1983, Ovarian follicle dynamics in mice: A comparative study of three inbred strains and an F1 hybrid, *J. Endocrinol.* **96:**23–33.

Federation CECOS, Schwartz, D., and Mayaux, M. J., 1982, Female fecundity as a function of age. Results of artificial insemination in 2193 nulliparous women azoospermic husbands, *N. Engl. J. Med.* **306:**404–406.

Felicio, L. S., Nelson, J. F., and Finch, C. E., 1980, Spontaneous pituitary tumorigenesis and plasma estradiol in aging female C57BL/6J mice *Exp. Gerontol.* **15:**139–142.

Felicio, L. S., Nelson, J. F., and Finch, C. E., 1984, Longitudinal studies of estrous cyclicity in aging C57BL/6J mice: II. Cessation of cyclicity and the duration of persistent vaginal cornification, *Biol. Reprod.* **31:**446–453.

Felicio, L. S., Nelson, J. F., Gosden, R. G., and Finch, C. E., 1983, Restoration of ovulatory cycles by young ovarian grafts in aging mice: Potentiation by long-term ovariectomy decreases with age, *Proc. Natl. Acad. Sci. (USA)* **80:**6076–6080.

Ferguson-Smith, M. A., 1983, Prenatal chromosome analysis and its impact on birth incidence of chromosome disorders, *Br. Med. Bull.* **39:**355–364.

Field, T. M., Dabiri, C., and Shuman, H. H., 1977, Developmental effects of prolonged pregnancy and the postmaturity syndrome, *J. Pediatr.* **90:**836–839.

Finch, C. E., 1976, The regulation of physiological changes during mammalian aging, *Q. Rev. Biol.* **51:**49–83.

Finch, C. E., 1978, Genotypic influences on female reproductive senescence in rodents, in: *Genetics and Aging (Birth Defects: Original Articles Series)*, Volume 14 (S. Bersura and D. E. Harrison, eds.), The National Foundation, New York, pp. 335–354.

Finch, C. E., Felicio, L. D., Flurkey, K., Gee, D. M., Mobbs, C., Nelson, J. F., and Osterburg, H. H., 1980, Studies on ovarian-hypothalamic-pituitary interactions during reproductive aging in C57BL/6J mice, *Peptides* **1** (Suppl. 1):163–176.

Finch, C. E., and Holinka, C. F., 1982, Aging and uterine growth during implantation in C57BL/6J mice, *Exp. Gerontol.* **17:**235–241.

Finch, C. E., Felicio, L. S., Mobbs, C. V., and Nelson, J. F., 1984, Ovarian and steroidal influences on neuroendocrine aging processes in mammals, *Endocrine Rev.* **5:**467–497.

Fink, G., 1979, Feedback actions of target hormones on hypothalamus and pituitary with special reference to gonadal steroids, *Annu. Rev. Physiol.* **41:**571–585.

Flint, M., 1976, Does the chimpanzee have a menopause? *Am. J. Phys. Anthropol.* **44:**178–179.

Flurkey, K., Gee, D. M., Sinha, Y. N., and Finch, C. E., 1982, Age effects on luteinizing hormone, progesterone, and prolactin in proestrous and acyclic C57BL/6J mice, *Biol. Reprod.* **26**:835–846.

Fraser, I., and Baird, D. T., 1974, Blood production and ovarian secretion rates of estradiol-17B and estrone in women with dysfunctional uterine bleeding, *J. Clin. Endocrinol. Metab.* **39**:564–570.

Fugo, N. W., and Butcher, R. L., 1971, Effects of prolonged estrous cycles on reproduction in aged rats, *Fertil. Steril.* **22**:98–101.

Garn, S. M., Rohmann, C. G., and Wagner, B., 1967, Bone loss as a general phenomenon in man, *Fed. Proc.* **26**:1729–1736.

Gee, D. M., Flurkey, K., and Finch, C. E., 1983, Aging and the regulation of luteinizing hormone in C57BL/6J mice: Impaired elevations after ovariectomy and spontaneous elevations at advanced ages, *Biol. Reprod.* **28**:598–607.

Gee, D. M., Flurkey, K., and Finch, C. E., 1984, The regulation of the LH surge by estradiol in ovariectomized C57BL/6J mice: Effects of estradiol implant size, postovariectiomy duration and aging, *Endocrinology* **114**:685–693.

German, J., 1968, Mongolism, delayed fertilization, and human sexual behavior, *Nature* **217**:516–518.

Gibson, D. C., Adelman, R. C., and Finch, C. E. (eds.), 1979, *Development of the Rodent as a Model System of Aging*, NIH Publications No. 79-161.

Golbus, M. S., 1983, Oocyte sensitivity to induced meiotic non-disjunction and its relationship to advanced maternal age, *Am. J. Obstet. Gynecol.* **146**:435–438.

Goldzieher, J. W., 1981, Polycystic ovarian disease, *Fertil. Steril.* **35**:371–381.

Gosden, R. C., 1973, Chromosomal anomalies of preimplantation mouse embryos in relation to maternal age, *J. Reprod. Fertil.* **35**:351–354.

Gosden, R. C., 1979, Effects of age and parity on the breeding potential of mice with one and two ovaries, *J. Reprod. Fertil.* **57**:477–487.

Gosden, R. C., 1985, *The Biology of Menopause: The causes and consequences of ovarian aging*, Academic Press, New York.

Gosden, R. C., 1985, Maternal age: a major factor affecting the prospects and outcome of pregnancy, *Ann. N.Y. Acad. Sci.* **442**:45–57.

Gosden, R. C., Laing, S. C., Felicio, L. S., Nelson, J. F., and Finch, C. E., 1983a, Imminent oocyte exhaustion and reduced follicular recruitment mark the transition to acyclicity in aging C57BL/6J mice, *Biol. Reprod.* **28**:255–260.

Gosden, R. C., Laing, S. C., Flurkey, K., and Finch, C. E., 1983b, Graafian follicle growth and replacement in anovulatory ovaries of aging C57BL/6J mice, *J. Reprod. Fertil.* **69**:453–462.

Gougeon, A., 1982, Rate of follicular growth in the human ovary, in: *Follicular Maturation and Ovulation* (R. Rolland, E. V. van Hall, S. C. Hillier, D. P. McNatty, and J. Schoemaker, eds.), Excerpta Medica, pp. 155–163.

Gould, K. C., Flint, M., and Graham, C. E., 1981, Chimpanzee reproductive senescence: A possible model for evolution of the menopause, *Maturitas* **3**:157–166.

Gould, S. F., Shannon, J. M., and Cusha, G. R., 1983, The autoradiographic demonstration of estrogen binding in the normal human cervix and vagina during the menstrual cycle, pregnancy, and menopause, *Am. J. Anat.* **168**:229–238.

Govoni, S., Pasinetti, G., Trabucci, M., Inzoli, M. R., and Rozzini, R., 1983, Plasma prolactin in a large population of healthy old people, *Brt. Med. J.* **287**:1107.

Graham, C. E., Kling, G. R., and Steiner, R. A., 1979, Reproductive senescence in female non human primates, in: *Aging in Non-human Primates* (D. M. Bowden, ed.), Van Nostrand Reinhold, New York, pp. 183–202.

Gray, G. D., and Wexler, B. C., 1980, Estrogen and testosterone sensitivity of middle-aged female rats in the regulation of LH, *Exp. Gerontol.* **15**:201–207.

Gray G. D., Tennent, B., Smith, E. R., and Davidson, J. M., 1980, Luteinizing hormone regulation and sexual behavior in middle-aged female rats *Endocrinology* **107**:187–194.

Gropp, A., 1976, Morphological consequences of trisomy in mammals, in: *Embryogenesis in Mammals*, Ciba Foundation Symp. 40 (new series), Elsevier-Excerpta Medica-North Holland, Amsterdam, pp. 155–171.

Gropp, A., Ciers, D., and Kolbus, U., 1974, Trisomy in the fetal backcross of male and female metacentric heterozygotes of the mouse, *Cytogenet. Cell Genet.* **13:**511–535.

Gudelsky, G. A., Nansel, D. D., and Porter, J. C., 1981, Dopaminergic control of prolactin secretion in the aging male rat, *Brain Res.* **204:**446–450.

Guyton, A. C., 1981, *Textbook of Medical Physiology*, 6th ed., Saunders, Philadelphia, p. 1016.

Hansmann, I., and El-Nahass, E., 1979, Incidence of nondisjunction in mouse oocytes, *Cytogenet. Cell Genet.* **24:**115–121.

Harman, S. M., and Talbert, C. B., 1985, Aging of the reproductive system, in: *Handbook of the Biology of Aging* (C. E. Finch and E. L. Schneider, eds.), Van Nostrand, New York, pp. 457–510.

Henderson, S. A., and Edwards, R. G., 1968, Chiasma frequency and maternal age in mammals, *Nature* **218:**22–28.

Hertig, A. T., 1967, The overall problem in man, in: *Comparative Aspects of Reproductive Failure* (K. Benirschke, ed.), Springer-Verlag, Berlin, pp. 11–41.

Hoffman, J. C., 1973, The influence of photoperiods on reproductive functions in female mammals, in: *Handbook of Physiology*, Volume 2 (R. O. Creep, ed.), American Physiological Society, Washington D.C., pp.57–79.

Holinka, C. F., and Finch, C. E., 1977, Age-related changes in the decidual response of the C57BL/6J mouse uterus, *Biol. Reprod.* **16:**385–393.

Holinka, C. F., and Finch, C. E., 1981, Efficiency of mating in C57BL/6J female mice as a function of age and previous parity, *Exp. Gerontol.* **16:**393–398.

Holinka, C. F., Tseng, Y. C., and Finch, C. E., 1978, Prolonged gestation, elevated preparturitional plasma progesterone and reproductive aging in C57BL/6J mice, *Biol. Reprod.* **19:**807–816.

Holinka, C. F., Tseng, Y. C., and Finch, C. E., 1979a, Reproductive aging in C57BL/6J mice: Plasma progesterone, viable embryos, and resorption frequency throughout pregnancy, *Biol. Reprod.* **20:**1201–1212.

Holinka, C. F, Tseng, Y. C., and Finch, C. E., 1979b, Impaired preparturitional rise of plasma estradiol in aging C57BL/6J mice, *Biol. Reprod.* **21:**1009–1014.

Hook, E. R., 1981, Rates of chromosome abnormalities at different maternal ages, *Obstet. Gynecol.* **58:**282–285.

Horsman, A., Jones, M., Francis, R., and Nordia, C., 1983, The effect of estrogen dose on postmenopausal bone loss, *N. Engl. J. Med.* **309:**1405–1407.

Huang, H. H., Marshall, S., and Meites, J., 1976, Capacity of old versus young female rats to secrete LH, FSH, and prolactin, *Biol. Reprod.* **14:**538–543.

Jacobs, P. A., and Hassold, T. J., 1980, in: *Human Embryonic and Fetal Death* (I. H. Porter and B. B. Hook, eds.), Academic Press, pp. 289–298.

Jick, H., Porter, J., and Morrison, A. S., 1977, Relation between smoking and the age of natural menopause, *Lancet* **1:**1354–1355.

Jones, E. C., and Krohn, P. L., 1961a, The relationships between age, numbers of oocytes, and fertility in virgin and multiparous mice, *J. Endocrinol.* **21:**469–496.

Jones, E. C., and Krohn, P. L., 1961b, Effect of hypophysectomy on age changes in the ovaries of mice, *J. Endocrinol.* **21:**469–496.

Judd, H. L., 1983, Pathophysiology of menopausal hot flushes, in: *Neuroendocrinology of Aging* (J. Meites, ed.), Plenum Press, New York, pp. 173–202.

Kennedy, T. G., and Kennedy, J. P., 1972, Effects of age and parity on reproduction in young female mice, *J. Reprod. Fertil.* **28:**77–84.

King, C. R., Magenis, B., and Bennett, S., 1978, Pregnancy and the Turner syndrome, *Obstet. Gynecol.* **52:**617–624.

Knobil, B., 1980, The neuroendocrine contol of the menstrual cycle, *Rec. Prog. Hormone Res.* **36:**53–88.

Kram, D., and Schneider, S. E., 1978, An effect of reproductive aging: Increased risks of genetically abnormal offspring, in: *The Aging Reproductive System* (E. L. Schneider, ed.), Raven Press, New York, pp. 237–270.

Krishna Rao, C. V. C., and Draper, H. H., 1969, Age-related changes in the bones of adult mice, *J. Gerontol.* **24:**149–151.

Krohn, P. L., 1962, Review lectures on senescence. II. Heterochronic transplantation in the study of aging, *Proc. R. Soc. Lond. (Ser. B)* **157:**128–147.

Laing, S. C., Gosden, R. G., and Fraser, H. M., 1984, Cytogenetic analysis of mouse oocytes after experimental induction of follicular overripening, *J. Reprod. Fertil.* **70:**387–393.

Landfield, P. W., Waymire, J. C., and Lynch, G., 1978, Hippocampal aging and adrenocorticoids: Quantitative considerations, *Science* **202:**1098–1102.

Leadem, C. A., and Kalra, S. P., 1984, Stimulation with estrogen and progesterone of LHRH release from perifused adult female rat hypothalami: Correlation with the LH surge, Endocrinology **114:**51–56.

Leipheimer, R. E., and Callo, R. V., 1983, Acute and long-term changes: Control and pituitary mechanisms regulating pulsatile luteinizing hormone secretion after ovariectomy in the rat, *Neuroendocrinology* **37:**421–426.

Lieblum, S., Bachmann, G., Kemmann, B., Colburn, D., and Swartzman, L., 1983, Vaginal atrophy in postmenopausal women, *J. Am. Med. Assoc.* **249:**2195–2190.

Lin, T. J., and So-Bosita, J. L., 1972, Pitfalls in the interpretation of estrogenic effect in postmenopausal women, *Am. J. Obstet. Gynecol.* **114:**929–931.

Lin, T. J., So-Bosita, J. L., Brar, H. K., and Roblete, B. V., 1973, Clinical and cytologic responses of postmenopausal women to estrogen, *Obstet. Gynecol.* **41:**97–107.

Lu, K. H., Hopper, B. R., Vargo, T. M., and Yen, S. S. C., 1979, Chronologic changes, in sex steroid, gonadotropin, and prolactin secretion in aging female rats displaying different reproductive states, *Biol. Reprod.* **21:**193–203.

Luthardt, F. W., Palmer, C. C., and Yu, P.-L., 1973, Chiasma and univalent frequencies in aging female mice, *Cytogenet. Cell Genet.* **12:**68–79.

McCoshen, J. A., 1982, In vivo sex differentiation of congeneic germinal cell aplastic gonads, *Am. J. Obstet. Gynecol.* **142:**83–88.

McDonough, P. G., Byrd, J. R., Tho, P. T., and Mahash, V. B., 1977, Phenotypic and cytogenetic findings in 82 patients with primary ovarian failure—changing trends, *Fertil. Steril.* **28:**638–641.

Malinin, G. I., and Malinin, I. M., 1972, Age-related spontaneous uterine lesions in mice, *J. Gerontol.* **27:**193–196.

Mandel, F. P., Davidson, B. J., Erlik, Y., Judd, H. L., and Meldrum, D. R., 1982, Effects of progestins on bone metabolism in postmenopausal women, *J. Reprod. Med.* **27**(suppl.):51–514.

Mandl, A. M., and Shelton, M., 1959, A quantitative study of oocytes in young and old nulliparous laboratory rats, *J. Endocrinol.* **18:**444–450.

Mandl, A. M., and Zuckerman, S., 1951, The relation of age to numbers of oocytes, *J. Endocrinol.* **7:**190–193.

Marcus, R., Kosek J., Pfefferbaum, A., and Horning, S., 1983, Age-related loss of trabecular bone in premenopausal women: A biopsy study, *Calif. Tissue Int.* **35:**406–409.

Masukawa, T., 1960, Vaginal smears in women past 40 years of age, with emphasis on their remaining hormonal activity, *Obstet. Gynecol.* **16:**407.

Mattison, D. R., and Thorgeirsson, S. S., 1979, Smoking and industrial pollution, and their effects on menopause and ovarian cancer, *Lancet* **1:**187–188.

Mattison, D. R., Evans, M. I., Schwimmer, W. B., White, B. J., Jensen, B. and Schulman, J. D., 1984, Familial premature ovarian failure, *Am. J. Human Genet.* **36:**1341–1348.

Maudlin, I., and Fraser, L. R., 1978, Maternal age and the incidence of aneuploidy in first-cleavage mouse embryos, *J. Reprod. Fertil.* **54:**423–426.

Max, C., 1977, Cytological investigation of embryos in low-dose X-irradiated young and old female inbred mice, *Hereditas* **85:**199–206.

Meisel, R. L., and Ward I. L., 1981, Fetal female rats are masculinized by male littermates located caudally in the uterus, *Science* **213:**239–242.

Meites, J., 1982, Changes in neuroendocrine control of anterior pituitary function during aging in *Neuroendocrinology* **34:**151.

Metcalf, M. G., Donald, R. A., and Livesey, J. H., 1981a, Pituitary-ovarian function in normal women during the menopaual transition, *Clin. Endocrinol.* **14:**245–255.

Metcalf, M. G., Donald, R. A., and Livesey, J. H., 1981b, Classification of menstrual cycles in pre- and postmenopausal women, *J. Endocrinol.* **91:**1–10.

Michael, S. D., Taguchi, O., and Nishizuka, Y., 1981, Changes in hypophyseal hormones associated with accelerated aging and tumorigenesis of the ovaries in meonatally thymectomized mice, *Endocrinology* **108:**2375–2380.

Mikkelsen, M., Poulsen, H., Grinsted, J., and Lange, A., 1980, Nondisjunction in trisomy-21: Study of chromosomal heteromorphisms in 110 families, *Ann. Hum. Genet.* **44:**17–28.

Miller, M. E., and Chatten, J., 1967, Ovarian changes in ataxia telangiectasea, *Acta Paediatr. Scand.* **56:**559–561.

Miller A. B., and Riegle, C. D., 1980, Temporal changes in serum progesterone in aging female rats, *Endocrinology* **106:**1579–1583.

Mintz, B., and Russell, B. S., 1957, Gene-induced embryological modifications of primordial germ cells in the mouse, *J. Exp. Zool.* **134:**207–237.

Mobbs, C. V., Flurkey, K. C., Gee, D. M., Yamamoto, K., Sinha, Y. N., and Finch C. B., 1984a, Estradiol-induced anovulatory syndrome in female C57BL/6J mice: Age-like neuroendocrine, but not ovarian impairments, *Biol. Reprod.* **30:**556–563.

Mobbs, C. V., Gee, D. M., and Finch, C. E., 1984b, Reproduction in senescence in female C57BL/6J mice: Ovarian impairments and neuroendocrine impairments that are partially reversible and delayable by ovariectomy, *Endocrinology* **116:**813–820.

Morrison, J. C., Gwens, J. R., Wiser, W. L., and Fish, S. A., 1975, Mumps oophoritis: A cause of premature menopause, *Fertil. Steril.* **26:**655–659.

Nass, T. E., LaPolt, P. J., Judd, H. L., and Lu, J. H. K., 1984, Alterations in ovarian steroid and gonadotrophin secretion preceding the cessation of regular oestrous cycles in aging female rats, *J. Endocrinol.* **100:**43–50.

Nelson, J. F., Felicio, L. S., Sinha, Y. N., and Finch, C. E., 1980, An ovarian role in the spontaneous pituitary tumorigenesis and hyperprolactinemia of aging female mice, *Gerontologist* **20** (No. 5, Part II):171 (abstract).

Nelson, J. F., Felicio, L. S., Osterburg, H. H., and Finch, C. E., 1981, Altered profiles of estradiol and progesterone associated with prolonged estrous cycles and persistent vaginal cornification in aging C57BL/6J mice, *Biol. Reprod.* **24:**784–794.

Nelson, J. F., Felicio, L. S., Randall, P. K., Simms, C., and Finch, C. E., 1982, A longitudinal study of estrous cyclicity in aging C57BL/6J mice. I. Cycle frequency, length, and vaginal cytology, *Biol. Reprod.* **27:**327–339.

Nelson, J. F., and Felicio, L. S., 1985, Reproductive aging in the female: an etiological perspective, *Rev. Biol. Res. Aging* **2:**251–314.

Nordin, B. E. C., Marshall, D. H., Francis, R. M., and Crilly, R. C., 1981, The effects of sex steroids and corticosteroid hormones on bone, *J. Steroid. Biochem.* **15:**171–174.

Novak, E. R., 1970, Ovulation after 50, *Obstet. Gynecol.* **36:**903–910.

Novak, E. R., Goldberg, B., Jones, G. S., O'Toole, R. V., 1965, Enzyme histochemistry of the menopausal ovary associated with normal and abnormal endometrium, *Am. J. Obstet. Gynecol.* **93:**669–680.

Odell, W. D., and Swerdloff, R. S., 1968, Progesterone-induced luteinizing and follicle-stimulating hormone surge in post-menopausal women: A simulated ovulatory peak, *Proc. Natl. Acad. Sci. (USA)* **61:**529-536

Padmanabhan, V., Keech, C., and Convey, E. M., 1983, Cortisol inhibits and adrenocorticotropin has no effect on luteinizing hormone-releasing hormnone-induced release of luteinizing hormone from bovine pituitary cells in vitro, *Endocrinology* **112:**1782-1787.

Page, R. D., and Butcher, R. L., 1982, Follicular and plasma patterns of steriods in young and old rats during normal and prolonged estrous cycles, *Biol. Reprod.* **27:**383-392.

Papadaki, L., Beilby, J. O. W., Chowaniec, J., Coulson, W. F., Darby, A. J., Newman, J., O'Shea, A., and Wykes, J. R., 1979, Hormone replacement therapy in the menopause: A suitable animal model, *J. Endocrinol.* **83:**67-77.

Parfitt, A. M., Mathews, C. H. E., Villanueva, A. R., and Kleerekoper, M., 1983, Relationships between surface, volume, and thickness of iliac trabecular bone in aging and in osteoporosis, *J. Clin. Invest.* **72:**1396-1409.

Parkening, T. A., Collins, T. J., and Smith, E. R., 1980, Plasma and pituitary concentrations of LH, FSH, and prolactin in aged C57BL/6 mice, *Neurobiol. Aging* **3:**31-36.

Parsons, P. A., 1964, Parental age and the offspring, *Q. Rev. Biol.* **39:**258-275.

Pedersen, T., 1972, Follicle growth in the mouse ovary, in: *Oogenesis* (J. D. Biggers and A. W. Schuetz, eds.), University Park Press, Baltimore, pp. 361-376.

Pellika, P. A., Sullivan, W. P., Coulam, C. B., and Toft, D. O., 1983, Comparison of estrogen receptors in human premenopausal and postmenopausal uteri using isoelectric focusing, *Obstet. Gynecol.* **62:**430-434.

Peppler, R. D., 1971, Effects of unilateral ovariectomy on folicular development and ovulation in cycling, aged rats, *Am. J. Anat.* **132:**423-428.

Pohl, C. R., Richardson, D. W., Hutchinson, J. J., Germark, J. A., and Knobil, E., 1983, Hypophysiotropic signal frequency and the functioning of the pituitary-ovarian system in the rhesus monkey, *Endocrinology* **12:**2076-2080.

Polani, P. E., and Jagiello, C. M., 1976, Chiasmata, meiotic univalents and age in relation to aneuploid imbalance in mice *Cytogenet. Cell Genet.* **16:**505-529.

Raisz, L. G., 1982, Osteoporosis, *J. Am. Geriatr. Soc.* **30:**127-138.

Ramirez, V. D., Feder, H. H., and Sawyer, C. H., 1984, The role of brain catecholamines in the regulation of LH secretion: A critical inquiry, in: *Frontiers in Neuroendocrinology*, Volume 8 (L. Martini and W. F. Canong, eds.), Raven Press, New York, pp. 27-84.

Reyes, F. I., Koh, K. S., and Faiman, C., 1976, Fertility in women with gonadal dysgenesis, *Am. J. Obstet. Gynecol.* **126:**668-670.

Reymond, M. J., and Porter, J. C., 1981, Secretion of hypothalamic dopamine into pituitary stalk blood of aged female rats, *Brain Res. Bull.* **7:**69-73.

Riggs, B. L., Worhner, H. W., Dunn, W. L., Mazess, R. B., Offord, K. P., and Melton, L. J., III, 1981, Differential changes in bone mineral density of the appendicular and axial skeleton with aging: Relationship to spinal osteoporosis, *J. Clin. Invest.* **67:**328-355.

Roth, G. S., and Hess, G. D., 1982, Changes in the mechanisms of hormone and neurotransmitter action during aging: Current status of the role of receptor and post-receptor alterations—A review, *Mech. Aging Devl.* **20:**175-194.

Sapolsky, R. M., Krey, L. C., and McEwen, B. S., 1983, The adrenocortical stress-response in the aged male rat: Impairment in the recovery from stress, *Exp. Gerontol.* **18:**55-64.

Schroder, R., 1954, Endometrial hyperplasia in relation to genital function, *Am. J. Obstet. Gynecol.* **68:**294-309.

Shah, B. G., Krishna Rao, C. V. C., and Draper, H. H., 1967, The relationship of Ca and P nutrition during adult life and osteoporosis in aged mice, *J. Nutrition* **92:**30-42.

Shelton, M., 1959, A comparison of the population of oocytes in nulliparous and multiparous senile laboratory rats, *J. Endocrinol.* **18:**451-455.

Sherman, B. M., and Korenman, S. G., 1975, Hormonal charactristics of the human menstrual cycle throughout reproductive life, *J. Clin. Invest.* **55:**699-706.

Sherman, B. M., West, J. H., and Korenmen, S. C., 1976, The menopausal transition: Analysis of LH, FSH, estradiol and progesterone concentrations during menstrual cycles of older women, *J. Clin. Endocrinol. Metab.* **42:**629–636.

Silberberg, M., and Silberberg, R., 1941, Age changes of bones and joints in various strains of mice, *Am. J. Anat.* **68:**69–90.

Silberberg, M., and Silberberg, R., 1962, Osteoarthrosis and osteoporosis in senile mice, *Gerontologia* **6:**19–101.

Sopelak, V. M., and Butcher, R. L., 1982a, Contribution of the ovary versus hypothalamus-pituitary to termination of estrous cycles in aging rats using ovarian transplants, *Biol. Reprod.* **27:**29–37.

Sopelak, V. M., and Butcher, R. L., 1982b, Decreased amount of ovarian tissue and maternal age affect embryonic development in old rats, *Biol. Reprod.* **27:**449–455.

Speed, R. M., 1977, The effects of aging on the meiotic chromosomes of male and female mice, *Chromosome* **64:**241–254.

Speed, R. M., and Chandley, A. C., 1981, The response of germ cells of the mouse to the induction of non-disjunction by X-rays, *Mut. Res.* **84:**409–418.

Speed, R. M., and Chandley, A. C., 1983, Meiosis in the foetal mouse ovary. II. Oocyte development and age-related aneuploidy. Does a production line exist? *Chromosoma* **88:**184–189.

Staats, J., 1976, Standardized nomenclature for inbred strains of mice: Sixth listing, *Cancer Res.* **36:**4333–4377.

Sugawara, S., and Mikamo, K., 1983, Absence of correlation between univalent formation and meiotic non-disjunction in aged female Chinese hamsters, *Cytogenet. Cell Genet.* **35:**34–40.

Syftestad, G. T., and Urist, M. R., 1982, Bone aging, *Clin. Orthoped.* **162:**288–296.

Tang, V. W., and Faimen, C., 1983, Premature ovarian failure: A search for circulating factors against gonadotropin receptors, *Am. J. Obstet. Gynecol.* **146:**816–821.

Tease, C., 1982, Similar does-related chromosome non-disjunction in young and old female after X-irradiation, *Mut. Res.* **95:**287–296.

Thompson, B., Hart, S. A., and Durno, D., 1973, Menopausal age and symptomatology in general practice, *J. Biosocial. Sci.* **5:**71–82.

Thung, P. L., Boot, D. M., and Muhlbock, O., 1956, Senile changes in the estrous cycles and in ovarian structure in some inbred strains of mice, *Acta Endocrinol.* **23:**8–23.

Treloar, A. E., 1974, Menarche, menopause, and intervening fecundability, *Hum. Biol.* **46:**89–107.

Treloar, A. E., 1981, Menstrual cyclicity and the pre-menopause, *Maturitas* **3:**249–264.

Treloar, A. E., Boyntonk, F. E., Behn, D. G., and Brown, B. W., 1967, Varation of the human menstrual cycle through reproductive life, *Int. J. Fertil.* **12:**77–126.

Trentin, J. J., 1950, Vaginal sensitivity to estrogen as related to mammary tumor incidence in mice, *Cancer Res.* **10:**580–583.

Tucker, H. St. G., Lankford, H. V., Gardner, D. F., and Blackard, W. C., 1980, Persistent defect in regulation of prolactin secretion after persistent pituitary tumor removal in women with the galactorrhea-amenorrhea syndrome, *J. Clin. Endocrinol. Metab.* **51:**968–971.

Tulandi, T., and Kinch, R. A. H. 1981, Premature ovarian failure, *Obstet. Gynecol. Surv.* **36:**521–527.

Turgeon, J. L., 1979, Estradiol-luteinizing hormone relationship during the proestrous gonadotropin surge, *Endocrinology* **105:**731–736.

Urist, M. R., 1962, Osteoporosis, *Annu. Rev. Med.* **13:**273–286.

Van Look, P. F. A., Lothian, H., Hunter, W. H., Michie, E. A., and Baird, D. T., 1977, Hypothalamic-pituitary-ovarian function in perimemopausal women, *Clin. Endocrinol.* **7:**13–31.

Van Loon, G. R., 1978, A defect in catecholamine neruons in patients with prolactin-secreting pituitary adenoma, *Lancet* **2:**868–871.

Van Wagenen, G., 1972, Vital statistics from a breeding colony: Reproduction and pregnancy outcome in *Macaca mulatta, J. Med. Primatol.* **1:**3–28.

Van der Schoot, P., 1976, Changing pro-estrous surges of luteininzing hormone in aging 5-day cyclic rats, *J. Endocrinol.* **69:**287–288.

Vom Saal, F. S., 1983, The interactions of circulating oestrogens and androgens in regulating mammalian sexual differentiation, in: *Hormones and Behavior in Higher Vertebrates* (J. Balthazart, E. Prove, and R. Gilles, eds.), Springer-Verlag, Berlin, pp. 159–177.

Vom Saal, F. S., and Moyer, C. L., 1985, Parental effects on reproductive capacity during aging in female mice, *Biol. Reprod.* **32:**1116–1126.

Vom Saal, F. S., Pryor, S., and Bronson, F. H., 1981, Effects of prior intrauterine position and housing on oestrous cycle length in adolescent mice, *J. Reprod. Fertil.* **62:**33–37.

Weir, B. J., 1971, The reproductive organs of the female plains viscacha, *Lagestomus maximus, J. Reprod. Fertil.* **25:**365–373.

Wilkes, M. M., Lu, K. H., Hopper, B. R., and Yen, S. S. C., 1979, Altered neuroendocrine status of middle-aged rats prior to the onset of senescent anovulation, *Neuroendocrinology* **29:**255–261.

Wink, C. S., and Felts, W. J. L., 1980, Effects of castration on the bone structure of male rats: A model of osteoporosis, *Calcif. Tissue Int.* **32:**77–82.

Wise, P. M., 1983, Aging of the female reproductive system, *Rev. Biol. Res. Aging* **1:**195–222.

Wronski, T. J., Lowry, P. L., Walsh, C. C., and Ignaszewski, L. A., 1985, Skeletal alterations in ovariectomized rats. *Calcif. Tissue Int.* **37:**324–328.

Yamaji, T., Shimanto, K., Ishibashi, M., Kosaka, K., and Crimo, H., 1976, Effect of age and sex on circulating and pituitary prolactin levels in humans, *Acta Endocrinol.* **83:**711–719.

Yamamoto, M., Endo, A., and Watanabe, G., 1973, Maternal age dependence of chromosome anomalies, *Nature New Biol.* **241:**141–142.

Zerwekh, J. E., Sakhaec, K., Glass K., and Pak, C. Y. C., 1983, Long term 25-hydroxyvitamin D3 therapy in postmenopausal osteoporosis: Demonstration of responsive and nonresponsive subgroups, *J. Clin. Endocrinol. Metab.* **56:**410–413.

2

METHODOLOGICAL ISSUES IN CLINICAL RESEARCH ON THE AGING REPRODUCTIVE SYSTEM

KENNETH L. MINAKER and JOHN W. ROWE

Introduction

Aging of the reproductive system in both men and women results in symptoms that frequently require physician assistance, and changes in sexual function with advancing age represent a "hidden" concern of many older patients. Physiologic studies of aging including studies of reproductive aging require special attention to subject selection, characterization, and study design (Rowe, 1977). Failure to consider these factors explains much of the inconsistency in the literature, and careful adherence to study design and subject selection will lead to clarification of the mechanisms accounting for reproductive aging.

Cross-Sectional and Longitudinal Studies

Two general study designs are available to the clinical gerontologist. In cross-sectional studies, groups of various ages are observed, and age-related differences are sought. In longitudinal studies, serial prospective measurements are obtained in one group of subjects at specified intervals, and age-related changes are determined. Since the human life span is long, most longitudinal studies follow subjects in several age cohorts throughout the adult age range

KENNETH L. MINAKER and JOHN W. ROWE • Division on Aging, Harvard Medical School; Gerontology Division, Joint Department of Medicine, Beth Israel and Brigham and Women's Hospitals, Boston, Massachusetts 02215; Geriatric Research Education Clinical Center, West Roxbury/Brockton Veterans Administration Medical Center, West Roxbury, Massachusetts 02132.

concurrently (cross-sequential design), so that the rates of change for the age-related variables under study can be compared across the adult age range (Rowe, 1977).

Cross-sectional studies must be interpreted with caution, since there are several ways in which they may not reflect true age-related changes. One error in design is based on a common misconception of the human life cycle. It is often assumed that the growth-and-development phase ends before the age of 20 years and is followed by a prolonged "plateau phase," during which the biological or physiological variable under study is stable, and then, at about the age of 60 years, by the onset of a fairly rapid decline. However, for most variables that change with age, the growth-and-development phase ends near the age of 20–30 years and is followed by a gradual, often linear, decline. Several studies of sympathetic activity and aging in man have included individuals as young as 10 years old in the young group, thus introducing possible confusion between the effects of growth-and-development and senescence. A dynamic change in a landmark event in the human life cycle can further alter the age selection of subjects for study. The progressively decreasing age of onset of the menarche is one such change in the reproductive system.

In the interpretation of cross-sectional studies, it is important to remember that subjects older than approximately 74 years represent a sample of biologically superior survivors from a cohort that has experienced at least a 50% mortality. If the variable under study is a risk factor for shortened survival (elevations in cholesterol or blood pressure), a cross-sectional study will seem to show age-related declines that do not represent true aging but rather reflect the progressive loss of individuals with high values. This effect is termed selective mortality and is a serious methodological obstacle that can be avoided by use of prospective, longitudinal study design, in which each subject is followed over time and the rate of change of each variable is calculated for each age group.

Despite their advantages over cross-sectional studies, longitudinal studies also have major drawbacks, including the need to observe a stable population over a long period and a particular sensitivity to alterations in methods. Subtle changes in laboratory techniques over several years may introduce "laboratory drifts" that are difficult to separate from age-related changes. An example is the "aging" of a sphygmomanometer. Unfamiliarity with the study situation may also introduce error into results, and as subjects return at regular intervals and become increasingly familiar with the testing environment, the "learning" or "stress" effect lessens. For example, in the Framingham Study, blood pressure measurements for all groups fell between the first and second biannual visit and again between the second and third; thereafter, they increased with each successive visit. Thus, it was necessary to exclude the first two measurements from the analysis of the impact of age on blood pressure (Gordon and Shurtleff, 1973).

The confidence in characterizing a study population is greatest for recent clinical status. This will underestimate the long-term influence of environmental factors on physiologic regulation because of the failure to recognize, appreciate, or quantify past or cumulative physiologic stressors. The processes of reproductive aging are highly integrated over a period of years and could be expected to be profoundly influenced by past events occurring during "vulnerable" periods of time. The menopause itself may be viewed as an adaptive response to failing reproductive capacity for the human species and as such may be influenced by a series of environmental and intrinsic event over a span of years (Grimley Evans, 1981, 1984). Current efforts to follow patients longitudinally through reproductive landmarks such as the menopause will at least standardize environmental influences and clarify mechanisms responsible for this aspect of physiologic aging.

To date only two longitudinal studies have been reported that follow sexual function in aging men. Nieschlag *et al.* (1982) followed sexual activity and hormonal and semen analysis in healthy fathers and grandfathers over a period of 3 years, and reported a decline in seminal fluid fructose and sperm motility but preservation of volume of ejaculate, sperm count, and percent of normally formed spermatozoa. Verwoerdt *et al.* (1969) followed sexual activity in healthy men for 8 years and confirmed prior cross-sectional studies indicating a decline in sexual activity over time. One longitudinal study of sexual activity in females shows a decline in female activity only after age 55 years (George and Weiler, 1981).

"Normal" versus "Usual" Aging

Gerontologists strive to differentiate the physiologic consequences of aging from those of concomitant disease. In the area of reproductive physiology, which involves multiple organs and complex, integrated responses, this age-versus-disease dichotomy poses a special challenge, since exclusion of all potentially contaminating illnesses would require many, varied and often invasive procedures. Clearly, the greater the effort expended in excluding patients with relevant disease the more likely "normal" aging is being studied, and conversely, studies bypassing detailed clinical screening are likely to result in data characteristic of the "usual" individual, who in advanced old age harbors chronic incremental illnesses.

Aging processes in man produce a spectrum of physiologic changes with varied clinical relevance (Rowe and Minaker, 1985). At one end of the spectrum measurable aging changes lack functional significance and at the other aging creates symptoms that have direct, predictable, adverse clinical sequelae. Investigators must be aware of normative changes in physiologic systems that have an impact on reproductive physiology and select subjects for study that reflect these aging changes. "Superselection" (inclusion of subjects who do not demonstrate normative age changes) is an important pitfall in clinical aging studies. An example of "superselection" in studies of male reproductive aging after 65 years is exclusion of men with any evidence of benign prostatic hypertrophy (BPH). BPH is so prevalent in elderly men that it is considered a "normal" component of aging and exclusion of these individuals yields data that are not generalizable. Similarly, in studies of female reproductive aging, screening of hundreds of 60-year-old women in order to identify a handful who are still premenopausal is unlikely to provide a proper group for a study of normal aging.

The available studies of plasma testosterone levels in aging men provide a clear example of "usual" intermixed with "normal." Most early studies included patients from clinics and hospitals and reported a decline in plasma testosterone during aging. In more recent studies extreme care has been taken to select men of good health, with social and environmental stability, and plasma testosterone does not decline (Swerdloff and Heber, 1982). It must also be appreciated that a disease included in a study population may have direct or subtle indirect impacts on reproductive function. For example, studies of male potency that include subjects with diabetes, which increases in prevalence with age from less than 1% to age 20 to 7.2% at age 60 and up to 17% at age 85, may yield data that are influenced by several different pathologic effects of this chronic disease. Direct local circulatory or neurologic effects may be present, or more indirect effects, such as impaired cardiac function or urinary tract infection, may adversely influence potency.

Both the "normal" and the "usual" older individual are worthy of our study, the former as a goal that more disabled groups may hope to reach once therapeutic strategies develop and are applied. The responsibility of the clinical investigator is to adequately characterize his/her study population so that a clear judgement regarding its disease contamination can be made.

Physiologic Covariables of Aging and Their Impact on Reproductive Physiology

It is now recognized that numerous covariables have a major impact on the results of gerontologic endocrine studies (Minaker, et al., 1984). They include diet, exercise, anthropometric changes, psychosocial factors, and personal habits, including medication use. Failure to consider the impact of these variables in study design no doubt contributes to the increased variance in physiologic performance observed in many studies of aging. At the present time, some of these factors are considered to be aging processes themselves; the relationships of others to physiologic aging remains to be clarified.

Diet

The influence of antecedent diet on physiologic events such as the menopause is significant. Calcium intake and absorption has been shown to progressively decrease with age and in many postmenopausal and perimenopausal females is an almost certain partial contributor to postmenopausal osteopenia. Gradual lowering of the age at menarche is felt by many to be related to societal improvements in general nutrition; consequent increases in body fat play an important role in aromatization of estrogen precursors to more potent estrogenic compounds, which in turn promotes the gonadotrophic pulses associated with the initiation of the menarche. Starvation is also a known modulator of the menstrual cycle. While dietary intake of healthy older men varies mildly during normal aging (Elahi et al., 1983), dietary intake in older individuals with chronic disease is often impaired. Thus dietary factors must be characterized in studies of reproductive aging.

Exercise

Recent interest in exercise has brought recognition of its effect on central nervous system alterations in many physiology systems including the regulation of the menstrual cycle. As in the case of diet, exercise may exert its effect on menstrual physiology via substantial lowering of body fat which results in menstrual irregularity. In contrast with extreme exercise that results in the anthropometric changes described above, less rigorous exercise programs that do not result in changes in body fat content lead to improvement in menstrual regularity.

Body Composition

A variety of factors, including exercise, may influence the qualitative and quantitative anthropometric changes that occur during aging (Rossman, 1977). With age, lean body mass declines and relative adiposity increases. The magnitude of these reasonably balanced changes (i.e., overall body weight is generally maintained) appears to be less than previously estimated as healthier populations are studied and may be in the range of 10% over a span of 50 years (Rowe et al., 1983). In addition to these quantitative changes, it is becoming increasingly apparent that the distribution of fat, in particular, is altered during aging, with increased centripetal fat distribution (Borkan et al., 1984). It has been clearly established that endocrine regulation of fat by both cortisol and insulin differs in central versus peripheral fat depots (Rudman and Girolamo, 1971). The impact of anthropometric differences on the reproductive system have already been documented. Numerous investigators have revealed a positive correlation between production rates and plasma levels of estradiol and estrone and body fat content. The suggested mechanism is potentiation of the conversion of androstendione to estradiol by fat tissues. There is an increasing effort to quantitate fatness in physiologic and clinical research, particularly in studies of protein metabolism and glucose homeostasis. The effort has resulted in methodologies that accurately and noninvasively determine these anthropometric changes. Where relevant in reproductive aging, these new techniques should be utilized to characterize anthropometric changes for subjects recruited for physiologic studies of the reproductive system.

Psychosocial Factors

The psychosocial milieu has a great impact on reproductive capacity and sexual opportunity. Population demographics alone underscore the magnitude of this phenomenon. Life expectancy for females is presently almost 7 years greater than that for men and the ratio of elderly females to males approaches $3:2$. The average married American woman can now anticipate at least a decade of widowhood. In addition remarriage rates are relatively low in late life. Cultural and religious attitudes influence sexual activity and changes in the expression of these values have led to a profound difference in common attitudes towards premarital sex between present teenagers and their counterparts of eight decades ago.

It is widely appreciated that the more complex, integrated, and less homeostatically crucial a physiologic system is, the more sensitive it is to be subject to major disruption (Shock, 1979). This has been widely recognized in evaluations for impotence which can be caused by a broad spectrum of disorders. Twenty to forty percent of cases are psychogenic, indicating the significant impact of psychologic factors. With advancing age, social and economic status usually decline, altering mobility and opportunity, which sustain the relationships that lead to sexual expression.

Personal Habits

Cigarette smoking has recently been widely recognized as altering fetal growth, hastening the onset of the menopause, and modifying the metabolism of drugs, some of which have relevance to reproductive physiology.

Another personal habit, alcohol intake, has significant effects, both short and long term, on reproductive physiology. Studies in healthy nonalcoholic men have provided evidence that short-term high rates of alcohol intake depresses testosterone secretion (Gordon *et al.*, 1976). Long-term high rates of alcohol consumption are associated with premature death and disability from cancer, cardiovascular disease, accidents and suicide, all of which complicate normative evaluation of reproductive physiology. Alcohol consumption patterns change with advancing age, introducing another variable to be considered when evaluating reproductive function. Only 45% of the population over age 65 drink alcohol, whereas 71% of those aged 30–49 regularly consume alcohol (Atkinson and Schuckit, 1981). While alcoholism is associated with premature mortality, its prevalence in the elderly remains high at between 5–15%. New alcoholics in late life appear to supplement the numbers of young alcoholics surviving into old age. In recent studies, low rates of alcohol intake do not appear to influence serum testosterone in otherwise very healthy old men. (Sparrow *et al.*, 1980). However, casual intake of alcohol may have a significant influence if the population under study regularly consume a variety of other medications and are suffering from several chronic diseases.

Medications

While it is commonly incorrectly assumed that only individuals with physician supervised illness take medications that might influence studies of reproductive aging, in the elderly, 50% of medication is consumed without a physician's prescription, and often without his knowledge. As medications can exert major effects on reproductive functioning, careful inquiry and detailed reporting is necessary to characterize study populations. For example, studies of spermatozoal chromosomal profiles of individuals taking cimetidine or smoking cannabis without the investigator's knowledge are flawed. Studies of sexual activity that do not quantify medication intake may well lead to the conclusion that an effect is age related when in fact a significant impact is made by medication. In one recent study (Slag *et al.*, 1983), 25% of impotence in an outpatient population of mean age 59 years was due to medication effects. Drugs implicated were diuretics, antihypertensive agents, major tranquillizers, and in 7%, significant alcohol intake. The use of all of these common drug groups increases dramatically with advancing age.

Technical Aspects of Study Design

Measurement of circulating hormone levels under both basal and stimulated conditions represents the most common method of evaluating endocrine aging. While, organ response is often assumed to reflect ambient hormonal stimulation it is now recognized that response to hormones also depend on binding hormone levels, circadian rhythms, pulsatile hormonal secretion, and conversion of the hormone to another form at the end organ. Study design and data analysis must be altered to accommodate possible aging effects in these factors.

The binding of testosterone to sex hormone-binding globulin (SHBG) is a major determinant of free hormone level and androgenic stimulation of tissues. In aging men SHBG increases presumably because of anthropometric changes that accompany aging; obesity increases peripheral conversion of androgens to estrogen, which increases hepatic synthesis of SHBG (Kley et al., 1974). This results in a decrease in plasma free testosterone. A circadian rhythm has been demonstrated for testosterone with a morning peak and lower afternoon plateau. The literature varies regarding the influence of age on diurnal variation in plasma testosterone (some reports suggesting maintenance, others alterations) (Bremner et al., 1983). Since there may be a circadian rhythm in plasma testosterone, studies using single measures should study patients at the same time of day. With the increased capacity to accurately measure hormone levels on small amounts of plasma, frequent sampling techniques have demonstrated pulsatile secretion of many hormones, including testosterone. This contributes to the increased variance in studies where single samples are obtained. The relevance of pulsatile secretion to aging physiology is established. The appearance of episodic and pulsatile secretion of luteinizing hormone (LH) has a profound influence on the initiation of sexual maturation, and the menopausal hot flash is associated with hormonal pulses as well.

Although not well studied in reproductive physiology, blood sampling technique is known to influence basal hormonal values in the elderly. The effect of the length of recumbency on plasma catecholamines has been studied with respect to age. Thirty minutes after recumbency, plasma catecholamines are higher in the elderly than in the young. However, 9 hr after, recumbency catecholamine levels fall to similar levels in both groups (Saar and Gordon, 1979). Standardization of sampling technique is important for certain hormones with short half-lives and may have relevance for reproductive physiology.

Evaluation of reproductive physiology also includes dynamic testing to defined stimuli that can be influenced by age. Reproductive capacity of females has been difficult to evaluate across the age span simply due to the decline in frequency of the necessary stimulus, sexual intercourse. While this example of altered stimulus is obvious, more subtle variants likely occur in other areas of reproductive physiology. Positive and negative feedbacks of estrogen on the hypothalamic-pituitary axis regulate FSH and LH secretion. Present evidence suggests that the midcycle surge of estrogen initiates the ovulatory surge of LH. Judd et al. (1976) have suggested that with advancing age plasma estrogen levels decrease in the late follicular and luteal portions of the menstrual cycle and that this may have implications for the irregular cycles characteristic of the perimenopausal female.

The investigative milieu may provide differential stimulation for elderly and young individuals. The stress effect of admission to a clinical research center (CRC) is almost certainly different for young and old volunteers due to differences in lifestyle, ability to adapt to CRC regimentation, and variation in the perception of the significance of being admitted to hospital.

Summary

Subject selection is perhaps the most crucial element of clinical studies of reproductive aging. Increasingly it is becoming possible to recognize and quantify important covariables of aging that will decrease the variance of data gathered and lead to a clearer elucidation of mechanisms of reproductive aging. A judicious adoption of longitudinal study design will further enhance the assessment of reproductive aging.

References

Atkinson, J. H., and Schuckit, M. A., Alcoholism and over-the-counter and prescription drug misuse in the elderly, in: *Annual Review of Gerontology and Geriatrics* (C. Eisdorfer, ed.), Springer, New York, pp. 255–284.

Borkan, G. A., Hults, D. E., Gerzof, S. G., Robbins, A. H., Silbert, C. A., 1984, Age changes in body composition revealed by computed tomography, *J. Gerontol.* 38(6):673–677.

Bremner, W. J., Vitiello, M. V., and Prinz, P. N., 1983, Loss of circadian rhythmicity in blood testosterone levels with aging in normal men, *J. Clin. Endocrinol. Metab.* 56(6):1278–1281.

Elahi, V. K., Elahi, D., Andres, R., Tobin, J. D., Butler, M. S., and Norris, A. H., 1983, A longitudinal study of nutritional intake in men, *J. Gerontol.* 38(2):162–180.

George, L. K., and Weiler, S. J., 1981, Sexuality in middle and late life. The effects of age, cohort, and gender, *Arch. Gen. Psychiatry* 38(8):919–923.

Gordon, G. G., Altman, K., Southern, A. L., Rubin, E., and Lieber, G. S., 1976, Effect of alcohol on sex hormone metabolism in normal men, *N. Engl. J. Med.* 295:793.

Gordon, T., and Shurtleff, D., 1973, Means at each examination and interexamination variation of specified characteristics: Framingham study, exam 1 to exam 10, The Framingham study: An epidemiologic investigation of cardiovascular disease (DHEW Publication No. NIH 74-478) (W. B. Kannel, T. Gordon, eds.), Washington, D.C., U.S. Government Printing Office.

Grimley Evans, J., 1981, The biology of human aging, in: *Recent Advances in Medicine* (A. M. Dawson, N. Compston, and G. M. Besser, eds.), Churchill Livingston, London, pp. 17–38.

Grimley Evans, J., 1984, Prevention of age-associated loss of autonomy: Epidemiologic approaches, *J. Chronic Dis.* 37(5):353–363.

Judd, H. D., Lucas, W. E., and Yen, S. S. C., 1976, 17 beta estradiol and estrone levels in postmenopausal women with and without endometrial cancer, *J. Clin. Endocrinol. Metab.* 43:272–278.

Kley, H. K., Nieschlag, E., Bidlingmaier, F., and Kruskemper, H. L., 1974, Possible age dependent influence of estrogens on the binding of testosterone in plasma of adult men, *Horm. Metab. Res.* 6:213–215.

Minaker, K. L., Rowe, J. W., Young, J. B., Pallotta, J., Landsberg, L., 1982, Effect of age on insulin stimulation of sympathetic nervous system activity in man, *Metabolism* 31:1181–1184.

Minaker, K. L., Meneilly, G. S., and Rowe, J. W., 1985, Endocrine systems, in: *Handbook of the Biology of Aging*, 2nd edition (C. Finch and E. Schneider, eds.), Van Nostrand Reinhold, New York, pp. 433–456.

Nieschlag, E., Lammers, U., Freischem, C. W., Langer, K., Wickings, E. J., 1982, Reproductive function in young fathers and grandfathers, *J. Clin. Endocrinol. Metab.* 55(4):676–681.

Rossman, I., 1977, Anatomic and body composition changes with aging, in: *Handbook of the Biology of Aging* (C. Finch and L. Hayflick, eds.), Van Nostrand Reinhold, New York, pp. 189–221.

Rowe, J. W., 1977, Clinical research on aging: Strategies and directions, *N. Engl. J. Med.* **297**:1332–1336.

Rowe, J. W., and Minaker, K. L., 1985, Geriatric medicine, in: *Handbook of the Biology of Aging*, 2nd edition (C. Finch and E. Schneider, eds.), Van Nostrand Reinhold, New York pp. 932–959.

Rowe, J. W., Minaker, K. L., Pallotta, J., and Flier, J. S., 1983, Characterization of the insulin resistance of aging, *J. Clin. Invest.* **71**:1581–1587.

Rudman, D. D., and Girolamo, M., 1971, Effects of adrenal cortical steroids on lipid metabolism, in: *The Human Adrenal Cortex* (N. P. Christy, ed.), Harper and Row, New York, pp. 241–256.

Saar, N., and Gordon, R. D., 1979, Variability of plasma catecholamine levels: age, duration of posture and time of day. *Br. J. Clin. Pharmacol.* **8**:353–358.

Shock, N., 1979, Systems physiology and aging, *Fed. Proc. Fed. Am. Soc. Exp. Biol.* **38**:161–162.

Slag, M. F., Morley, J. E., Elson, M. K., Trence, D. L., Nelson, C. V., Nelson, A. E., Kinlow, W. B., Beyer, H. S., Nuttall, F. Q., Shafer, and R. B., 1983, Impotence in medical clinic outpatients, *JAMA* **249**(13):1736–1740.

Sparrow, D., Bosse, R., and Rowe, J. W., 1980, The influence of age, alcohol consumption, and body build on gonadal function in men, *J. Clin. Endocrinol. Metab.* **51**(3):508–512.

Swerdloff, R. S., and Heber, D., 1982, Effects of aging on male reproductive function, in: *Endocrine Aspects of Aging* (S. G. Korenman, ed.), Elsevier Biomedical, New York, pp. 119–136.

Verwoerdt, A., Pfeiffer, E., and Wang, H. S., 1969, Sexual behavior in senescence II. Patterns of sexual activity and interest, *Geriatrics* **24**:137–154.

II

AGE-RELATED CHANGES IN REGULATORY SYSTEMS

3

NEUROENDOCRINE CORRELATES OF AGING IN THE MALE

WILLIAM J. BREMNER, ALVIN M. MATSUMOTO, ROBERT A. STEINER, DONALD K. CLIFTON, and DANIEL M. DORSA

Prominent age-associated changes occur in the male reproductive system. With aging, normal men experience a gradual decrease in both sexual activity (Figure 1) and in the interest in sexual performance (Kinsey *et al.*, 1948). Despite the fact that occasional cases of fertility persisting in men in their nineties have been reported, fertility rates decline in men during the normal aging process (Harman, 1978). Similarly, in other mammals such as rats and nonhuman primates, various measures of male sexual activity and fertility decline with aging. Rates of contact with females, mounting, and ejaculation all decrease in normal old animals (Larsson and Essberg, 1962; Phoenix and Chambers, 1982).

In men, a series of studies have assessed whether testicular androgen production declines with age and, if so, whether the decline is due to decreased gonadotropin production or to a progressive impairment in testicular responsiveness (Vermeulen *et al.*, 1972; Stearns *et al.*, 1974; Baker *et al.*, 1976). Results of early studies involving single samples of blood or urine from small groups of nomal young and old men were about evenly divided on whether there is an age-related decline in testosterone levels. These inconclusive results

WILLIAM J. BREMNER and ALVIN M. MATSUMOTO • Endocrinology Section, Veterans Administration Medical Center, Seattle, Washington 98108. **ROBERT A. STEINER and DONALD K. CLIFTON** • Departments of Obstetrics and Gynecology, and Physiology and Biophysics, University of Washington School of Medicine, Seattle, Washington 98195. **DANIEL M. DORSA** • Gerontology Research, Education, and Clinical Center, Veterans Administration Medical Center, Seattle, Washington 98108.

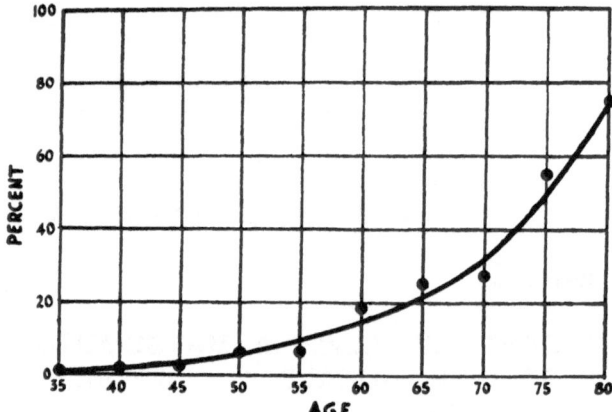

Figure 1. Percentage of total male population that is impotent in each age group. From Kinsey *et al.* (1948).

prompted the study of larger number of subjects, often with more than one sample per subject. Several large studies demonstrated a clear tendency for testosterone levels to decrease and luteinizing hormone (LH) levels to increase with increasing age of the subjects (Vermeulen *et al.*, 1972; Stearns *et al.*, 1974; Baker *et al.*, 1976; Harman, 1978). The fraction of testosterone, unbound to proteins, that circulates in plasma decreased even more than total testosterone levels since the capacity of plasma-binding proteins increased with age (Vermeulen *et al.*, 1972). Since the unbound fraction is thought to represent the physiologically active steroid, these studies seemed to establish that testicular androgen production does decrease with age. The fact that gonadotropin levels increased in older men implied that failure of the testes, rather than of the central nervous system or pituitary, was the cause of the decreased androgen levels.

More recently, these studies have been criticized on the grounds that many of the elderly subjects were relatively inactive nursing home residents or had coexistent illnesses and that the low androgen levels could have been due to inanition or illness. In one study in which the old men were carefully identified as healthy, there was no decline in testosterone levels with age (Harman and Tsitouras, 1980). But this study differed from previous ones in that the blood samples were obtained in the afternoon, near the low point in the normal circadian rhythm of blood testosterone levels. This difference suggested that the major change that occurs with aging could be a loss or circadian rhythmicity in serum testosterone levels, so that the morning rise found in normal young men is smaller or absent in healthy elderly subjects.

We studied the circadian rhythm of serum testosterone levels in 29 normal men, 17 young (age range 23–28 years, 25.2 ± 1.8, mean ± S.D.) and 12 old (age range 58–82 years, 71.0 ± 7.8, mean ± S.D.) (Bremner *et al.*, 1983). The subjects were paid volunteers, recruited through advertisements for normal men. The men were all within 15% of ideal body weight. Each was a non-smoker and nonabuser of alcohol and was receiving no medication. Normality was confirmed by medical history, physical examination, electrocardiogram, complete blood count, and urinalysis.

Each man was admitted to a clinical research center and allowed to adjust for 24 hr. On the second morning, hourly blood sampling through an indwelling peripheral venous cannula was begun at 0800 and continued until 0700 the next morning. Electroencephalographic monitoring was performed to confirm normal sleep patterns.

Total testosterone levels were measured in each serum sample by radioimmunoassay (Bremner *et al.*, 1981). Assay sensitivity was less than 10 pg/tube (less than 0.1 ng/ml). Intraassay and interassay variabilities were 5.1% and 9.8%, respectively. All specimens from each subject were measured in a single assay run. Analysis of variance and Duncan's multiple range test were used to determine differences between the groups at each time. Student's *t*-test and cosinor analysis (Nelson *et al.*, 1979) were used for other comparisons.

A clear circadian rhythm was seen in the serum testosterone levels of young men (Figure 2). Maximal levels occured at approximately 0800, minimal levels between 1900 and 2100 hr. In old men, the circadian rhythm was much less apparent (Figure 2). Mean testosterone levels were significantly higher in young men than in old men at each time point studied between 0200 and 1300. From 1400 to 0100, testosterone levels of the two age groups did not differ significantly, but the younger men tended to have higher levels. The circadian excursion of serum testosterone levels (highest point minus lowest point) was greater in the young men (3.5 ± 0.3 ng/ml, mean ± S.E.M.) than in the old men (2.1 ± 0.1, $p < 0.001$ by unpaired *t*-test). The amplitudes of the circadian rhythms determined by cosinor analysis were also significantly greater in the young men (1.6 ± 0.2 ng/ml, mean ± S.E.M.) than in the old men (0.7 ± 0.1, $p < 0.001$). The amplitude of a slight circadian rhythm ($p < 0.05$ by cosinor analysis), detectable in six of the elderly men, was markedly lower than in the young men.

Mean testosterone levels in the two groups of men did not differ significantly between midafternoon (1400) and late evening (0100), but testosterone values averaged throughout the 24-hr period were significantly higher in the young men (6.2 ± 1.3 ng/ml) than in the old men (5.1 ± 0.4, $p < 0.05$) by unpaired *t*-test.

These data demonstrate distinct differences between circulating testosterone levels in normal young and old men. Since the old men we recruited were physically and mentally active and had no coexistent disease, their lower testosterone levels cannot be attributed to inanition or illness. The differences in

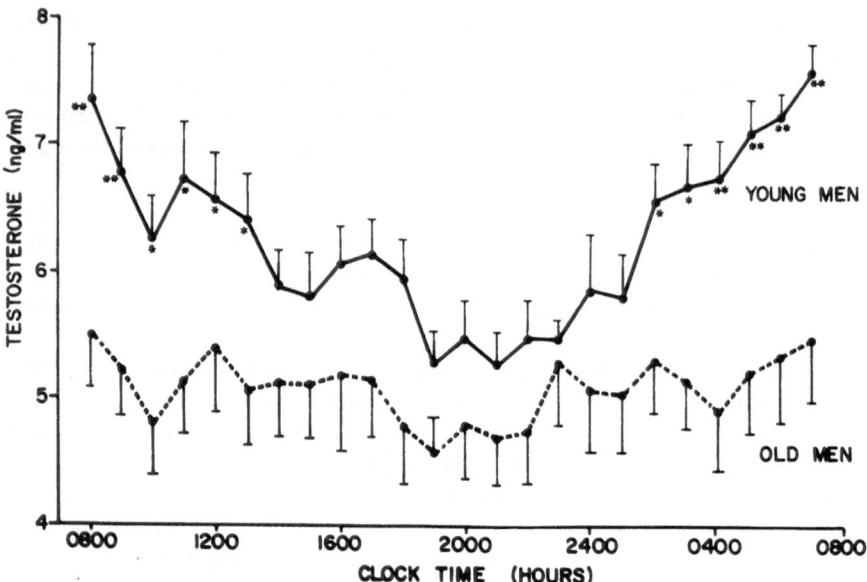

Figure 2. Hourly serum testosterone levels (mean ± S.E.M.) in normal young ($n = 17$) and old ($n = 12$) men. Blood samples were obtained using an indwelling peripheral venous cannula which allowed free movement and normal sleep. (*) $P < 0.05$ and (**) $P < 0.01$ for the significance levels of the differences between young and old men at each time point. Absence of an asterisk denotes that there was no significant difference at that time point. From Bremner et al. (1983).

mean testosterone levels and circadian patterns observed by us are most likely the consequences of normal aging in men.

Our results imply that the ability to demonstrate an age-related decrease in testosterone levels in men depends in part on the time of day blood sampling is performed. Samples obtained in the morning are most likely to demonstrate an age effect; those obtained in the afternoon or evening are least likely to differ between young and old men.

Marrama et al. (1982) on the basis of single blood samples obtained at 0800 and 1800 in normal young and older men also reported a loss of the circadian variation in testosterone levels with aging. Murono et al. (1982), however, reported persistence of a significant circadian rhythm in their elderly men based on single morning and afternoon blood samples. Other recent studies (Sparrow et al., 1980; Nieschlag et al., 1982) were unable to demonstrate a difference in testosterone levels between young and old men even though the single samples were obtained in the morning. But, even in these studies, serum LH and follicle-stimulating hormone (FSH) levels were higher in elderly men than in younger ones. Furthermore, the testosterone responsiveness to human chorionic gonadotropin (hCG) stimulation was less in older men (Nieschlag et *al.*, 1982). These findings appear to be consistent with a decrease in testicular endocrine function with age despite the fact that these two studies were unable to demonstrate lower basal testosterone levels in older men. Frequent blood sampling from each man over longer periods of time might have allowed demonstration of age-related changes in testosterone levels. Two large studies using morning blood samples obtained from healthy men over the entire age range from 20 to 88 years both demonstrated significant declines in testosterone levels and more marked decreases in free testosterone levels with aging (Purifoy *et al.*, 1981; Royer *et al.*, 1984). The decreasing testosterone levels with age are consistent with earlier observations of decreasing numbers of Leydig cells in the testes of older men (Kaler and Neaves, 1978).

Our finding of a loss in circadian testosterone rhythmicity with aging raises the question of whether this loss is due to age-related changes in gonadotropin production or to decreased testicular responsiveness to gonadotropins. Careful studies on circadian patterns of gonadotropin secretion and on testicular responses to gonadotropins with normal aging will be required to answer this question. The fact that gonadotropin levels are high suggests strongly that a decrease in testicular responsiveness to gonadotropins is an important part of aging in men. But our results demonstrating a loss of circadian rhythmicity imply that aging could also be associated with important changes in the mechanisms of the central nervous system that control circadian patterns in gonadotropin secretion.

We have recently found an important circadian variation in the pattern of LH secretion in normal young men. We studied seven healthy men, age range 21–44 years. Each man was admitted to a Clinical Research Center for 48 hr allowing 24 hr of adaptation prior to the initiation of studies. Electroencephalographic monitoring was performed to demonstrate normality of sleep. Serum samples were obtained from indwelling peripheral venous cannulas at 10-min intervals during the second 24 hr of the admission. LH levels were measured by radioimmunoassay (Bremner *et al.*, 1981). LH pulses were defined as increments in LH level of greater than 4 times the intraassay coefficient of variability. Interpulse intervals were defined as the time in minutes between LH pulses.

Mean LH interpulse intervals varied significantly in a circadian rhythm (statistical assessment by cosinor methodology; Nelson *et al.*, 1979). Interpulse intervals were usually greater in the evening and while the subjects were asleep than during the day (Figures 3 and 4). However, there was considerable variability among the subjects (as shown by the large standard error) with some men not showing a significant circadian variation in interpulse interval. There was also a tendency for the LH pulses to be of greater amplitude when the interpulse interval was greater. It is possible that the circadian pattern of LH secretion is important in controlling the circadian rhythm of testosterone. We are presently comparing the circadian pattern of LH secretion in normal young

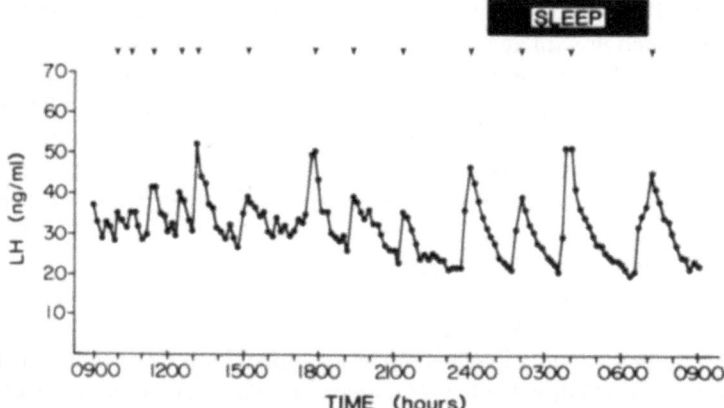

Figure 3. Example of LH levels measured every 10 min for 24 hr in a normal young man. Episodic LH secretion is less frequent in evening and during sleep than in the morning. Time is clock time. Arrows denote LH pulses defined as increments of greater than 4 times the intraassay coefficient of variation.

and old men to determine whether there may be an age-associated alteration in the pattern of gonadotropin secretion that could explain the loss of the testosterone circadian rhythm with aging.

An alternate explanation of the increased LH levels and decreased testosterone levels found in most studies of elderly men is that a different molecular

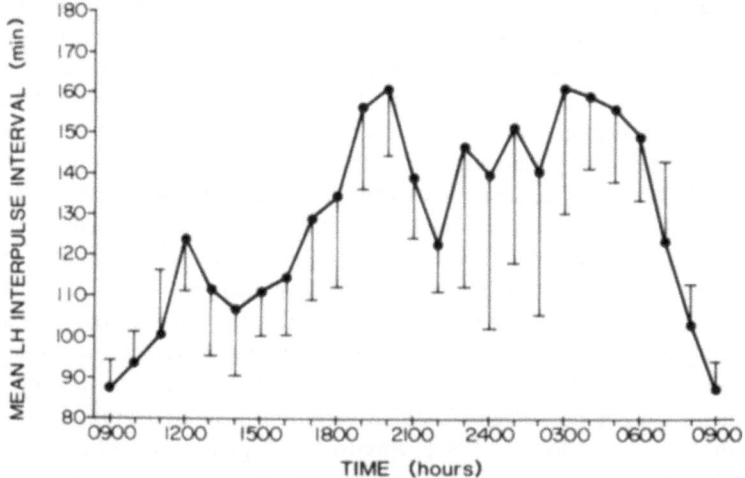

Figure 4. Mean ± S.E. LH interpulse intervals (interpulse interval is the time in minutes between LH pulses) for 24 hr in seven normal young men. Time is clock time.

form of LH is present in older men than younger ones. This different form of LH could have relatively low bioactivity compared to its immunoactivity and thereby not stimulate testosterone levels adequately despite being in high level by immunoassay. This possibility has been suggested by studies from two groups (Marrama et al., 1984; Warner et al., 1985). In preliminary studies of morning serum samples obtained from 11 healthy old men and 13 healthy young men, we have not found a significant difference in the ratio of bioactivity to immunoactivity. This ratio was 9.9 ± 1.2 (mean $\pm$ S.E.M.) in the young men and 8.5 ± 1.2 in the old men ($p = 0.42$; assay methodology in Bremner et al., 1981). However, further studies of this important issue are needed.

To investigate further the possible role of neuroendocrine events in aging of the mammalian reproductive system, we have initiated studies in rats. Rodent studies offer the advantage of easy access to adequate numbers of animals of advanced age and good health. We questioned whether an alteration in the pattern of LH secretion attends the aging process and is responsible for the decrease in testosterone levels found in normal old rats. In the first study (Steiner et al., 1984), we assessed the patterns of spontaneous LH and testosterone secretion in normal young (3 months) and old (22 months) Sprague–Dawley rats. Unanesthetized, freely mobile animals were bled at 10-min intervals for 8 hours through indwelling venous catheters. A blood replacement mixture was infused after each sample was obtained to maintain circulating blood volume. LH and testosterone levels were measured by radioimmunoassay. Mean testosterone levels were 50% lower in old animals compared to young ones ($p < 0.001$) The amplitude of LH pulses in older animals was markedly dereased compared to younger animals and the pulse frequency was also decreased.

In a second study (Karpas et al., 1983), we investigated the pattern of spontaneous LH secretion in castrated young (3 months), middle-aged (18 months) and old (26 months) Sprague–Dawley rats. Blood sampling was performed at 4-min intervals for 4 hr in unanesthetized, freely mobile animals. A distinct pulsatile pattern of blood LH levels was seen in all animals, and this varied with the age of the animals. Figure 5 depicts results in representative animals from the three different age groups. The mean blood LH levels were lower in the middle-aged and old animals than in the young. The LH interpulse interval was similar in the young and middle-aged animals, but considerably longer in the old rats. Figure 6 depicts the results of grouping data from all the animals of each age group. Mean LH level and LH pulse amplitude in the old and middle-aged groups were reduced to approximately 25 and 50%, respectively, of those observed in the young group. The mean interpulse interval was significantly lower in the old group than in the young (i.e., there was a 33% decrease in frequency of LH pulses with aging).

These studies demonstrate that LH pulse frequency and amplitude are markedly lower in old male rats than in younger animals. These results suggest

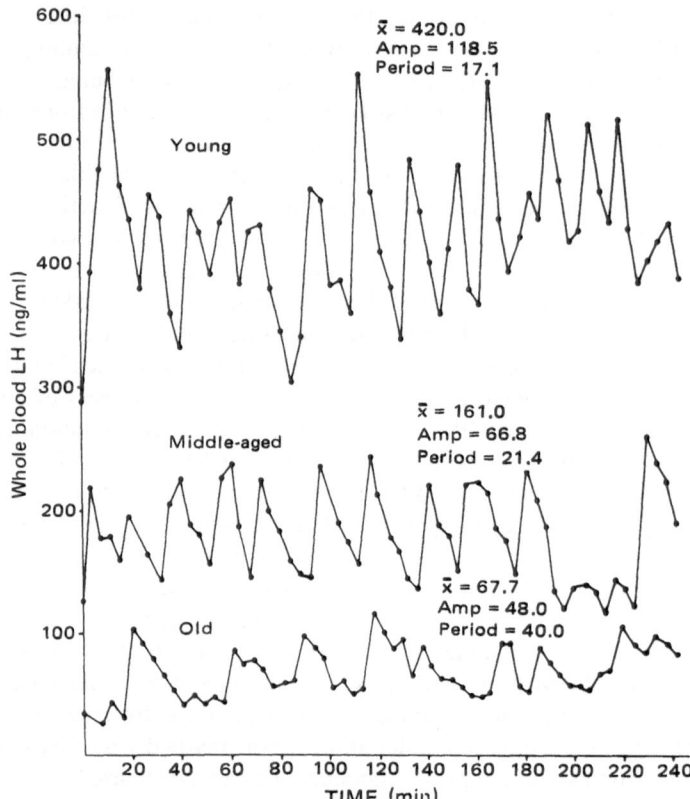

Figure 5. Typical patterns of LH levels in individual castrate male rats of three different ages: young (3 months), middle-aged (18 months) and old (26 months). From Karpas *et al.* (1983)

that the functional capacity of the luteinizing hormone-releasing hormone (LHRH) pulse generator in the central nervous system is compromised during the aging process. Coquelin and Desjardins (1982) have also found reduced pulsatile LH secretion in intact old male mice. Because in our studies this difference in LH secretory patterns between young and old rats was shown in the absence of gonadal steroid negative feedback, we believe that the age-dependent change is not simply the result of a differential testosterone negative feedback sensitivity between the groups. We have deduced that, in the rat, aging is accompanied by a progressive degeneration of the mechanisms governing pulsatile LH secretion, and we suggest that this accounts in part for the age-associated decrease in testicular function. Finding a lower LH pulse frequency in old animals implies the development of a major defect in the generating capacity of hypothalamic LHRH neurons. The reduced LH pulse amplitude found in old

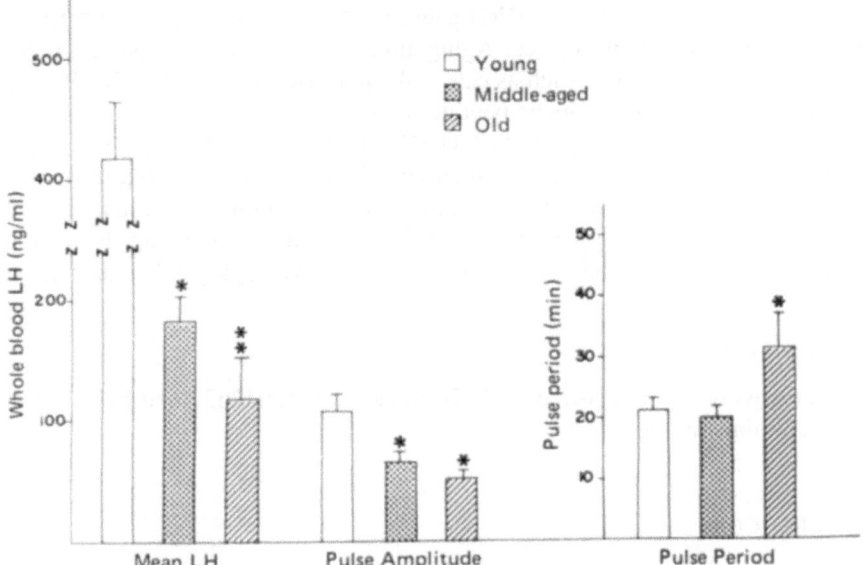

Figure 6. Mean LH levels, LH pulse amplitude, and interpulse interval in castrate male rats from the three age groups. Bars represent mean ± S.E.M. for young ($n = 8$), middle-aged ($n = 7$) and old ($n = 5$) age groups. (*) $p < 0.05$ compared to the young group; (**) $p < 0.01$ compared to the young group. From Karpas *et al.* (1983)

animals could be due either to diminution of LHRH secretion or to a reduction in pituitary responsiveness to LHRH (or to both).

The possibility that a defect in hypothalamic LHRH secretion is responsible for the degenerative changes in the patterns of LH secretion observed in old rats is reinforced by direct evidence provided by Miller and Reigle (1978), who demonstrated decreased levels of LHRH in the hypothalami of aged rats. Immunohistochemical evidence of LHRH neuronal deterioration with aging also bolsters the belief that LHRH secretion is compromised with aging (Hoffman and Sladek, 1980). However, Dorsa *et al.* (1984) have not confirmed a decrease in hypothalamic LHRH levels in certain groups of aging rats.

The degeneration of the LHRH pulse-generating system with aging may be related to a declining number of LHRH-secreting neurons. A report by Enright (1980) suggests, on theoretical grounds, that the inherent stability and performance of a neural oscillator depend on the number of pacemaker cells within the network. Therefore, the degenerative changes that we observed— markedly reduced frequency and amplitude—in the operating characteristics of the LHRH pulse generator of aged rats may have resulted from a loss of pacemaker cells within the LHRH oscillator.

The secretory activity of LHRH neurons is affected by various other transmitter and modulator substances. Aging affects the activity of important regulators of LHRH secretion, such as catecholamines (Meites et al., 1978; Ponzio et al., 1982), endogenous opiate peptides (Forman et al., 1981), and melatonin (Iguchi et al., 1982). Simpkins et al. (1981) and Miller and Reigle (1982) have reported that normal male rats lose their circadian rhythm in serum testosterone levels with aging; these results are similar to our data in men. Further study of the changes in the central nervous system that underlie rodent aging should add to our understanding of the neuroendocrine changes with aging in human beings.

ACKNOWLEDGMENTS

This work was supported by NIH grant P-50-HD 12629 and by the Veterans Administration.

References

Baker, H. W. G., Bremner, W. J., Burger, H. G., deKretser, D. M., Dulmanis, A., Eddie, L. W., Hudson, B., Keogh, E. J., Lee, V. W. K., and Rennie, G. C., 1976, Testicular control of follicle-stimulating hormone secretion, Rec. Prog. Horm. Res. 32:429–476.

Bremner, W. J., Matsumoto, A. M., Sussman, A. M., and Paulsen, C. A., 1981, Follicle stimulating hormone and human spermatogenesis, J. Clin. Investig. 68:1044–1052.

Bremner, W. J., Vitiello, M. V., and Prinz, P. N., 1983, A loss of circadian rhythmicity in blood testosterone levels with aging in normal men, J. Clin. Endocrinol. Metab. 56:1278–1281.

Coquelin, A., and Desjardins, C., 1982, Luteinizing hormone and testosterone secretion in young and old male mice, Am. J. Physiol. 243:E257–E263.

Dorsa, D. M., Smith, E. R., and Davidson, J. M., 1984, Immunoreative β-endorphin and LHRH levels in the brains of aged male rats with diminished sex behavior, Neurobiol. Aging 5:115–120.

Enright, J. J., 1980, Temporal precision in circadian system: A reliable plot from unreliable components, Science 209:1542–1545.

Forman, L. J., Sontag, W. E., VanVugt, D. A., and Meites, J., 1981, Immuno-reactive B-endorphin in the plasma, pituitary and hypothalamus of young and old male rats, Neurobiol. Aging 2:281.

Harman, S. M., 1978, clinical aspects of aging in the male reproductive system, in: The Aging Reproductive System (E. L. Schneider, ed.) Raven Press, New York, pp. 29–58.

Harman, S. M., and Tsitouras, P.D., 1980, Measurement of sex steroids, basal luteinizing hormone and Leydig cell response to human chorionic gonadotropin, J. Clin. Endocrinol. Metab. 51:35–40.

Hoffman, G. E., and Sladek, J. R., 1980, Age-related changes in dopamine, LHRH and somatostatin in the rat hypothalamus, Neurobiol. Aging 1:27.

Iguchi, H., Kato, K-I., and Ibayashi, H., 1982, Age-dependent reduction in serum melatonin concentrations in healthy human subjects, J. Clin. Endocrinol. Metab. 55:27–29.

Kaler, L. W., and Neaves, W. B., 1978, Attrition of the human Leydig cell population with advancing age, Anat. Rec. 192:513–518.

Karpas, A. E., Bremner, W. J., Clifton, D. K., Steiner, R. A., and Dorsa, D. M., 1983, Diminished LH pulse frequency and amplitude with aging in the male rat, Endocrinology 112:788–792.

Kinsey, A. C., Pomeroy, W. B., and Martin, C. E., 1948, *Sexual Behavior in the Human Male*, W. B. Saunders, Philadelphia.

Larsson, K., and Essberg, L., 1962, Effect of age on the sexual behavior of the rat, *Gerontologia* **6:**133–143.

Marrama, P., Carani, C., Baraghini, G. F., Volpe, A., Zini, D., Celani, M. F., and Montanini, V., 1982, Circadian rhythm of testosterone and prolactin in the aging man, *Maturitas* **4:**131–138.

Marrama, P., Montanini, V., Celani, M. F., Carani, C., Cioni, K., Bazzani, M., Cavani, D., and Baraghini, G. F., 1984, Decrease in luteinizing hormone biological activity/immunoreactivity ratio in elderly men, *Maturitas* **5:**223–231.

Meites, J., Huang, H. H., and Simpkins, J. W., 1978, Recent studies on neuroendocrine control of reproductive senescence in rats, in: *The Aging Reproductive System* (E. L. Schneider, ed.), Ravan Press, New York, pp. 213–235.

Miller, A. E., and Reigle, G. D., 1978, Hypothalamic LH releasing activity in young and aged intact and gonadectomized rats, *Exp. Aging Res.* **4:**145.

Miller, A. E., and Reigle, G. D., 1982, Temporal patterns of serum luteinizing hormone and testosterone and endocrine responses to luteinizing hormone releasing hormone in aging male rats, *J. Gerontol.* **37:**522–528.

Murono, E. P., Nankin, H. R., Lin, T., and Osterman, J., 1982, The aging Leydig cell. V. Diurnal rhythms in aged men, *Acta Endocrinol.* **99:**619–623.

Nelson, W., Tong, Y. L, Lee, J-K., and Malberg, F., 1979, Methods for cosinor-rhythmometry, *Chronobiologia* **6:**305–383.

Nieschlag, E., Lammers, U., Freischem, C. W., Langer, K., and Wickings, E. J., 1982, Reproductive function in young fathers and grandfathers, *J. Clin. Endocrinol. Metab.* **55:**676–681.

Phoenix, C. H., and Chambers, K. C., 1982, Sexual behavior in aging male rhesus monkeys, in: *Advanced Views in Primate Biology* (A. B. Chiarelli and R. S. Corrucini, eds.), Springer, Berlin, pp. 95–104.

Ponzio, F., Calderini, G., Lanusao, G., Vantim, G., Toffano, G., and Algieri, S., 1982, Changes in monoamines and their metabolite levels in some brain regions of aged rats, *Neurobiol. Aging* **3:**23.

Purifoy, F. E., Koopmans, L. H., and Mayes, D. M., 1981, Age differences in serum androgen levels in normal adult males. *Hum. Biol.* **53:**499–511.

Royer, G. L., Seckman, C. E., Schwartz, J. H., Bennett, K. P., and Hendrix, J. W., 1984, Relationship between age and levels of total, free, and bound testosterone in healthy subjects, *Curr. Ther. Res.* **35:**345–353.

Simpkins, J. W., Kalra, P. S., and Kalra, S. P., 1981, Alterations in daily rhythms of testosterone and progesterone in old male rats, *Exp. Aging Res.* **7:**25–32.

Sparrow, D., Basse, R., and Rowe, J. W., 1980, The influence of age, alcohol consumption and body build on gonadal function in man, *J. Clin. Endocrinol. Metab.* **51:**508–512.

Stearns, E. L., MacDonnell, J. A. Kaufman, B. J., Padua, R., Lucman, T. S., Winter, J. S. D., and Faiman, C., 1974, Declining testicular function with age: Hormonal and clinical correlates, *Am. J. Med.* **57:**761–766.

Steiner, R. A., Bremner, W. J., and Clifton, D. K., 1982, Regulation of luteinizing hormone pulse frequency and amplitude by testosterone in the male rat, *Endocrinology* **111:**2055–2061.

Steiner, R. A., Bremner, W. J., Clifton, D. K., and Dorsa, D. M., 1984, Reduced pulsatile luteinizing hormone and testosterone secretion with aging in the male rat, *Biol. Reprod.* **31:**251–258.

Vermeulen, A., Rubens, R., and Verdonck, L., 1972, Testosterone secretion and metabolism in male senescence, *J. Clin. Endocrinol. Metab.* **34:**730–735.

Warner, B. A., Dufau, M. L., and Santen, R. J., 1985, Effects of aging and illness on the pituitary testicular axis in men: Qualitative as well as quantitative changes in luteinizing hormone, *J. Clin. Endocrinol. Metab.* **60:**263–268.

4

REPRODUCTIVE FUNCTIONS IN GRANDFATHERS

EBERHARD NIESCHLAG and ERIK MICHEL

Introduction

Information on the reproductive functions of older men is scanty. The average age of men attending infertility clinics is around 30 years and only a few of these patients are older than 40 years. This indicates that the desire to procreate children is not very pronounced in advanced age and is, rather, restricted to younger ages. Moreover, only few children are born to marriages in which the husband is beyond the sixth decade of life; for example of all 568,000 children born in 1982 in the Federal Republic of Germany, only 142 were registered in marriages with husbands over 60 years old. This might be interpreted as an indication of reduced reproductive capacities, not only on the side of the female partner, but also on the male side. In addition, our society very often assumes that sexual activity is neither possible nor necessary for elderly men. ''From a psychosocial point of view the male over age 50 has to contend with one of the great fallacies of our culture. Every man in this age group is arbitrarily identified by both public and professional alike as sexually impaired'' (Masters and Johnson, 1981). Does this bias reflect physiology?

Although sexual activity in men decreases with advancing age, it by no means ceases (Kinsey *et al.*, 1948; Pfeiffer *et al.*, 1968; Nieschlag *et al.*, 1982; Tsitouras *et al.*, 1982; Davidson *et al.*, 1983). Testicular androgen and pituitary gonadotrophin secretion are not or only marginally influenced by old age (see

EBERHARD NIESCHLAG and ERIK MICHEL • Max Planck Clinical Research Unit for Reproductive Medicine and Department of Experimental Endocrinology, University Women's Hospital, D-4400 Münster, Federal Republic of Germany.

below), in any case not to an extent that would account for a cessation of sperm production and make infertility a physiological corollary of old age.

While hormone levels provided indirect evidence, firm data on reproductive functions of older men including seminal parameters were not available. MacLeod and Gold (1953) in their frequently cited study on "Semen quality in relation to age and sexual activity" only investigated men up to the age of 51 years. This dearth of data prompted us to investigate testicular functions in older men encompassing semen analysis as well as endocrine parameters. Special care was exerted in the selection of study subjects. In the first part of the study results from older men were compared with those from younger men (Nieschlag *et al.*, 1982); in the second part some of the older men were followed over 3 years.

Study Design and Volunteers

Initial Study

When investigating ejaculates and hormone levels from old men in order to establish representative values it would be desirable to exclude individuals with preexisting fertility disturbances. For our purposes old men who had just fathered children would have been best suited as volunteer subjects. As mentioned above, however, such cases are rare. To minimize the chance of preexisting infertility we recruited (with the help of a newspaper advertisement) 23 "vigorous grandfathers," i.e., men who had proven their fertility earlier in life. Although all men had children and all but one grandchildren, we were well aware that many years had passed between procreation and the present investigation and that fertility disturbances may have arisen during this interval. The 23 grandfathers were 60–88 years old (67 ± 7.8, S.D.). Twenty men were still married, two widowers, and one was separated from his wife. This distribution agrees approximately with that in the general population over 60 years in this country, of whom about 80% live with a spouse.

Twenty men, 24–37 years old (29.2 ± 3.2 S.D.), served as the control group. Each of them had fathered at least one child; most wives were still pregnant, and in the remaining cases wives had conceived not more than 4 years ago.

Although these grandfathers were deliberately not recruited from a hospital environment in order to exclude serious acute or chronic diseases, they were not entirely disease-free. Two had myocardial infarctions, 12 and 30 years previously; one had a stroke 11 years ago, and one had had a laryngectomy 19 years ago because of carcinoma. Eight had benign prostatic hypertrophy; seven were obese (at least 10% above normal weight); and four had coronary insufficiency.

All grandfathers claimed continued interest in sexual relations. On average, a frequency of intercourse episodes of 4.3 + 3.6 per month was reported compared with a frequency of 9.0 + 3.2 in the younger control group.

In young and old men basal blood samples were collected for hormone determinations. This was followed by a GnRH test (25 μ i.v.) and a human chorionic gonadotropin (hCG) stimulation test with 5000 I.U. injected on the same and the following day and blood samples being collected on the following and the 3rd day. All grandfathers and 11 young fathers followed this protocol.

All except one grandfather and all young fathers provided semen samples which were evaluated according to WHO guidelines (Belsey *et al.*, 1980). Sperm function of all young men and 16 grandfathers was assessed by the heterologous ovum penetration (HOP) test using zona pellucida-free hamster oocytes (Wickings *et al.*, 1983).

Follow-Up Study

Thirteen of the grandfathers were followed for 3 further years. In the 3rd year all 13 underwent the same endocrinological investigations as in the initial study. Basal and gonadotropin-releasing hormone (GnRH)-stimulated luteinizing hormone (LH) and follicle-stimulating hormone (FSH) serum concentrations of the young fathers as well as of the grandfathers were reassayed since the radioimmunoassay procedure for these hormones had been changed in our laboratory. Eight of the 13 older men also provided semen samples in the 2nd year of the follow-up period.

The general health of these 13 older men had deteriorated in so far as two had undergone prostatectomy, one had a myocardial infarction, and four now suffered from diabetes mellitus.

Spermatogenesis and Seminal Parameters

Testicular Volumes

Upon initial investigation, testicular volumes of grandfathers and young fathers were not different. The testicular volumes of the eight grandfathers had also not changed when reinvestigated three years later (Figure 1). This is in agreement with results from autopsies indicating no significant change of testicular volume with advancing age (Roessle and Roulet, 1932; Harbitz, 1973). Since the tubuli seminiferi constitute over 90% of the testis volume, the unaltered testis size can be considered as an indication that the seminiferous epithelium may remain intact in older men.

Semen Analysis

Overall, the seminal parameters from the older men were in good agreement with those from the younger men. Ejaculate volume and percentage of

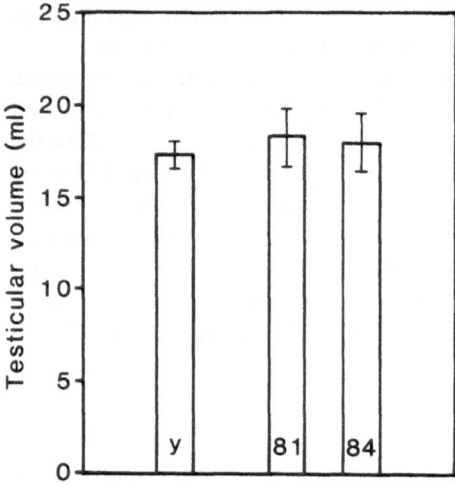

Figure 1. Testicular volumes of eight grandfathers followed up after 3 years (1981–1984) and providing semen samples compared with values of 20 young fathers (y) ($\bar{x}$ + S.E.M.).

normally formed sperm were not significantly different between younger and older men. Sperm density (but not total sperm count) was even higher in the older group. However, sperm motility and seminal fructose were significantly lower in the older men (Table I).

Table I

Testicular Volumes and Seminal Parameters in Young Fathers and Grandfathers (Mean ± S.D.)[a]

	Young men ($n = 20$)	Old men ($n = 22$)	P
Age (years)	29.2 ± 3.2	67.0 ± 7.8	
Age range (years)	24–37	60–88	
Testicular volume (ml)	17.3 ± 4.0	19.0 ± 5.7	NS[b]
Ejaculate volume (ml)	4.0 ± 1.7	3.2 ± 1.9	NS
Sperm density (million/ml)	78 ± 51	120 ± 101	0.05
Percentage progressively motile sperm	68 ± 14	50 ± 19	0.0005
Percentage normally formed sperm	52 ± 13	48 ± 9	NS
Fructose (mg/ml)	3.2 ± 1.1	1.7 ± 1.1	0.0005
pH	7.28 ± 0.09	7.25 ± 0.15	NS
HOP test (% penetrated ova)	54 ± 20	54 ± 18[c]	NS

[a] From Nieschlag et al. (1982).
[b] NS, not significant.
[c] $n = 16$.

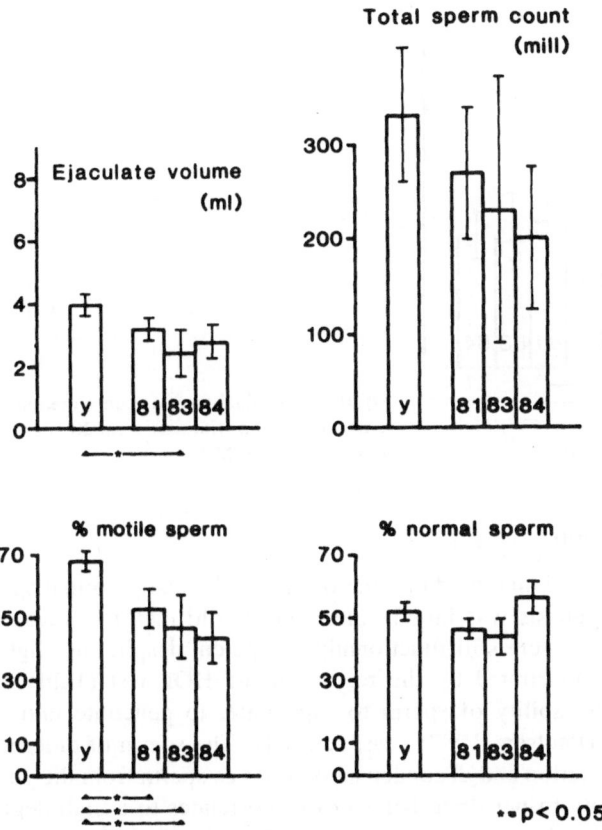

Figure 2. Seminal variables of eight grandfathers followed up after 2 and 3 years (1981, 1983, and 1984) compared with data from 20 young fathers (y) (1983: $n = 6$) ($\bar{x}$ + S.E.M.).

In those eight men who were reinvestigated 2 and 3 years later, no further deterioration occurred except for the ejaculate volume, which was at the time of the second investigation significantly smaller than in the young fathers (Figures 2 and 3).

The reduced percentage of progressively motile sperm can be interpreted as being a result of the lowered ejaculatory frequency in the older men, since sperm evaluated after prolonged periods of sexual abstinence show a gradual loss of motility. Lowered motility together with decreased seminal fructose concentration and decreased ejaculate volume could also be interpreted as a sign of lowered activity of the accessory sex glands, in particular of epidydimis and seminal vesicles (for which fructose is a functional indicator).

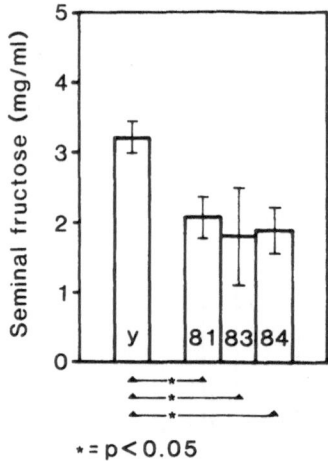

Figure 3. Seminal fructose concentration of eight grand-fathers compared with data from 20 young fathers (1983: $n = 6$) ($\bar{x}$ + S.E.M.).

* = p < 0.05

Sperm Function

However, the unaltered number of morphologically normal sperm indicate that spermatogenesis was largely intact in the old men investigated. The fact that these sperm were still functionally competent despite the slightly lowered motility is demonstrated by the results of the HOP test (Table I). This test investigates the ability of sperm to capacitate, to penetrate into ova, and to decondensate (Paulsen, 1983; Prasad, 1984). The sperm of older men showed the same penetration and decondensation rates as sperm from the young fathers. Since the ability to penetrate hamster ova correlates to a high degree with the ability to fertilize human eggs *in vitro* (Wolf *et al.*, 1983; Belkien *et al.*, 1985), it can be concluded from these results that not only number and morphological appearance but also functional integrity of sperm is maintained in relatively healthy old men.

Histology

Although the grossly unaltered seminal parameters in the investigated volunteers allowed us to conclude that spermatogenesis must have been normal, it would still have been of interest to investigate testicular histology. However, it is not justifiable to subject normal volunteers to testicular biopsies. In general, the interpretation of histological findings of testes from older men is hampered by the fact that it is almost impossible to obtain testes or biopsies from healthy subjects. Knowledge about the morphology of testes in advanced age is therefore based on autopsy material (Bürgi and Hedinger, 1956; Suoranta, 1971; Schlüter, 1978) or on testes removed in the course of treatment of prostatic carcinoma (Honoré, 1978; Holstein and Hubmann, 1981; Schulze, 1982) or inguinal hernia repair (Honoré, 1978). Reproductive history and possible fertility disturbances in such cases are not known.

In many cases the histological appearance of testes obtained from old men even under such pathological conditions is remarkably similar to that obtained from younger subjects. In 27 of 102 men over 90 years old testis morphology was perfectly normal (Schlüter, 1978). In other cases, however, alterations exist ranging from isolated anomalies to full testicular atrophy. On occasions, both normal and atrophic tubuli seminiferi are seen side by side in one and the same testis (Bürgi and Hedinger, 1956; Schlüter, 1978). Signs of atrophy and hyalinization may be found along with arteriosclerotic alterations of the blood vessels (Suoranta, 1971). In some cases abnormal variants of type A pale spermatogonia and an unusual multilayered arrangement of spermatogonia have been observed, a phenomenon rarely seen in younger men (Holstein and Hubmann, 1981). Schulze (1982) found generally decreased spermatogenic activity in older men. This allowed waves of spermatogenesis to be recognized, as seen in other mammalian species, but not in younger men.

In summary, although various histological alterations have been observed in testes of older men, none of these alterations appear to be typical of old age and these alterations can also be found in younger men. Since these findings derived predominantly from diseased men whose fertility status and reproductive history are unknown, it remains unclear whether these alterations are a consequence of the diseased status or a sign of old age per se. It should be emphasized that in quite a number of cases no histological alterations were found at all. From the histology and the seminal parameters available it is fair to conclude that testicular involution does not occur physiologically as a corollary of old age. Thus, in contrast to the gonadal atrophy occurring in women around menopause, there is no morphological basis for a male climacteric.

Endocrine Findings

High intratesticular testosterone concentrations are a prerequisite for normal spermatogenesis, and testosterone secreted from the testes is responsible for the development and maintenance of secondary male sex characteristics and for the function of the accessory reproductive glands. Therefore, evaluation of the endocrine testicular function and of the pituitary gland in control of the testis is of importance when assessing reproductive capacities of elderly men. With the advent of radioimmunoassays for steroids in the early 1970's, sensitive methods for the measurement of sex hormones in serum became available, and they were swiftly applied to the investigation of possible age-dependent alterations. As later studies showed, the selection of study subjects was of prime importance. In the first series of investigations, reconvalescent older patients,

patients from geriatric wards, and men from homes for the elderly were investigated. Our own group was among the first to perform such a study on men in a home for senior citizens (Nieschlag *et al.*, 1973). The conclusions drawn from this and other studies were later modified when the importance of selecting study subjects became apparent (Harman and Tsitouras, 1980; Sparrow *et al.*, 1980; Nieschlag *et al.*, 1982).

Basal Serum Testosterone

The first studies yielded the following picture. Serum testetosrone concentrations decrease slowly from the sixth decade on (Vermeulen *et al.*, 1972; Nieschlag *et al.*, 1973; Stearns *et al.*, 1974). In parallel, a slow decrease in the testosterone concentration in the spermatic vein was found (Serio *et al.*, 1977). A decrease in the testosterone production rate was also observed (Vermeulen *et al.*, 1972) as well as a decrease in the biologically active, free serum testosterone concentration. The latter phenomenon was explained by an estrogen-caused increase in sex hormone-binding globulin (SHBG) (Pirke and Doerr, 1973; Kley *et al.*, 1974).

These findings are contrasted by more recent investigations performed not on hospitalized or institutionalized patients, but on relatively healthy men still dwelling in their usual home and family environment. Harman and Tsitouras (1980) and Sparrow *et al.* (1980) investigated old men with exceptionally good health and found no decrease in serum testosterone concentrations. Similarly, the grandfathers from our study (Nieschlag *et al.*, 1982), who were free of acute diseases but were not characterized by exceptionally good health, during the first investigation showed neither a decrease in serum testosterone nor an increase in serum estradiol compared with the young fathers. Thus it appears that the general health condition of the investigated subjects influences testicular function insofar as "healthier" men show no decrease in serum testosterone.

However, these results did not remain unchallenged. Some authors concluded that the differences between the various studies might have been caused by different times of blood sampling since they found a difference in the amplitude of the diurnal serum testosterone variations (Bremner *et al.*, 1983). Zumoff *et al.* (1982) concluded that the differences might be caused by too few samples not allowing for representative values.

Our follow-up study may be of importance in this controversy. The endocrine investigations were performed using the same protocol on two occasions 3 years apart. While the serum testosterone concentrations of the 13 men who were followed up were not different from the younger control group on the first occasion, the serum testosterone was significantly below that of the control group 3 years later (Figure 4). Their health condition had also deteriorated. Two now suffered from angina pectoris; one had a myocardial infarction; and two had had a prostatectomy. While on the first occasion only one older man suffered

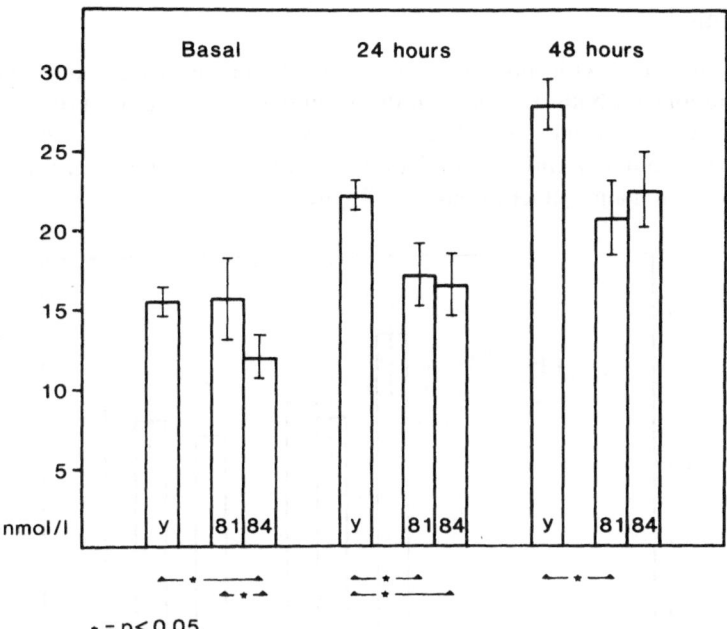

Figure 4. Serum testosterone response to hCG stimulation in 11 young fathers (y) and 13 grand-fathers on two occasions (1981 and 1984) ($\bar{x}$ + S.E.M.).

from diabetes mellitus, four did so 3 years later. Furthermore, while on the first occasion 22 of the 23 grandfathers could provide a semen sample, five of the 13 men followed up could not do so 3 years later. These findings indicate that the general health of these men had certainly deteriorated over the 3 years, and the decreased serum testosterone concentrations can be interpreted as a sign of this decline.

Endocrine Reserve Capacity of the Testis

Earlier and recent studies agree in the observation that the endocrine reserve capacity of the testes under hCG stimulation is reduced in older men (Nieschlag *et al.*, 1973; Rubens *et al.*, 1974; Harman and Tsitouras, 1980; Nankin *et al.*, 1981; Nieschlag *et al.*, 1982). Although basal serum testosterone levels decreased in the grandfathers in our study 3 years, the levels obtained after 2 days of hCG stimulation were identical on both occasions; yet in both instances they remained below the maximum values achieved by younger men (Figure 4). This agrees with the morphological finding that although the volume of the existing Leydig cells does not change, a decrease in total Leydig cell number occurs as a function of age (Kaler and Neaves, 1978). The decreased capacity appears to be a characteristic of old age.

Gonadotropins

In agreement with most other studies we found an increase in basal FSH concentrations on both occasions in the 13 grandfathers followed over 2 years compared to the younger men. The values stayed significantly above those from the young men under GnRH stimulation, indicating that this partial function of the pituitary remains intact in old age (Figure 5).

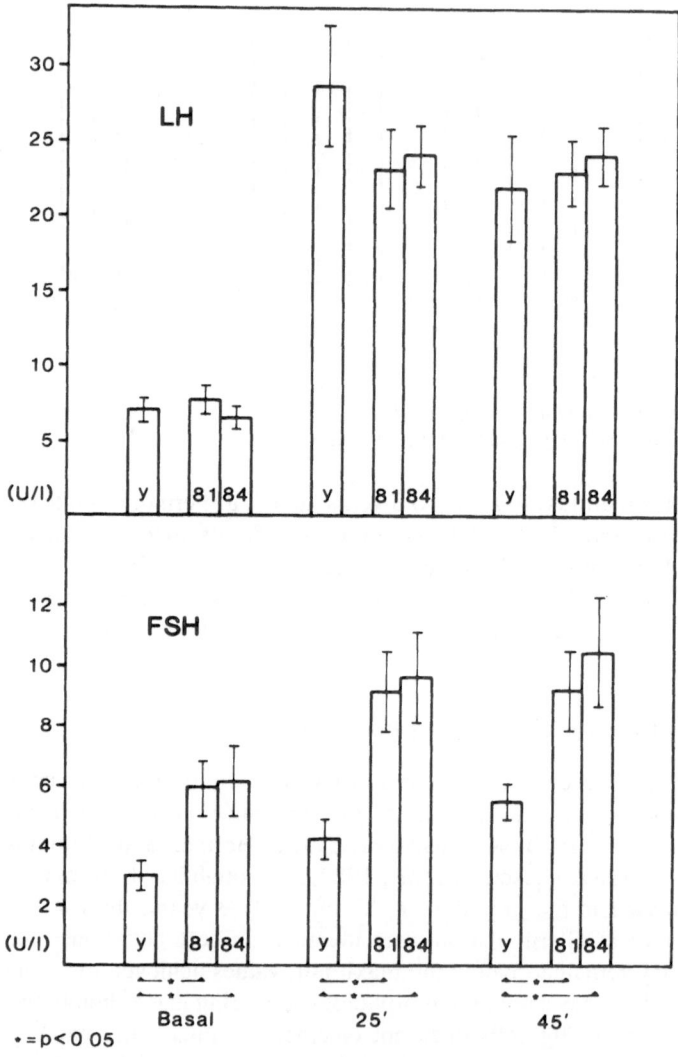

Figure 5. Pituitary response to 25 μg GnRH stimulation in 11 young fathers (y) and 13 grandfathers on two occasions (1981 and 1984) ($\bar{x}$ + S.E.M.).

Since during the day LH shows considerable variations in a pulsatile fashion, single LH determinations are of limited value. In the GnRH stimulation test a marginal yet significant decrease in the LH response has been observed in three studies including our own (Harman *et al.*, 1982; Nieschlag *et al.*, 1982; Winters and Troen, 1982). When 13 grandfathers were reinvestigated the LH responses were not different from the first occasion (Figure 5). Thus a slight decrease in the pituitary reserve capacity also appears to be a typical sign of old age.

Correlation between Endocrine Findings and Sexual Activity

As far as reported in the studies on endocrine testicular function in old age, a decrease of sexual activity was noted with advancing age. However, a correlation between serum testosterone levels and sexual activity could not be found (Nieschlag *et al.*, 1982; Tsitouras *et al.*, 1982). Such a correlation could not be found in our follow-up study, nor was there a correlation with the endocrine reserve capacity of the testes. The lack of such a correlation may be due to the relatively small sample size, since Davidson *et al.* (1983), who studied 193 men between the ages of 50 and 93, found a correlation between serum testosterone levels and decreasing sexual activity. For the individual patient, however, it appears that such a correlation cannot be established, as long as the serum testosterone levels remain above the range seen in hypogonadal patients.

Conclusion

A review of the available studies on testicular functions in old age provides neither biochemical nor morphological evidence for a physiological process in men that could be compared to the female climacteric. Although discrepancies exist between the studies, all investigators agree that the alterations in testicular and pituitary function observed in senescence occur over long periods of time and remain subtle compared to the sudden and profound changes in gonadal function during female menopause. The *physiological* changes seen in male senescence never reach the extent seen in clinically manifest hypogonadism (which does not exclude the possibility that hypogonadism may occur in old age as a *pathological* finding). While women lose their reproductive capacities during the menopause, sustained androgen and sperm production indicate that impotence and infertility are not a corollary of advancing age in men.

References

Belkien, L., Bordt, J., Freischem, C. W., Knuth, U. A., Schneider, H. P. G., and Nieschlag, E., 1985, Prognostic value of the heterologous ovum penetration test for human in vitro fertilization, *Int. J. Androl.* **8**:270–280.

Belsey, M. A., Eliasson, R., Gallegos, A. J., Moghissi, K. S., Paulsen, C. A., and Prasad, M.

R. N., 1980, *Laboratory Manual for the Examination of Human Semen and Semen-Cervical Mucus Interaction*. Press Concern, Singapore, 43 pp.

Bremner, W. J., Vitiello, M. V., and Prinz, P. N., 1983, Loss of circadian rhythmicity in blood testosterone levels with aging in normal men, *J. Clin. Endocrinol. Metab.* **56**:1278–1281.

Bürgi, H., and Hedinger, Chr., 1956, Histologische Hodenveränderungen im hohen Alter, *Schweiz. Med. Wschr.* **47**:1236–1239.

Davidson, J. M., Chen, J. J., Crapo, L., Gray, G. D., Greenleaf, W., and Catania, J. A., 1983, Hormonal changes and sexual function in aging men, *J. Clin. Endocrinol. Metab.* **57**:71–77.

Harbitz, T. B., 1973, Testis weight and the histology of the prostate in elderly men, *Acta Path. Microbial. Scand.* Section A **81**:148–158.

Harman, S. M., and Tsitouras, P. D., 1980, Reproductive hormones in aging men. I. Measurement of sex steroids, basal luteinizing hormone, and Leydig cell response to human chorionic gonadotropin, *J. Clin. Endocrinol. Metab.* **51**:35–40.

Harman, S. M., Tsitouras, P. D., Costa, P. T., and Blackman, M. R., 1982, Reproductive hormones in aging men. II. Basal pituitary gonadotropins and gonadotropin response to luteinizing hormone-releasing hormone, *J. Clin. Endocrinol. Metab.* **54**:547–551.

Holstein, A. F., and Hubmann, R., 1981, Spermatogonia in old age, *Fortschr. d. Androl.* **7**:77–88.

Honoré, L. H., 1978, Aging changes in the human testis: A light-microscopic study. *Gerontology* **24**:58–65.

Kaler, L. W., and Neaves, W. B., 1978, Attrition of the human Leydig cell population with advancing age. *Anat. Rec.* **192**:513–518.

Kinsey, A. C., Pommeroy, W. B., Martin, C. E., and Gebhardt, P. H., 1948, *Sexual Behavior in the Human Male*, Saunders, Philadelphia.

Kley, H. K., Nieschlag, E., Bidlingmaier, F., and Krüskemper, H. L., 1974, Possible age-dependent influence of estrogens on the percentage binding of testosterone in adult men, *Horm. Metab. Res.* **6**:213–215.

MacLeod, J., and Gold, R. Z., 1953, The male factor in fertility and infertility. VII. Semen quality in relation to age and sexual activity, *Fertil. Steril.* **4**:194–209.

Masters, W. H., and Johnson, V. E., 1981, Sex and the aging process, *J. Am. Geriatr. Soc.* **29**:385–390.

Nankin, H. R., Lin, T., Murono, E. P., and Osterman, J., 1981, The aging Leydig cell. III. Gonadotropin stimulation in men, *J. Androl.* **2**:181–189.

Nieschlag, U., Kley, H. K., Wiegelmann, W., Solbach, H. G., and Krüskemper, H. L., 1973, Lebensalter und endokrine Funktion der Testes des erwachsenen Mannes, *Dtsch. Med. Wschr.* **26**:1281–1284.

Nieschlag, U., Lammers, E., Freischem, C. W., Langer, K., and Wickings, E. J., 1982, Reproductive function in young fathers and grandfathers, *J. Clin. Endocrinol. Metab.* **55**:676–681.

Paulsen, C. A., 1983, Another look at the sperm penetration assay, *Fertil. Steril.* **3**:302–304.

Pfeiffer, E., Verwoerdt, A., and Wang, H. S., 1968, Sexual behavior in aged men and women, *Arch. Gen. Psychiatry* **19**:753–758.

Pirke, K. M., and Doerr, P., 1973, Age related changes and interrelationships between plasma testosterone, oestradiol and testosterone-binding globulin in normal adult males, *Acta Endocrinol.* **74**:792–800.

Prasad, M. R. N., 1984, The in vitro sperm penetration test: A review, *Int. J. Androl.* **7**:5–22.

Roessle, R., and Roulet, F., 1932, Mass und Zahl in der Pathologie, Springer, Berlin.

Rubens, R., Dhont, M., and Vermeulen, A., 1974, Further studies on Leydig cell function in old age, *J. Clin. Endocrinol. Metab.* **39**:40–43.

Serio, M., Cattaneo, S., Borrelli, D., Gonelli, P., Pazzagli, M., Forti, G., Fiorelli, G., Giannotti, P., and Giusti, G., 1977, Age-related changes in androgenic hormones in human spermatic venous blood, in: *Androgens and Antiandrogens* (L. Martini and M. Motta, eds.), Raven Press, New York, pp. 67–75.

Sparrow, D., Bosse, R., and Rowe, J. W., 1980, The influence of age, alcohol consumption, and body build on gonadal function in men, *J. Clin. Endocrinol. Metab.* **51**:508–512.

Schlüter, D., 1978, *Die endokrinen Organe der Über-Neunzigjährigen*, Dissertation, Universität Hamburg.

Schulze, W., 1982, Evidence of a wave of spermatogenesis in human testis, *Andrologia* **14**:200–207.

Stearns, E. L., MacDonnell, J. A., Kaufmann, B. J., Padua, R., Lucman, T. S., Winter, J. S. D., and Faiman, C., 1974, Declining testicular function with age, *Am. J. Med.* **57**:761–766.

Suoranta, H., 1971, Changes in the small blood vessels of the adult human testis in relation to age and to some pathological conditions, *Virchows Arch. Abt. A Path. Anat.* **352**:165–181.

Tsitouras, P. D., Martin, C. E., and Harman, S. M., 1982, Relationship of serum testosterone to sexual activity in healthy elderly men, *J. Gerontol.* **37**:288–293.

Vermeulen, A., Rubens, R., and Verdonck, L., 1972, Testosterone secretion and metabolism in male senescence, *J. Clin. Endocrinol Metab.* **34**:730–735.

Wickings, E. J., Freischem, C. W., Langer, K., and Nieschlag, E., 1983, Heterologous ovum penetration test and seminal parameters in fertile and infertile men, *J. Androl.* **4**:730–735.

Winters, S. J., and Troen, P., 1982, Episodic luteinizing hormone (LH) secretion and the response of LH and FSH to LH-releasing hormone in aged men: Evidence for coexistent primary testicular insufficiency and an impairment in gonadotropin secretion, *J. Clin. Endocrinol. Metab.* **55**:560–565.

Wolf, D. P., Sokoloski, J. E., and Quigley, M. M., 1983, Correlation of human in vitro fertilisation with the hamster egg bioassay, *Fertil. Steril.* **40**:53–59.

Zumoff, B., Strain, G. W., Kream, J., O'Connor, J., Rosenfeld, R. S., Levin, J., and Fukushima, D. K., 1982, Age variation of the 24-hour mean plasma concentrations of androgens, estrogens, and gonadotropins in normal adult men, *J. Clin. Endocrinol. Metab.* **54**:534–538.

5

AGING OF ADRENAL CORTEX AND OVARY

MORTIMER B. LIPSETT

Introduction

The early impetus to study hormone secretion in aging resulted from attempts to understand the failing libido noted in many older men. The topic has been investigated and reinvestigated as methods improved, from crude bioassays to crude steroid measurements (the 17-ketosteroids) arriving finally at our present level of sophistication in hormone analysis. Nevertheless, even the semiquantitative bioassays and the gross group measurements of steroids revealed the basic outlines of aging in the adrenal cortex and ovary.

Adrenal Cortex

In 1955, Pincus *et al.* (1955) summarized cross-sectional studies of 17-ketosteroid (17-KS) excretion with age. Although the origins of the 17-KS could not be appreciated then, they drew the correct conclusion that adrenal androgens must be involved. They demonstrated that there was a linear decline in 17-KS excretion with age, the rate of fall being about the same in men and women. They also showed that urinary corticosteroid excretion changes very little with age.

These data are unexceptional and may now be interpreted as follows: Cortisol secretion is maintained at a level necessary to support a normal plasma cortisol, but the secretion of dehydroepiandrosterone sulfate (DHAS) dehydroepiandrosterone (DHA), the principal 17-KS, shows an incommensurate fall

MORTIMER B. LIPSETT • National Institute of Child Health and Human Development, National Institutes of Health, Bethesda, Maryland 20205.

with age leading to increased cortisol/DHA ratios as shown by Zumoff *et al.* (1980) (Figure 1). Since the negative feedback system for cortisol remains intact throughout life, is the regulation of secretion of DHA an intraadrenal phenomenon or are there additional adrenal androgen stimulating factors?

This is not an easy problem to study since only man and the great apes secrete large amounts of DHA. Nevertheless, Parker and Odell (1979) produced evidence that there was an androgen-stimulating hormone and subsequently Parker *et al.* (1983) isolated from the pituitary gland a glycopeptide of molecular weight $\sim 60,000$ that stimulated DHA synthesis in the dispersed dog adrenal cell system with lesser changes in cortisol synthesis. Albertson *et al.* (1984) confirmed this hypothesis by examining DHA/cortisol ratios in sham-operated chimpanzees and in hypophysectomized chimpanzees replaced with ACTH. Although plasma cortisol levels were maintained at approximately equal levels in both groups of animals, the ACTH-treated animals had significantly lower DHA outputs, again signifiying the presence of something other than ACTH in the pituitary (Figure 2). Recently Hauffa *et al.* (1984) compared the plasma DHA, DHAS, and androstenedione levels in children with Cushing's disease with those of normal children. In spite of putatively increased ACTH levels and other proopiomelanocorticotropin-related peptides, the adrenal androgen levels remained appropriate for the patients' chronologic age, again pointing to a non-ACTH factor as responsible for the andrenal androgen secretion. Thus it seems possible that the phenomenon of decreased DHA production in aging may be aging of the pituitary-hypothalamic system rather than of the adrenal cortex.

There are two pathways to cortisol synthesis, one via the Δ^5-steroid, 17-hydroxyprogesterone. In general, plasma levels of the Δ^5-steroids decrease more with age than the Δ^4-steroids (Meldrum *et al.*, 1981) and are stimulated less by

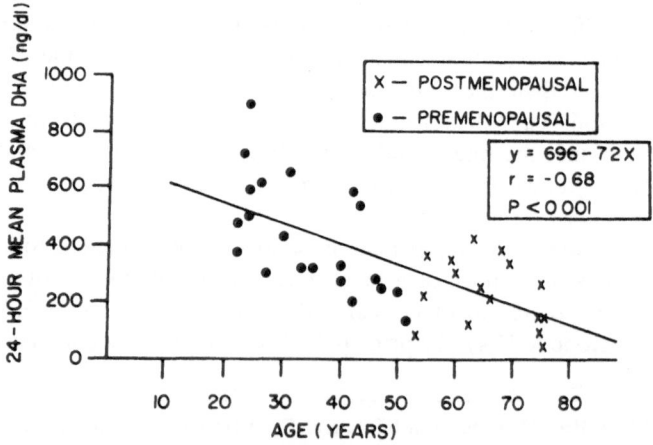

Figure 1. Twenty-four-hr mean plasma DHA concentration in normal women. From Zumoff *et al.* (1980), with permission.

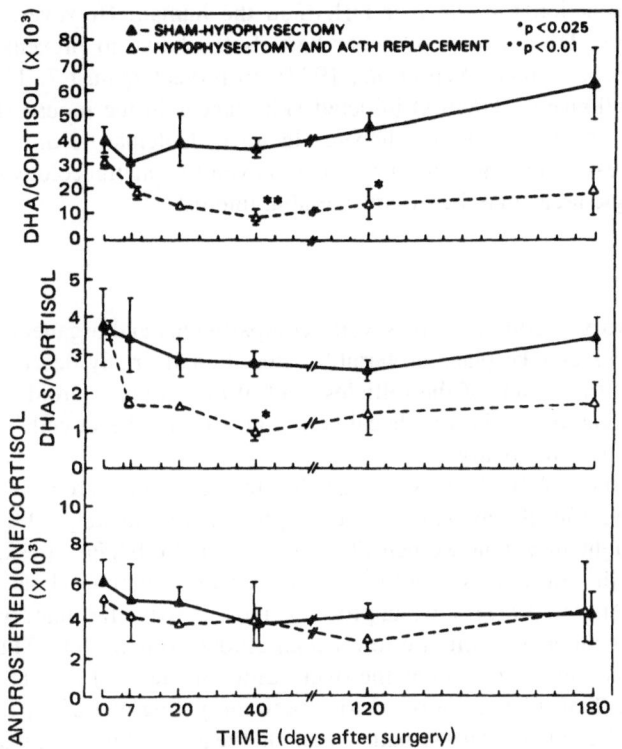

Figure 2. Plasma DHA to F, DHAS to F, and Δ^4 to F ratios during 3-hr ACTH infusions in hypophysectomized, ACTH-replaced chimpanzees and sham-hypophysectomized controls.* $P <$ 0.01 (compared to control animals). From Albertson *et al.* (1984), with permission.

ACTH than are the Δ^4-steroids (Vermeulen *et al.*, 1982). Thus the putative adrenal androgen stimulating factor is apparently necessary for regulation of the Δ^5-pathway.

Cumming *et al.* (1982) suggested that early oophorectomy led to a premature decline in DHAS levels and Lucky *et al.* (1979) presented evidence of an effect of estrogens on plasma DHA. In a more comprehensive study, Lobo *et al.* (1982) showed that estrogen replacement therapy that brought the plasma free estradiol to normal raised the plasma concentration of DHAS. From their measurements of other adrenal androgens, it was suggested that this effect of estradiol was greater with the Δ^5-3β-ol androgens as had been previously suggested. Thus there is still another possible regulatory effect that may contribute to the aging of the adrenal cortex as measured by DHAS.

It might be well to ask the question—why bother about DHA? Is it just an index of aging either of the adrenal cortex or of a still incompletely characterized function of the pituitary-hypothalamic unit? No one has been able to demonstrate a function for DHA or DHAS in the human. However, in the last several years, DHA and a congener have been shown to prevent obesity in genetically obese mice (Yen *et al.*, 1977), to protect against 7, 12-demethyl-benz [α] anthracene (DMBA)-initiated skin cancer in the mouse (Paschko *et al.*, 1984), and to ameliorate diabetes in mice (Coleman *et al.*, 1983). The doses of DHA were large, but nevertheless even the pharmacology of DHA in these biologic areas may be of considerable interest.

Ovary

The ovary is composed of several compartments and organelles, and age-related processes take place at variable rates. I shall briefly describe evidence for aging of the ovum, of the follicles, and of the various steroid-synthesizing cells, the granulosa cells, the thecal cells, the stromal thecal cells, and the interstitial cells of the ovary.

The ovary itself changes in size due to the progressive loss of ovarian follicles, and after the menopause the weight is approximately 50% that of the ovarian weight in young women (Talbert, 1978). Follicular atresia begins in fetal life with a rapid loss of follicles from several million to about 700,000 at birth (Zuckerman, 1951). Atresia proceeds more slowly after that, but 10 years before the menopause there are fewer than 20,000 follicles left. Thus, ovarian aging begins almost as soon as the ovary differentiates and continues past the menopause. Transplantation of ovaries between young and aging rats demonstrated that the young ovary remains young in the aged rat proving that the aging of the ovary is primary and not due to hypothalamic-pituitary malfunction (Blaha, 1970) (Krohn, 1962).

The follicle itself ages and the process of atresia may be considered as aging and death of the follicle. Atresia has been the subject of extensive investigations (Schwartz and Hunzicker-Dunn, 1981). When enzymes and other activities associated with growth are measured during atresia, all are decreased. However, lysosomal enzyme activity is increased as one might predict with all dissolution. Some have speculated that follicular atresia is programmed and others that androgens or other substances positively induce the processes of atresia.

Just before the menopause, plasma FSH levels show inappropriate elevations although plasma steroid concentrations are within the normal range (Sherman *et al.*, 1976). This may be attributed to a decreasing secretion of follicular "inhibin" and is perhaps another indication of follicular aging.

Estrogen

The contribution of the ovary to the hormonal milieu after the menopause has been clarified during the past decade. Catherization of ovarian veins has permitted direct assessment of ovarian secretions. However, the uncertainties of estimating blood flow makes quantitation difficult. Adrenal suppression with exogenous glucocorticoids has also been used to give meaningful data on ovarian secretions. Finally, the changes in blood hormone levels after ovariectomy in the postmenoupausal woman give a good estimate of ovarian activity.

After the last functional follicle becomes atretic (a few apparently nonfunctional follicles may be seen after the menopause) (Costoff and Mahesh, 1975), there is an abrupt fall in plasma estrogen concentration. Estradiol (E_2) decreased from 100–150 pg/ml to about 13 pg/ml. Estrone (E_1) becomes the prominent extrogen averaging about 35 pg/ml as compared with 50 to 100 pg/ml before the menopause.

Catheterization of the ovarian veins on postmenopausal women shows a still detectable secretion of E_1 and E_2 (Judd, 1976). However, when these women were castrated the fall in E_1 and E_2 plasma levels was questionably measurable. Thus, there is little direct secretion of estrogens from the postmenopausal ovary. Similarly Longcope (1971) found that the gradients for E_2 and E_1 were so low that using an estimated blood flow, direct secretion of E_2 and E_1 could not be more than 20 and 8 μg per day, respectively.

An unequivocal estimate of ovarian contribution to plasma estrogens may be gained by measuring plasma E_1 and E_2 levels following oophorectomy. Judd (1976) summarized his results in 16 postmenopausal women before and after oophorectomy and found essentially no differences in plasma E_1 and E_2 concentrations.

The source of estrogens after the menopause then becomes of interest. The adrenal cortex secretes little or no estrone or estradiol. Rather, these steroids enter the blood after peripheral aromatization of androstenedione (Grodin et al., 1973) (Longcope et al., 1980). The estradiol is almost certainly derived from the estrone since the limited conversion of testosterone to E_2 could not account for plasma estradiol levels.

Aromatization of androstenedione to E_1 occurs in many tissues, muscle, fat, kidney, liver, and brain. There is clear evidence that the rate of conversion of androstenedione to E_1 increases with body weight (Judd et al., 1980) (Casey and MacDonald, 1983), presumably as a function of fat cell activity (Figure 3). Thus, the exposure of a postmenopausal woman to estrogens will depend on the secretion of androstenedione by the ovaries and adrenal glands and the rate of aromatization of androstenedione.

Androstenedione

The interstitial cells of the ovary and stromal thecal cells remain active after the menopause. There is an easily measurable gradient for testosterone between the effluent ovarian veins and the systemic circulation (Judd, 1976). Androstenedione is secreted to a lesser extent (Judd, 1976).

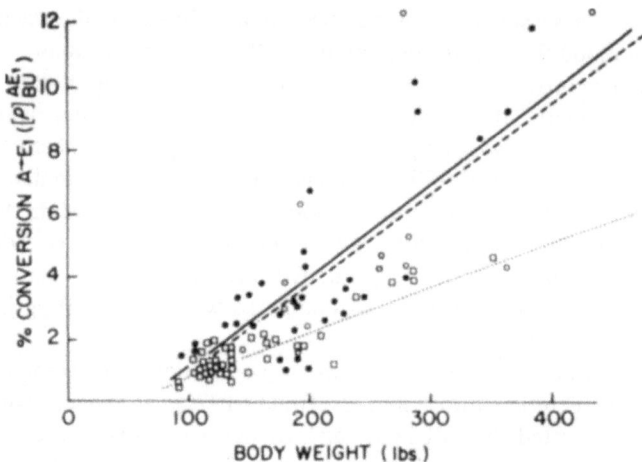

Figure 3. Conversion of androstenedione to estrone in relation to body weight. (□) Premenopausal women; (○) Postmenopausal women, normal; (●) with endometrial cancer. From Casey and MacDonald (1983), with permission.

The contribution of the ovary to the androstenedione production rate is small since ovariectomy effects very little change in plasma androstenedione levels (Judd, 1976). When the adrenal cortex is either stimulated by ACTH or suppressed with dexamethasone (Judd, 1976; Vermeulen, 1976), the androstenedione levels drop by about 70%.

Plasma testosterone is low after the menopause, values being in the range of 150 pg/dl as compared to about 400 pg/dl in the menstruating woman. The ovarian secretion of testosterone is higher than that of androstenedione and probably accounts for most of the testosterone in blood (Vermeulen, 1976). Conversion of androstenedione to testosterone in peripheral tissue would be of little significance.

None of these ovarian activities has been followed through the aging process. The recent reports on the efficacy of estrogens in slowing bone loss after the menopause emphasize the importance of the continued supply of estrogen precursors primarily from the adrenals.

References

Albertson, B. D., Hobson, W. C., Burnett, B. S., Turner, P. T., Clark, R. V., Schiebinger, R. J., Loriaux, D. L., and Cutler, G. B., Jr., 1984, Dissociation of cortisol and adrenal androgen secretion in the hypophysectomized, adreno-corticotropin-replaced chimpanzee, *J. Clin. Endrocrinol. Metab.* **59:**13–18.

Blaha, G. C., 1970, The influence of ovarian grafts from young donors on the development of transferred ova in aged golden hamsters, *Fertil. Steril.* **21:**268–273.

Casey, M. L. and MacDonald, P. C., 1983, Origin of estrogen and regulation of its formation in postmenopausal women, in *The Menopause* (H. J. Buchsbaum, ed.), Springer-Verlag, New York, pp. 1-8.

Coleman, D. L., Leiter, E. H., and Schwizer, R. W., 1983, Therapeutic effects of dehydroepiandrosterone (DHEA) in diabetic mice, *Diabetes* **31**:830-833.

Costoff, A., and Mahesh, V. B., 1975, Primordial follicles with normal oocytes in the ovaries of postmenopausal women, *J. Am. Geriatr. Soc.* **23**:193-197.

Cumming, D. C., Rebar, R. W., Hopper, B. R., and Yen, S. S. C., 1982, Evidence for an influence of the ovary on circulating dehydroepiandrosterone sulfate levels, *J. Clin. Endocrinol. Metab.* **54**:1069-1071.

Grodin, J. M., Siiteri, P. K., and MacDonald, P. C., 1973, Source of estrogen production in postmenopausal women, *J. Clin. Endocrinol. Metab.* **36**:207-214.

Hauffa, B. P., Kaplan, S. L., and Grumbach, M. M., 1984, Dissociation between plasma adrenal androgens and cortisol in Cushing's disease and ectopic ACTH-producing tumor: Relation to adrenarche, *Lancet* **1**:1373-1376.

Judd, H. L., 1976, Hormonal dynamics associated with the menopause, *Clin. Obstet. Gynecol.* **19**:775-788.

Judd, H. L., Davidson, B. J., Freeman, A. M., Shamonk, I. M., Lagasse, L. D., and Ballon, S. C., 1980, Serum androgens and estrogens in postmenopausal women with and without endometrial cancer, *Am. J. Obstet. Gynecol.* **136**:859-871.

Krohn, P. L., 1962, Heterochronic transplantation in the study of ageing, *Proc. R. Soc. Lond. (Biol.)* **157**:128-147.

Lobo, R. A., March, C. M., Goebelsmann, O., and Mishell, D. R., Jr., 1982, The modulating role of obesity and 17 β-estradiol (E_2) on bound and unbound E_2 and adrenal androgens in oophorectomized women, *J. Clin. Endocrinol. Metab.* **54**:320-324.

Longcope, C., 1971, Metabolic clearance rates of estrogens in postmenopausal women, *Am. J. Obstet. Gynecol.* **111**:778-781.

Longcope, C., Hunter, R., and Franz, C., 1980, Steroid secretion by the postmenopausal ovary, *Am. J. Obstet. Gynecol.* **138**:564-568.

Lucky, A. W., Marynick, S. P., Rebar, R. W., Cutler, G. B., Glen, M., Johnsonbaugh, E., and Loriaux, D. L., 1979, Replacement oral ethynylestradiol therapy for gonadal dysgenesis: Growth and adrenal androgen studies, *Acta Endocrinol. (Copenh)* **91**:519-524.

Meldrum, D. R., Davidson, B. J., Tataryn, I. V., and Judd, H. L., 1981, Changes in circulating steroids with aging in postmenopausal women, *Obstet. Gynecol.* **57**:624-628.

Parker, L. N., and Odell, W. D., 1979, Evidence for existence of cortical androgen-stimulating hormone, *Am. J. Physiol.* E616-E620.

Parker, L. N., Lifrak, E. T., and Odell, W. D., 1983, A 60,000 molecular weight human pituitary glycopeptide stimulate adrenal androgen secretion, *Endocrinology* **113**:2092-2096.

Paschko, L. L., Rovito, R. J., Williams, J. R., Sobel, E. L., and Schwartz, A. G., 1984, Dehydroepiandrosterone (DHEA) and 3 β-methylandrost-5-en-17-one: inhibitors of 7,12-demethylbenz [a]-anthracene (DMBA)-initiated and 12-O-tetradecanoylphorbol-13-acetate (TPA)-promoted skin papilloma formation in mice, *Carcinogenesis* **5**:463-466.

Pincus, G., Dorfman, R. I., Romanoff, L. P., Rubin, B. L., Bloch, E., Carlo, J., and Freeman, H., 1955, Steroid metabolism in aging men and women, *Recent Prog. Hormone Res.* **11**:307-334.

Schwartz, N. B., and Hunzicker-Dunn, M., 1981, *Dynamics of Ovarian Function*, Raven Press, New York, Chapters 1-7.

Sherman, M. B., West, J. H., and Korenman, S. G., 1976, The menopausal transition: Analysis of LH, FSH, estradiol, and progesterone concentrations during menstrual cycles of older women, *J. Clin. Endocrinol. Metab.* **42**:629-636.

Talbert, G. B., 1978, Effect of aging of the ovaries and female gametes on reproductive capacity, in *The Aging Reproductive System*, Vol. 4 (E. L. Schneider, ed.), Raven Press, New York.

Vermeulen, A., 1976, The hormonal activity of the postmenopausal ovary, *J. Clin. Endocrinol. Metab.* **42:**247–253.

Vermeulen, A., Deslypere, J. P., Schelfhrout, W., Verdonck, L., and Rubens, R., 1982, Adrenocortical function in old age. Response to acute adrenocorticotropin stimulation, *J. Clin. Endocrinol. Metab.* **54:**187–191.

Yen, T. T., Allan, J. V., Pearson, D. V., Acton, J. M., and Greenberg, M., 1977, Prevention of obesity in A^{vy}/a mice by dehydroepiandrosterone, *Lipids* **12:**409–413.

Zuckerman, S., 1951, The number of oocytes in the mature ovary, *Recent Prog. Hormone Res.* **6:**63–84.

Zumoff, B., Rosenfeld, R. S., Strain, G. W., Levin, J., and Fukushima, D. K., 1980, Sex differences in the twenty-four-hour mean plasma concentrations of dehydroisoandrosterone and dehydroisoandrosterone sulfate (DHAS) and the DHA to DHAS ratio in normal adults, *J. Clin. Endocrinol. Metab.* **51:**330–333.

6

CHANGES IN THE CENTRAL NERVOUS SYSTEM AND NEUROENDOCRINE CONTROL OF REPRODUCTION IN MALES AND FEMALES

PHYLLIS M. WISE

Introduction

Normal reproductive function in male and female mammals depends upon complex interactions among the brain, the anterior pituitary gland, and the gonad. Aging of this reproductive axis is associated with changes in each component. At this time, there is no general agreement among reproductive endocrinologists as to which of these changes are of primary importance and which are secondary effects of changes at another level of the reproductive axis. I will discuss the results of studies that demonstrate that alterations occur in the central nervous system (CNS), particularly in the hypothalamus, early during the aging process. These studies support the hypothesis that neurochemical and neuroendocrine changes may have important cascading repercussions at other points in the reproductive axis that ultimately result in infertility.

Investigations into the role of CNS in the aging of the reproductive system are only relevant if it is clearly established that the CNS plays an important regulatory role in normal reproductive function when animals are young. A large body of literature supports the concept that (1) the CNS, and in particular the hypothalamus, provides the neurochemical signals that regulate the synthesis and release of luteinizing hormone-releasing hormone (LHRH) and ulti-

PHYLLIS M. WISE • Department of Physiology, School of Medicine, University of Maryland, Baltimore, Maryland 21201.

mately the secretion of luteinizing hormone (LH) and follicle-stimulating hormone (FSH); (2) neural signals are influenced by steroidal milieu and therefore feedback control occurs via the CNS; and (3) cyclic neural inputs are the basis of cyclic gonadotropin release in females. Several neurotransmitters have been implicated in the control of gonadotropin release in young animals. The majority of work has concentrated on the effects of norepinephrine, dopamine, and serotonin on gonadotropin release and steroid feedback to these neurotransmitters. Our research has concentrated on the neurochemical and neuroendocrine regulation of cyclic release of gonadotropins and aging females. Therefore I will focus primarily on the role of these three neurotransmitters in the aging female reproductive system.

Role of Norepinephrine, Dopamine, and Serotonin in Gonadotropin Release in Young Animals

Norepinephrine (NE) plays an important stimulatory role in cyclic LH release. The following lines of evidence demonstrate that NE stimulates LH release. First, NE, when infused intraventricularly, triggers an ovulatory surge of LH (Krieg and Sawyer, 1976; Sawyer et al., 1974; Vijayan and McCann, 1978). Second, progesterone-induced LH surges are blocked by diethyldithiocarbamate, an inhibitor of NE synthesis (Kalra et al., 1972). Finally, NE turnover rates increase in specific hypothalamic nuclei coincident with LH surges on proestrus (Honma and Wuttke, 1980; Lofstrom, 1977; Rance et al., 1981a) and in estradiol- and estradiol–progesterone-treated rats (Honma and Wuttke, 1980; Wise et al., 1981). In contrast, NE turnover rates are low when LH surges are suppressed in phentobarbital-blocked (Rance and Barraclough, 1981), prolactin-treated (Wise, 1983a), or androgen-sterilized rats (Lookingland et al., 1982). Steroids feed back and influence NE concentrations and activity (Lofstrom et al., 1977; Wise et al., 1981).

α- and β-adrenergic receptors are thought to mediate the stimulatory and inhibitory effects of NE, respectively. Clonidine and phenylephrine, α agonists, stimulate LH release in ovariectomized steroid-primed rats (Leung et al., 1982a) and cause the reappearance of pulsatile LH release in ovariectomized aged rats (Estes and Simpkins, 1982). Isoproterenol, a β agonist, has no effect or inhibits steroid-induced LH surges (Caceres and Taleisnik, 1980; Leung et al., 1982b). Steroids influence the ratio of α/β receptors in several brain areas (Greenberg and Weiss, 1983) and thereby could influence neuroendocrine events. Noradrenergic activity may directly influence synthesis and transport and/or release of LHRH into hypophysial portal blood (Sarkar and Fink, 1981). Alternatively, NE may modulate the action of other neurotransmitters (Baraban and Aghajanian, 1981; Mansky et al., 1982) known to affect LHRH release.

Dopamine (DA) may inhibit LH release. Elevated DA neuronal activity in the median eminence and/or the medial preoptic nucleus has been associated with the lack of LH surges in 5-day cycling rats (Lofstrom, 1977), steroid-treated animals (Rance et al., 1981b) and constant-light illuminated or andro-gen-sterilized acyclic rats (Fuxe et al., 1972). DA turnover rates (Lofstrom, 1977; Rance et al., 1981a) and DA concentrations in hypophysial portal blood (Ben-Jonathan et al., 1977) decrease at the time of the preovulatory LH surge. DA antagonists increase LHRH concentrations in portal blood; this effect is reversed by the presence of apomorphine, a DA receptor agonist (Sarkar and Fink, 1981).

Considerable controversy exists concerning the role of serotonin (5-HT) in cyclic LH release. Initial data suggested that 5-HT may inhibit LH secretion. Direct infusion into the third ventricle or drug-induced increases in 5-HT levels (Kordon and Glowinski, 1972; Lippman, 1968; Schneider and McCann, 1970) were correlated with inhibition of LH release and/or ovulation. More recently investigators have come to appreciate the potential importance of the circadian rhythm in hypothalamic 5-HT concentration (Quay, 1964; 1968; Hery et al., 1972) and the role that this rhythm may play in the effect of 5-HT on LH release. Thus, 5-HT may enhance or block LH surges depending upon whether it amplifies or blunts its circadian rhythm (Walker, 1980, 1983). Depletion of 5-HT blocks proestrous (Walker, 1980) and steroid-induced LH surges (Coen and MacKinnon, 1979; Hery et al., 1976, 1978). 5-hydroxytryptophan, which temporarily restores 5-HT synthesis, reinstates the LH surge if administered at certain times of day (Coen and MacKinnon, 1979; Hery et al., 1976). Serotonin may regulate LH secretion by directly affecting LHRH metabolism and release directly (Jennes et al., 1982; Leonardelli et al., 1974).

Effects of Aging on Norepinephrine, Dopamine, and Serotonin Dynamics: Correlations with Changes in Reproduction

Norepinephrine

Alterations in NE activity are apparent using a variety of experimental methods in old animals. First, norepinephrine concentrations are diminished in old rodents in several brain areas, including the hypothalamus (Estes and Simp-kins, 1980; Finch, 1973; Simpkins et al., 1977; Sun, 1976), the organum vas-culosum of the lamina terminalis and the preoptic area (Estes and Simpkins, 1980), and the brain stem and spinal cord (Algeri et al., 1983). Sun (1976) reported that catecholamine-storing vesicles appear to be less dense in nerve terminals of old males. Second, uptake of NE in the hypothalamus (Sun, 1976) and turnover rates are diminished in old males (Finch, 1973; Simpkins et al., 1977) and are altered in acyclic females (Estes and Simpkins, 1984). Third, NE-induced cAMP accumulation in the cerebral cortex is diminished in old males (Berg and Zimmerman, 1975). This correlates well with the decreased β-adrenergic receptor concentrations observed by Greenberg and Weiss (1983) in the cerebral cortex as well as in the pineal gland and hippocampus of old constant diestrus females. Finally, administration of NE or precursors systemically (Huang *et al.*, 1976; Quadri *et al.*, 1973) or directly into the preoptic area (Cooper *et al.*, 1979) reinitiates cyclicity or ovulation in old, previously acyclic females. Drugs that elevate catecholamine concentrations (Clemens and Bennett, 1977; Clemens *et al.*, 1969; Quadri *et al.*, 1973) or stimuli thought to act via catecholaminergic mechanisms (Clemens *et al.*, 1969; Everett, 1940, 1943; Huang *et al.*, 1976) can reinitiate cyclicity in old female rats. Electrical stimulation of the preoptic area, which may bypass any catecholaminergic deficiency and directly stimulate LHRH release, results in ovulation in old constant estrus rats (Clemens *et al.*, 1969).

The animals used in the above studies were between 18–30 months old. The data provide compelling evidence that changes in NE activity exist in old animals in which reproductive abilities have deteriorated. However, these studies do not indicate whether such neurochemical changes initiate the cascade of events that result in infertility or whether they occur after the transition to infertility is complete and, therefore, are secondary to changes in other components of the reproductive axis.

We have recently examined NE turnover rates in middle-aged rats that are entering the transition to estrous acyclicity under a variety of controlled endocrine conditions. We observed that preovulatory LH surges are delayed in onset and attenuated in amplitude when rats are between 7–9 months of age (Figure 1) (Wise, 1982a). To determine whether the changes in the pattern of proestrous LH release are due to changes at the hypothalamic neurotransmitter level, we measured NE turnover rates in key brain areas involved in the regulation of LHRH dynamics (Wise, 1982b). We examined neurotransmitter dynamics in the suprachiasmatic nucleus and the medial preoptic nucleus because both of these are thought to be involved in cyclic LH release (Brown-Grant and Raisman, 1977). In particular, the suprachiasmatic nucleus is considered a putative circadian oscillator that influences rhythms in a variety of physiological systems (Rusak and Zucker, 1979). In addition, these areas are among the rostral areas that contain cell bodies of LHRH-containing neurons; therefore, they are possible sites of contact between NE and LHRH. We also monitored NE turnover rates in the arcuate nucleus and the median eminence because these areas are primarily involved with the basal release of LHRH. Terminals of LHRH-containing neurons reside in the median eminence, where the releasing hormone is secreted and transported via hypophysial portal blood vessels to the anterior pituitary gland. We examined turnover rates (1) in the morning prior to the anticipated LH surge, (2) during the "critical period" when neural signals for LH surges would be expected, (3) during the afternoon when peak LH concen-

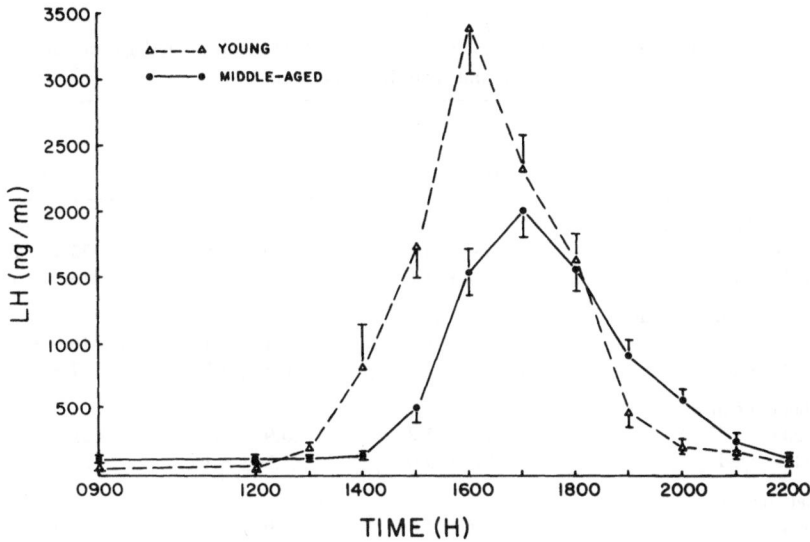

Figure 1. Proestrous LH concentrations in young (△- - -△) and middle-aged (●———●) rats. Animals were bled sequentially from right atrial cannulae. Points represent mean ±S.E. Baseline LH concentrations prior to the surge were the same in middle-aged and young rats. In young rats, the first significant increase in plasma LH was at 1300 hr, peak concentrations were attained between 1600–1700 hr and returned to baseline by 2200 hr. In contrast, the surge in middle-aged animals began 1 hr later and peaked between 1600–1900 hr. Hormone concentrations were significantly lower in middle-aged animals from 1300–1700 hr but remained elevated longer and were significantly higher than in young rats at 2000 hr. From Wise (1982a).

trations occur, and (4) during the evening when LH concentrations return to baseline concentrations. In addition, these hypothalamic areas and times were chosen because previous studies have demonstrated that in young rats, NE turnover rates increase in all of these areas during the afternoon concomitant with the LH surge (Rance *et al.*, 1981a).

The results of these studies demonstrate that selective changes in NE turnover rates occur in the suprachiasmatic and medial preoptic nuclei of middle-aged proestrus rats (Table I). The expected increase in NE turnover rates in the median eminence and arcuate nucleus occurred in middle-aged rats at the same time during the afternoon as in young rats. In contrast, we observed no increase during the afternoon in either of the two hypothalamic areas, which are thought to be involved in cyclic LH release. We believe these data demonstrate that changes in NE dynamics occur during the middle-aged period, before the establishment of the acyclic state. This strongly suggests that such changes are not merely symptoms of the loss of reproductive competence. Hypothalamic alterations may, in fact, contribute to the continuing deterioration of the repro-

Table I

Norepinephrine Turnover Rates (pg/μg Protein/hr) in Microdissected Brain Areas of Young and Middle-Aged Rats[a]

	Time (hr)		
	0900–1100	1200–1400	1500–1700
Medial preoptic nucleus			
Young	16.4 ± 4.1	28.0 ± 3.6[b]	33.9 ± 6.6[b]
Middle-aged	14.5 ± 2.5	12.7 ± 3.3	13.3 ± 1.8
Suprachiasmatic nucleus			
Young	1.3 ± 1.7	7.1 ± 4.1	16.0 ± 3.4[b]
Middle-aged	3.6 ± 2.7	10.3 ± 2.2	7.5 ± 2.5
Median eminence			
Young	8.6 ± 3.2	20.9 ± 2.4[b]	35.7 ± 4.9[b]
Middle-aged	1.6 ± 3.0	23.2 ± 3.1[b]	23.9 ± 4.1[b]
Arcuate nucleus			
Young	10.2 ± 2.8	11.9 ± 1.7	20.2 ± 2.0[b]
Middle-aged	0.5 ± 2.5	10.7 ± 2.9[b]	5.6 ± 2.7

[a] Taken from Wise (1982b).
[b] Significantly different from turnover rate in the morning (0900–1100 hr).

ductive system. Furthermore, because the changes are limited to the two hypothalamic nuclei involved in cyclic LH release and biological rhythms, they raise the intriguing possibility that changes in the integrity of the biological clock may initiate changes in the ability of the female reproductive system to cycle in the precisely timed manner characteristic of the young animal. Many systems exhibit altered or suppressed circadian rhythms in old age. Our data suggest that changes in the biological clock may begin early during the aging process and may cause gradual loss of temporal integration of multiple physiological events. The female reproductive system is highly dependent upon the close timing of complex and interactive neuroendocrine events. Alterations in the biological clock or its ability to be entrained by environmental cues would be expected to influence deleteriously the female estrous cycle within a short period of time.

Several other aspects of noradrenergic activity are altered in middle-aged rodents. In the hypothalamus of young mice, the activity of dopamine-β-hydroxylase, the enzyme that converts dopamine to norepinephrine, increases during proestrus. In contrast, in 16- to 20-month old animals, enzyme activity does not show any significant proestrous rise (Banerji *et al.*, 1982). Greenberg and Weiss (1983) reported age-related changes in the preovulatory pattern of hypothalamic α-adrenergic binding sites. Using the α-adrenergic receptor antagonist prazosin they found that in young rats, α-receptor concentrations increase late on proestrus and during early estrus. No such increase could be detected in middle-aged rats exhibiting prolonged or irregular cycles.

The ability of estradiol to feed back positively is compromised in middle-aged females (Lu *et al.*, 1981; Mobbs *et al.*, 1984; Steger *et al.*, 1980; Wise, 1984). We found that physiological concentrations of estradiol, administered to ovariectomized young and middle-aged rats, stimulate significantly different LH surges (Wise, 1984). In young rats, estradiol induces maximal LH surges in all rats within 2 days of estradiol administration. Middle-aged rats require 3 days of exposure before a maximal LH response can be elicited. Even at these times, the LH surge is delayed and attenuated compared to that in young rats (Figure 2). We have correlated the change in the pattern of LH release with alterations in the pattern of NE turnover rates in hypothalamic nuclei. In young rats, the estradiol-induced LH surge is accompanied by an increase in NE turnover rates in the medial preoptic nucleus, the suprachiasmatic nucleus, and the median eminence. Middle-aged rats exhibited no diurnal change in the medial preoptic nucleus or suprachiasmatic nucleus. The age-related alteration in NE turnover rates during estradiol-induced LH surges (Wise, 1984) is strikingly similar to that observed in proestrous rats (Wise, 1982b). Thus, the altered ability of estradiol to facilitate proestrous or steroid-induced LH surges may be related to its inability to induce a specific pattern of NE turnover rates prerequisite to normal LH release.

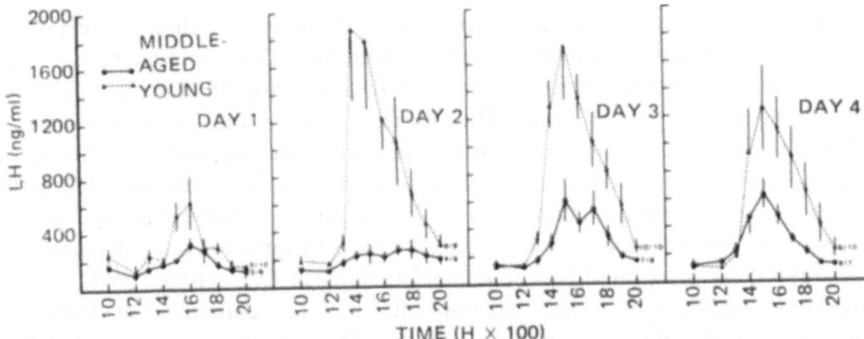

Figure 2. Plasma LH concentrations in ovariectomized estradiol-treated young (▲- - -▲) and middle-aged (●———●) rats on days 1–4 after implantation of estradiol-containing Silastic capsules on day 0. The fraction at the end of each LH profile is the number of rats that displayed an LH surge per total number of rats bled. Virtually all young rats exhibited an LH surge in response to the presence of estradiol on days 2, 3, and 4. LH concentrations were significantly elevated by 1300 hr and peaked between 1500–1600 hr each day. In contrast, middle-aged rats required the presence of estradiol for 3 days before maximal LH surges were observed in the majority of animals. LH surges were delayed and attenuated in this age group. From Wise (1984).

Dopamine

Several aspects of DA metabolism are altered in old animals. Concentrations of DA in various brain regions of old males and females are reduced compared to young animals. The decreases observed in DA concentrations in the arcuate nucleus and median eminence are of particular interest to reproductive neuroendocrinologists since this dopaminergic tract is involved with regulation of both prolactin and LH secretion. Decreases in DA concentrations in old males and females in the medial basal hypothalamus (Porter *et al.*, 1980; Simpkins *et al.*, 1977) and in the median eminence (Gudelsky *et al.*, 1981; Sarkar *et al.*, 1982) are as high as 50%. Median eminence DA turnover rates are significantly lower in old males and females (Demarest *et al.*, 1982; Gudelsky *et al.*, 1981; Osterburg *et al.*, 1981; Simpkins *et al.*, 1977). This is reflected in decreased DA concentrations in hypophysial portal plasma (Gudelsky *et al.*, 1981; Reymond and Porter, 1981).

How do these changes in tuberinfundibular DA neurons affect prolactin or LH secretion in the aged animal? Prolactin concentrations are dramatically increased in old male and female rodents (Flurkey *et al.*, 1982; Porter *et al.*, 1980). Hyperprolactinemia is known to suppress reproductive function by actions at several components of the reproductive axis (Evans *et al.*, 1981). Therefore, it is tempting to speculate that age-related decreases in reproductive capacity are causally related to decreasing tuberoinfundibular DA activity with age. Although virtually all aspects of DA metabolism are altered in old animals, the temporal pattern of these neurochemical changes does not suggest that changes at this level cause a loss of estrous cyclicity in females. We (Wise, 1982b) have reported that the proestrous pattern of DA turnover rates is virtually unchanged in middle-aged rats, with the exception of a transient decrease during the morning prior to the LH or prolactin surges. Furthermore, we found that estradiol-induced prolactin surges are normal in amplitude and timing in middle-aged ovariectomized rats at a time when LH surges are attenuated in the same group of animals (Wise, 1984). These animals showed no age-related changes in median eminence DA concentrations or turnover rates. Demarest *et al.* (1982) have reported that functional deficits in DA activity could be observed in middle-aged female rats only after they had completed the transition to acyclicity. No changes in median eminence DA concentrations, rate of DOPA accumulation, or DA content in the anterior pituitary gland could be demonstrated in rats that were still cycling. These data suggest that although tuberinfundibular DA neurons exhibit dramatic changes in the aged animal that may be related to the maintenance of high prolactin and suppressed LH concentrations, these changes do not occur prior to the transition to acyclicity and therefore do not cause the transition to infertility.

Tuberoinfundibular DA neurons appear to be autoregulated by circulating levels of prolactin. Thus, in young animals elevated concentrations of prolactin stimulate dopaminergic activity in the median eminence as measured by an increase in DA synthesis, turnover rates, and concentrations in hypophysial portal blood. In contrast, in old animals all of these indices of dopaminergic activity are diminished despite high concentrations of prolactin (Demarest *et al.*, 1982; Porter *et al.*, 1980), suggesting that prolactin does not feed back as effectively in old animals compared to young.

Serotonin

Serotonin's effects on LH release and its role in aging of the reproductive system remain less well understood. Serotonin concentrations in the hypothalamus (Quay, 1968) and the pineal gland (Quay, 1964) exhibit a circadian rhythm. Meyer and Quay (1976) observed a circadian rhythm in the uptake of 5-HT in the hypothalamus and suprachiasmatic nucleus *in vitro* in males and females. Serotonin content in the hypothalamus rises during the critical period when the neural signal for LH surges would be anticipated (Walker and Timiras, 1982) (Figure 3). 5-Hydroxyindoleacetic acid, the major oxidation product of 5-HT, increases as 5-HT content decreases. The pattern of LH secretion can be modified by drug-induced modifications in the amplitude of the 5-HT circadian rhythm (Walker, 1983). Together these data suggest that serotonin metabolism exhibits a circadian rhythm and that hypothalamic 5-HT activity modulates the timing and/or the amplitude of LH surges.

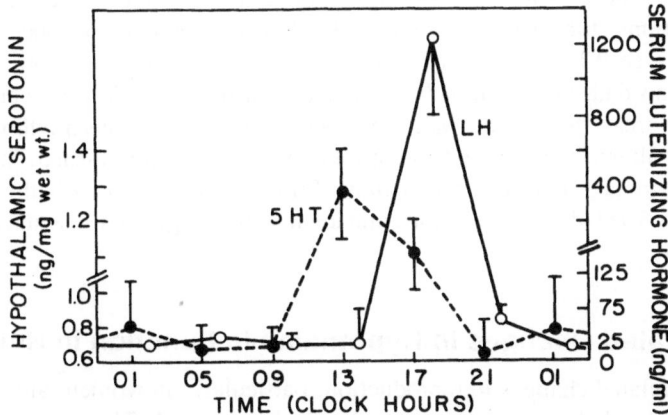

Figure 3. Temporal relationship between circadian rhythms in serum LH and hypothalamic 5-HT content. Ovariectomized rats were kept in a standard photoperiod (LD 14 : 10 with lights on at 0500 hr). Estradiol containing capsules were implanted to initiate LH surges. Individual animals were serially bled for LH determinations; groups of six rats were killed at each time point for 5-HT determinations. Values represent mean ±S.E. Peak and nadir values for 5-HT, 1300 hr vs. 2100 hr and LH, 1800 hr vs. 0100 hr were significantly different. From Walker and Timiras (1982).

Changes in serotonin activity have been reported in old animals. First, hypothalamic 5-HT concentrations remain constant or are slightly elevated in old males (Simpkins *et al.*, 1977). Likewise, in females, 5-HT concentrations do not decrease in various areas of the brain, including those which contain cell bodies and terminals (Timiras *et al.*, 1983). Since both NE and DA concentrations decrease in many of the same brain areas, aging involves an increase in the ratio of 5-HT/NE or 5-HT/DA. Timiras and her colleagues (1983) believe that this may affect the rate of aging. Second, turnover rates in both males (Simpkins *et al.*, 1977) and females (Timiras *et al.*, 1983) are elevated in the hypothalamus and cerebral hemispheres respectively. Third, restriction in the diet, from the time of weaning, of the precursor amino acid tryptophan results in decreased brain concentrations of 5-HT and delays the onset of infertility in females (Timiras *et al.*, 1983). Finally, the constant estrus syndrome, characteristic of acyclic aged female rats, can be induced experimentally by destroying the suprachiasmatic nucleus or altering the concentration of 5-HT (Walker *et al.*, 1980). Furthermore, LH surges can be restored in old constant estrous rats by pharmacologically increasing the serotonergic signal (Walker, 1982).

The data in young animals clearly indicates that the rhythm in serotonin activity, and not the absolute concentrations of this neurotransmitter, is an important factor to consider. Most studies to date have (1) examined 5-HT concentrations at one time during the day or (2) used pharmacological manipulations to determine whether amplification, suppression, or alteration in the timing of the serotonergic signal can change the pattern of LH release to mimic the changes that occur with age. The use of drugs to manipulate neurotransmitter concentrations, turnover rates or their binding to receptors are open to many criticisms (see Wise, 1983b, for more detailed discussion). At the present time, there are no data in old animals on the rhythm of serotonin turnover rates in key brain areas that would indicate whether the endogenous rhythm is suppressed or altered in timing when reproductive ability is diminished. Such studies are necessary before we can draw firm conclusions on whether critical changes in 5-HT dynamics occur coincident with the gradual loss in reproductive ability.

Age-Related Changes in Hypothalamic Function in Humans

Age-related changes in reproduction, particularly in women, are generally thought to result from changes at the level of the gonad. Thus, concentrations of estradiol secreted by a diminishing pool of follicles decrease during the perimenopausal period and are considered to be critical changes in the aging reproductive system. The possibility that concomitant neural changes may contribute to irregular menstrual cycles, and therefore to ultimate acyclicity, should be considered. The results of several studies make this an intriguing possibility.

First, hot flushes begin to appear in women during the perimenopausal transition. Such episodic increases in temperature have been correlated with episodic increases in plasma LH in menopausal women (Casper *et al.*, 1979; Tataryn *et al.*, 1979). Data support the concept that hypothalamic and not pituitary factors initiate menopausal hot flushes. Some of the cell bodies of LHRH-containing neurons exist in the medial preoptic area in close apposition to those involved in temperature regulation. Thus, it appears possible that both changes in the release pattern of LH and the altered ability to thermoregulate are indices of changes in hypothalamic function in humans that occur during middle age. Second, if changes in the biological clock occur early in the aging process, they may affect the timing of hypothalamic and pituitary secretions, resulting in suboptimal development of follicles during the menstrual cycle. Since the correct functioning of the reproductive system depends upon correct temporal and quantitative interactions and feedback of multiple neurotransmitters and hormones, changes in either gonadal or neural signal will rapidly affect other components of the reproductive axis to cause a cascading deterioration of reproductive abilities.

ACKNOWLEDGMENTS

This work was supported by NIH grant AG 02224 and NIH Research Career Development Award AG 00168.

References

Algeri, S., Calderini, G., Toffano, G., and Ponzio, F., 1983, Neurotransmitter alterations in aging rats, in: *Aging of the Brain*, Volume 22 (D. Samuel, S. Algeri, S. Gershon, V. E. Grimm, and G. Toffano, eds.), Raven Press, New York, pp. 227–243.

Banerji, T. K., Parkening, T. A., and Collins, T. J., 1982, Effects of aging on the activity of hypothalamic dopamine-beta-hydroxylase during various stages of the estrous cycle in C57BL/6 mice, *Neuroendocrinology* 34:14–19.

Baraban, J. M., and Aghajanian, G. K., 1981, Noradrenergic innervation of serotonergic neurons in the dorsal raphe: Demonstration by electron microscopic autoradiography, *Brain Res.* 204:1–11.

Ben-Jonathan, N., Oliver C., Weiner, H. J., Mical, R. S., and Porter, J.C., 1977, Dopamine in hypophysial portal plasma of the rat during the estrous cycle and throughout pregnancy, *Endocrinology* 100:452–458.

Berg, A., and Zimmerman, I. D., 1975, Effects of electrical stimulation and norepinephrine on cyclic-AMP levels in the cerebral cortex of the aging rat, *Mech. Ageing Develop.* 4:377–383.

Brown-Grant, K., and Raisman, G., 1977, Abnormalities in reproductive function associated with destruction of the suprachiasmatic nuclei in female rats. *Proc. R. Soc. Lond. B* 198:279–290.

Caceres, A., and Taleisnik S., 1980, Inhibition of secretion of LH induced by electrochemical stimulation of the anterior cingulate cortex mediated by a β-adrenergic mechanism, *J. Endocrinol.* 87:419–429.

Casper, R. F., Yen, S. S. C., and Wilkes, M. M., 1979, Menopausal flushes: A neuroendocrine link with pulsatile LH secretion, *Science* 205:823–825.

Clemens, J. A., and Bennett, D. R., 1977, Do aging changes in the preoptic area contribute to loss of cyclic endocrine function? *J. Gerontol.* **32**:19–24.

Clemens, J. A., Amenomori, Y., Jenkins, T., and Meites, J., 1969, Effects of hypothalamic stimulation, hormones, and drugs on ovarian function in old female rats, *Proc. Soc. Exp. Biol. Med.* **132**:561–563.

Coen, C. W., and MacKinnon, P. C. B., 1979, Serotonin involvement in the control of phasic LH release in the rat: evidence for a critical period, *J. Endocrinol.* **82**:105–113.

Cooper, R. L., Brandt, S. J., Linnoila, M., and Walker, R. F. 1979, Induced ovulation in aged female rats by L-dopa implants into the medial preoptic area, *Neuroendocrinology* **28**:234–240.

Demarest, K. T., Moore, K. E., and Riegle, G. D., 1982, Dopaminergic neuronal function, anterior pituitary dopamine content, and serum concentrations of prolactin, LH and progesterone in the aged female rat, *Brain Res.* **247**:347–354.

Estes, K. S., and Simpkins, J. W., 1980, Age-related alterations in catecholamine concentrations in discrete preoptic area and hypothalamic regions in the male rat, *Brain Res.* **194**:556–560.

Estes, K. S. and Simpkins, J. W., 1982, Resumption of pulsatile LH release after α-adrenergic stimulation in aging constant estrous rats, *Endocrinology* **111**:1778–1784.

Estes, K. S., and Simpkins, J. W., 1984, Age-related alterations in dopamine and norepinephrine activity within microdissected brain regions of ovariectomized Long Evans rats, *Brain Res.* **298**:209–218.

Evans, W. S., Cronin, M. J., and Thorner, M. D., 1981 Hypogonadism in hyperprolactinemia; proposed mechanism, *Frontiers Neuroendocrinol.* **7**:77–122.

Everett, J. W., 1940, The restoration of ovulatory cycles and corpus luteum formation in persistent estrous rats by progesterone, *Endocrinology* **27**:681–686.

Everett, J. W., 1943, Further studies on the relationship of progesterone to ovulation and luteinization in the persistent-estrous rat, *Endocrinology* **32**:285–292.

Finch, C. E., 1973, Catecholamine metabolism in the brains of ageing male mice, *Brain Res.* **52**:261–276.

Flurkey, K., Gee, D. M., Sinha, Y. N., Wisner, J. R., and Finch, C. E., 1982, Age effects on LH, progesterone and prolactin in proestrous and acyclic C57BL/6J mice, *Biol. Reprod.* **26**:835–846.

Fuxe, K., Hokfelt, T., and Nilsson, O., 1972, Effect of constant light and androgen sterilization on the amine turnover of the tuberoinfundibular dopamine neurons: Blockade of cyclic activity and induction of a persistent high dopamine turnover in the median eminence, *Acta Endocrinol.* **69**:625–639.

Greenberg, L. H., and Weiss, B., 1983, Neuroendocrine control of catecholaminergic receptors in aging brain, in: *Aging Brain and Ergot Alkaloids*, Volume 23 (A. Agnoli, G. Crepaldi, P. F. Spano, and M. Trabucchi, eds.), Raven Press, New York, pp. 37–52.

Gudelsky, G. A., Nansel, D. D., and Porter, J. C., 1981, Dopaminergic control of prolactin secretion in the aging male rat, *Brain Res.* **204**:446–450.

Hery, F., Rouer, E., and Glowinski, J., 1972, Daily variations of serotonin metabolism in the rat brain, *Brain Res.* **43**:445–465.

Hery, M., Laplante, E., and Kordon, C., 1976, Participation of serotonin in the phasic release of LH. I. Evidence from pharmacological studies, *Endocrinology* **99**:496–503.

Honma, K., and Wuttke, W. O., 1980, Norepinephrine and dopamine turnover rates in the medial preoptic area and the mediobasal hypothalamus of the rat brain after various endocrinological manipulations, *Endocrinology* **106**:1848–1853.

Huang, H. H., Marshall, S., and Meites, J., 1976, Induction of estrous cycles in old non-cyclic rats by progesterone, ACTH, ether stress or l-dopa, *Neuroendocrinology* **20**:21–34.

Jennes, L., Beckman, W. C., Stumpf, W. E., and Grzanna, R., 1982, Anatomical relationships of serotoninergic and noradrenalinergic projections with the GnRH system in septum and hypothalamus, *Exp. Brain Res.* **46**:331–338.

Kalra, P. S., Kalra, S. P., Krulich, L., Fawcett, C. P., and McCann, S. M., 1972, Involvement of norepinephrine in transmission of the stimulatory influence of progesterone on gonadotropin release, *Endocrinology* **90**:1168–1176.

Kordon, C. and Glowinski, J., 1972, Role of hypothalamic monoaminergic neurons in the gonadotropin release regulating mechanisms, *Neuropharmacology* **11**:153–162.

Krieg, R. J., and Sawyer, C. H., 1976, Effects of intraventricular infusion of catecholamines on LH release in ovariectomized steroid-primed rats, *Endocrinology* **99**:411–415.

Leonardelli, J., Dubois, M. P., and Poulain, P., 1974, Effect of exogenous serotonin on LHRH secreting neurons in the guinea pig hypothalamus revealed by immunofluorescence, *Neuroendocrinology* **15**:69–72.

Leung, P. C. K., Arendash, G. W., Whitmoyer, D. I., Gorski, R. A., and Sawyer, C. H., 1982a, Differential effects of central adrenoceptor agonists on LH release, *Neuroendocrinology* **34**:207–214.

Leung, P. C. K., Whitmoyer, D. I., Garland, K. E., and Sawyer, C. H., 1982b, β-adrenergic suppression of progesterone-induced LH surge in ovariectomized estrogen-primed rats, *Proc. Soc. Exp. Biol. Med.* **169**:161–164.

Lippman, W., 1968, Relationship between hypothalamic norepinephrine and serotonin and gonadotropin secretion in the hamster, *Nature* **218**:173–174.

Lofstrom, A., 1977 Catecholamine turnover alterations in discrete areas of the median eminence of the 4- and 5-day cyclic rat, *Brain Res.* **120**:113–130.

Lofstrom, A., Eneroth, P., Gustafsson, J. A., and Skett, P., 1977, Effects of estradiol benzoate on catecholamine levels and turnover in discrete areas of the median eminence and limbic forebrain, and on serum LH, FSH, and prolactin concentrations in the ovariectomized female rat, *Endocrinology* **101**:1559–1569.

Lookingland, K. J., Wise, P. M., and Barraclough, C. A., 1982, Failure of the hypothalamic noradrenergic system to function in androgen-sterilized rats, *Biol. Reprod.* **27**:268–281.

Lu, K. H., Gilman, D. P., Meldrum, D. R., Judd, H. L., and Sawyer, C. H., 1981, Relationship between circulating estrogens and the central mechanisms by which ovarian steroids stimulate LH secretion in aged and young female rats, *Endocrinology* **108**:836–841.

Mansky, T., Mestres-Ventura, P., and Wuttke, W., 1982, Involvement of GABA in the feedback action of estradiol on gonadotropin and prolactin release: hypothalamic GABA and catecholamine turnover rates, *Brain Res.* **231**:353–364.

Meyer, D. C., and Quay, W. B., 1976, Hypothalamic and suprachiasmatic uptake of serotonin *in vitro*: twenty-four hour changes in male and proestrous female rats, *Endocrinology* **98**:1160–1165.

Mobbs, C. V., Gee, D. M., and Finch, C. E., 1984, Reproductive senescence in female C57BL/6J mice: ovarian impairments and neuroendocrine impairments that are partially reversible and delayable by ovariectomy, *Endocrinology* **115**:1653–1662.

Osterburg, H. H., Donahue, H. G., Severson, J. A., and C. E. Finch, 1981, Catecholamine levels and turnover during aging in brain regions of male C57BL/6J mice, *Brain Res.* **224**:337–352.

Porter, J. C., Nansel, D. D., Gudelsky, G. A., Reymond, M. J., Pilotte, N. S., Foreman, M. M., and F. H. Tilders, 1980, Some aspects of hypothalamic and hypophysial secretion in aging rats, *Peptides* **1**:135–139.

Quadri, S. K., Kledzik, G. S., and Meites, J., 1973, Reinitiation of estrous cycles in old constant estrous rats by central-acting drugs, *Neuroendocrinology* **11**:248–255.

Quay, W. B., 1964, Circadian and estrous rhythms in pineal melatonin and 5-hydroxyindole acetic acid. *Progr. Brain Res.* **8**:61–63.

Quay, W. B., 1968, Difference in circadian rhythms in 5-hydroxytryptamine according to brain region. *Am. J. Physiol.* **215**:1448–1453.

Rance, N., and Barraclough, C. A., 1981, Effects of phenobarbital on hypothalamic LHRH and catecholamine turnover rates in proestrous rats, *Proc. Soc. Exp. Biol. Med.* **166**:425–431.

Rance, N., Wise, P. M., Selmanoff, M. K., and Barraclough, C. A., 1981a, Catecholamine turn-over rates in discrete hypothalamic areas associated changes in median eminence LHRH and serum gonadotropins on proestrus and diestrous day 1, *Endocrinology* **108**:1795–1802.

Rance, N., Wise, P. M., and Barraclough, C. A., 1981b, Negative feedback effects of progesterone correlated with changes in hypothalamic norepinephrine and dopamine turnover rates. *Endocrinology* **108**:2194–2199.

Reymond, M. J., and Porter, J. C., 1981, Secretions of hypothalamic dopamine into pituitary stalk blood of aged female rats, *Brain Res. Bull.* **7**:69–73.

Rusak, B., and Zucker, I., 1979, Neural regulation of circadian rhythms, *Physiol. Rev.* **59**:449–526.

Sarkar, D. K., and Fink, G, 1981, GnRH surge: possible modulation through postsynaptic α-adrenoreceptors and two pharmacologically distinct dopamine receptors, *Endocrinology* **108**:862–867.

Sarkar, D. K., Gottschall, P. E., and Meites, J., 1982, Damage to hypothalamic dopaminergic neurons is associated with development of prolactin-secreting pituitary tumors. *Science* **218**:684–686.

Sawyer, C. H., Hilliard, J., Kanematsu, S., Scaramuzzi, R., and Blake, C. A., 1974, Effects of intraventricular infusions of norepinephrine and dopamine on LH release and ovulation in the rabbit, *Neuroendocrinology* **15**:328–337.

Schneider, H. M. G., and McCann, S. M., 1970, Mono- and indoleamines and the control of LH secretion. *Endocrinology* **86**:1127–1133.

Simpkins, J. W., Mueller, G. P., Huang, H. H., and Meites, J., 1977, Evidence for depressed catecholamine and enhanced serotonin metabolism in aging male rats: Possible relation to gonadotropin secretion, *Endocrinology* **100**:1672–1678.

Steger, R. W., Huang, H. H., Chamberlain, D. S., and Meites, J., 1980, Changes in control of gonadotropin secretion in the transition period between regular cycles and constant estrus in aging female rats, *Biol. Reprod.* **22**:595–603.

Sun, A. Y., 1976, Aging and in vivo norepinephrine uptake in mammalian brain, *Exp. Brain Res.* **2**:207–219.

Tataryn, I. V., Meldrum, D. R., Lu, K. H., Frumar, A. M., Judd, H. L., 1979, LH, FSH and skin temperature during the menopausal hot flash, *J. Clin. Endocrinol. Metab.* **49**:152–154.

Timiras, P. S., Cole, G, Croteau, M., Hudson, D. B., Miller, C., and Segall, P. E., 1983, Changes in brain serotonin with aging and modification through precursor availability in: *Aging Brain and Ergot Alkaloids* Volume 23 (A. Agnoli, G. Crepaldi, P. F. Spano, and M. Trabucchi, ed.), Raven Press, New York, pp. 23–36.

Vijayan, E., and McCann, S. M., 1978, Re-evaluation of the role of catecholamines in control of gonadotropins and prolactin release, *Neuroendocrinology* **25**:150–165.

Walker, R. F., 1980, Serotonin neuroleptics change patterns of preovulatory secretion of LH in rats, *Life Sci.* **27**:1063–1068.

Walker, R. F., 1982, Reinstatement of LH surges by serotonin neuroleptics in aging constant estrous rats, *Neurobiol. Aging* **3**:253–257.

Walker, R. F., 1983, Quantitative and temporal aspects of serotonin's facilitatory action on phasic secretion of LH in female rats, *Neuroendocrinol.* **36**:468–474.

Walker, R. F., and Timiras, P. S., 1982, Pacemaker insufficiency and the onset of aging, in: *Cellular Pacemakers II* (D. Carpenter, ed.), Wiley Interscience, New York, pp. 234–265.

Walker, R. F., Cooper, R. L., and Timiras, P. S., 1980, Constant estrus; role of rostral hypothalamic monoamines in development of reproductive dysfunction in aging rats, *Endocrinology* **107**:249–255.

Wise, P. M., 1982a, Alterations in proestrus LH, FSH and prolactin surges in middle-aged rats, *Proc. Soc. Exp. Biol. Med.* **169**:348–354.

Wise, P. M., 1982b, Norepinephrine and dopamine dynamics in microdissected brain areas in middle-aged and young rats on proestrus, *Biol. Reprod.* **27**:562–574.

Wise, P. M., 1983a, Effect of hyperprolactinemia on estrous cyclicity, plasma LH, estradiol and progesterone and catecholamine turnover rates in microdissected brain areas, *Endocrine Soc. Abstr.* No. 347.

Wise, P. M., 1983b, Aging of the female reproductive system, in: *Review of Biological Research in Aging*, Volume 1 (M. Rothstein, ed.), Alan R. Liss, New York, pp. 195–224.

Wise, P. M., 1984, Estradiol-induced daily LH and prolactin surges in young and middle-aged rats: Correlations with age-related changes in pituitary responsiveness and catecholamine turnover rates in microdissected brain areas, *Endocrinology* **115**:801–809.

Wise, P. M., Rance, N., and Barraclough, C. A., 1981, Effects of estradiol and progesterone on catecholamine turnover rates in discrete hypothalamic regions in ovariectomized rats, *Endocrinology* **108**:2186–2193.

7

EFFECTS OF AGING ON THE MECHANISMS OF ESTROGEN ACTION IN RAT UTERUS

GEORGE S. ROTH

Introduction

The aging process is characterized by an altered ability to regulate various physiological functions (Shock, 1962; Roth, 1979; Masoro, 1981). Generally, such
control is exerted by certain external factors such as nutrition, temperature,
stress, and others, working through internal mechanisms involving hormones
and neurotransmitters (Aldelman and Roth, 1982).

Literally hundreds of reports dealing with changes in response to hormones
and neurotransmitters during aging have appeared. To attempt an exhaustive
compilation is beyond the scope of an article such as this; the reader is directed
to several recent reviews (Roth, 1979; Pradham, 1980; Gregerman and Bierman, 1981; Roth and Hess, 1982). Although a number of hormone and neurotransmitter responses do not appear to change with age and sensitivity and/or
magnitude of response may actually increase with age in a few cases, the majority of such studies report either decreased sensitivity and/or magnitude of
response with increasing age. It is this latter finding that has inspired considerable recent interest and attempts to explain such decrement in cellular and
molecular terms.

Changes in responsiveness during aging occur for a wide variety of hormone and neurotransmitter types, in a great number of target cell and tissue
types, in various animal species as well as in man, and for a wide variety of

GEORGE S. ROTH • Molecular Physiology and Genetics Section, Gerontology Research Center, National Institute on Aging, National Institutes of Health, Baltimore City Hospitals, Baltimore,
Maryland 21224.

physiological and biochemical processes. It should be pointed out that the time and rate of change with age may vary markedly depending upon the aforementioned variables. Such diversity might suggest that the causes of altered responsiveness are indeed multifactorial or that if a single primary aging mechanism exists it has been ramified manyfold before manifesting itself at the target cell and tissue level. Nevertheless, such apparent complexity has not discouraged scores of laboratories from attempting to elucidate the mechanisms responsible for age-related alterations in hormone and neurotransmitter sections.

In order to carry out this task, it is essential to understand those basic mechanisms by which hormones and neurotransmitters act at the cellular and molecular levels under normal conditions, independent of the aging process. A detailed discussion of these events can be found in several recent reviews (King and Mainwaring, 1974; Cuatrecases, 1974; O'Malley and Means, 1978).

Essentially all of the cellular and molecular events that mediate hormone/neurotransmitter action are candidates for age-related modification, possibly resulting in altered hormonal and neurotransmitter responsiveness. However, a major emphasis at present seems first to be the differentiation between changes in receptor and postreceptor events as causes of altered functional response during the aging process (Roth and Hess, 1982).

We have recently catalogued the various types of receptor alterations reported to occur during the aging process (Roth and Hess, 1982; Hess and Roth, 1984). Taking into account reports appearing up to the present time, about 200 different receptor systems have been studied as a function of age during the adult portion of the lifespan. In general, about 50% of the studies report reduced receptor concentrations with increased age; 35% report no changes whatsoever; 10% find increased receptor concentrations with increased age; and 5% report changes in affinity, usually decreases with increased age. A number of these reports have been independently confirmed by other laboratories while nearly as many are the subject of dispute. Reasons for possible discrepancies have been previously discussed (Roth and Hess, 1982), and these controversies need to be resolved by standardization of experimental models and methodology. However, reasonable good agreement on age-related receptor loss exists for at least the following systems (see Roth and Hess, 1982 and Hess and Roth, 1984, for exact citations): (1) striatal dopamine receptors in various species, (2) estrogen receptors in rodent uterus, (3) β-adrenergic, (4) estrogen, and (5) glucocorticoid receptors in several rat brain regions, (6) androgen receptors in rat prostrate, and (7) glucocorticoid receptor loss from human fibroblasts during aging *in vitro*.

Reasonably close, if not causal, relationships between receptor and response loss exist in about 35 cases. However, correlations alone do not establish causality and receptor loss may not necessarily be responsible for reduced responsiveless in some of these systems. In addition, disagreement exists between laboratories as to the extent and/or importance of receptor loss in several cases. Probably, the best agreement on receptor loss as a probable cause of response loss during aging exists for (1) dopamine receptors—rotational and stereotypic behavior, adenylate cyclase in various species; and (2) β-adrenergic receptors—adenylate cyclase in rat cerebellar cortex. However, despite some controversy, independent confirmation from different laboratories also exists for (1) glucocorticoid receptors—RNA synthesis in rat liver; and (2) insulin—glucose oxidation in human skin fibroblasts. Most of the remaining systems have only been examined in individual laboratories and thus await further confirmation.

Possible causal relationships between age changes in postreceptor events and altered responsiveness are even more tenuous. Only a few such cases have been reported and only from individual laboratories. Thus, considerably more work will be necessary to confirm these observations. Some general patterns may be emerging, however, even from the limited number of existing studies. For example, age changes within the cell membrane itself may lead to altered responsiveness to hormones and neurotransmitters. Examples include altered coupling of the adenylate cyclase subunits in aged human lymphocytes and impaired regulation of the aged rat adipocyte glucose transport system.

An even more widespread phenomenon appears to be an altered ability to mobilize calcium in aged cells. Five hormone/neurotransmitter response systems that require calcium movement can essentially be "rejuvenated" if sufficient calcium is made to enter aged cells. These are: (1) α-adrenergic stimulation of electrolyte secretion and glucose oxidation in parotid cells (Ito *et al.*, 1982; Gee *et al.*, 1983); (2) β-adrenergic stimulation of myocardial contraction (Guarnieri *et al.*, 1980); (3) stimulation of mast cell histamine release by compound 48/80 (Orida and Feldman, 1982); (4) α-adrenergic and serontonergic stimulation of aortic contraction (Cohen and Berkowitz, 1976); and (5) cholinergic regulation of motor function (Peterson and Gibson, 1983). So far, receptor loss per se has not been causally implicated in such reduced responsiveness, but, in all cases examined, calcium flux is impaired with aging.

Estrogen Receptors in the Aged Rat Uterus

Over the past 5 years, one project in our laboratory has focused on the effects of aging on estrogen action in the rat uterus. This system exhibits alterations at both the receptor and postreceptor levels that appear to be at least partially responsible for reductions in estrogenic responsiveness (Holinka *et al.*, 1975; Singhal *et al.*, 1969; Saidududdin and Zassenbas, 1979; Hseueh *et al.*, 1979; Jiang and Peng, 1981; Gesell and Roth, 1981; Haji and Roth, 1984; Haji *et al.*, 1985; Chuknyiska *et al.*, 1984). Initially, we became interested in elucidating the mechanisms of uterine estrogen receptor loss during aging since substantial agreement existed among various laboratories as to the loss of these receptors during rodent senescence (Holinka *et al.*, 1975; Nelson *et al.*, 1976; Saiduddin and Zassenhaus, 1979; Hseueh *et al.*, 1979; Gesell and Roth, 1981; Jiang and Perry, 1981; Belisle and Beaudry, 1982).

However, some controversy arose as to the appropriate experimental conditions for such analyses and the possible mechanisms by which receptor loss occurs. For example, since estrogen positively controls its own receptors (Pavlik and Coulson, 1976), age-related decreases in receptor might simply reflect lower circulating estrogen levels in senescent rats (Huang *et al.*, 1978; Steger *et al.*, 1979). On the other hand, Saiduddin and Zassenhaus (1979) used ovariectomized animals that had been primed with equivalent estrogen dosages per body weight and observed over 50% fewer receptors in the uteri of senescent rats than in those of their mature counterparts. Some early studies examined only the uptake of estrogen into the tissue rather than its specific binding to receptor (Larson *et al.*, 1972; Peng and Peng, 1973). Although uptake was also decreased with increasing age, it was not clear to what extent such measurements reflected the status of estrogen receptors.

Although it was clear that many (but not all) types of hormone, neurotransmitter, and other receptors change during the aging process, many fundamental questions regarding the mechanisms of receptor changes during aging in general remained unanswered. Some evidence existed to suggest that the biosynthetic rate of certain receptors may be reduced in aged cells (Chang *et al.*, 1981; Joseph *et al.*, 1981; Rosner and Cristofalo, 1981; Pitha *et al.*, 1982; Henry and Roth, 1985), although it was also possible that only the percentage of receptors that are functional, i.e., able to bind hormones with specificity and high affinity, and not the receptor concentrations, changed with age.

This last possibility was tested for rat uterine estrogen receptors, since specific antisera became available to immunochemically complement various physiochemical analyses of receptor functionality (Greene *et al.*, 1977). In light of the controversy regarding the specifics of age-related uterine estrogen receptor changes, we therefore attempted to analyze these receptors in mature and senescent rats at various times after ovariectomy. In addition, physiochemical and especially immunochemical properties of these receptors were studied in order to determine whether altered or nonfunctional uterine estrogen receptors was present in the uteri of senescent rats.

Table I summarizes the results of Scatchard analyses of $[2,4,6,7\text{-}(n)\text{-}^3H]$-estradiol binding to uterine cytosol receptors in mature (6–12 months) and senescent (22–24 months) rats 1–4 weeks after ovariectomy. It is clear that specific binding was reduced (by 46%) in the senescent group, maximum binding was 468.0 ± 63.4 fmoles/mg protein in the mature animals and 251.2 ± 31.6 fmoles/mg protein in the senescent animals. Also, binding affinity (K_d) was not altered as a function of age, remaining 1.5–1.6 nM. We were unable to detect any effect of time after ovariectomy between 2–50 days on the specific binding

Table I

Effect of Age on Rat Uterine Estrogen Receptor Concentrations
(B_{max}) and Affinities $(K_d)^a$

	B_{max} (fmoles/mg protein)	K_d (nM)
Mature	468.0 ± 63.4 (15)	1.5 ± 0.2 (15)
Senescent	251.2 ± 31.6 (13)	1.6 ± 0.2 (13)
P value	<0.006	NS

^{a} Uterine extracts were prepared and cytoplasmic estrogen receptors were assayed as described in Gesell and Roth (1981). Values represent the mean $\pm$ S.E. for the number of experiments indicated in parentheses.

of 1 nM [2,4,6,7-(N)-^{3}H]estradiol in either age group (data not shown). In agreement with previous reports (Zava *et al.*, 1976; Jiang and Peng, 1981), specific binding was minimal and not significantly different in nuclei from avariectomized animals of both ages, ranging from 0–10% and 5–20% of the total uterine receptors in the mature and senescent groups, respectively.

Since possible age differences in uterine protein or cell content could influence interpretation of results expressed per milligram protein, DNA and cytosol protein determinations were performed on mature and senescent uteri. Wet weight and cytosol protein content per uterus were followed for 4 weeks after ovariectomy. Mean values decreased about 25% in both groups, but were not found to be statistically significantly different between ages (wet weight ranged from 150–300 mg; cytosol protein contents ranged from 3–6 mg; $n = 36$). The DNA content per uterus was changed after ovariectomy and appeared to be slightly lower (23%) in the senescent group (887 ± 47 and 684 ± 66 μg for the mature and senescent uteri, respectively). Thus, the 46% reduction in receptor concentration could not be accounted for simply by cell loss in the senescent uteri, and certainly not by an increased cytosol protein content.

We next examined various physiochemical properties of uterine estrogen receptors in mature and senescent rats. No significant age differences were detected in receptor stability, both at various temperatures and in the presence or absence of bound estradiol (Gesell and Roth, 1981). In addition, no age differences were seen in the potency of various steroids to compete for binding to the receptor (Gesell and Roth, 1981).

Although no functional or physiochemical property, except apparent receptor concentration, differed between mature and senescent uteri, the possibility remained that the latter difference was due to nonfunctional receptors in the senescent preparations that were undetectable by all of the above analyses. It was, therefore, decided to immunochemically titrate these receptors to determine whether specific binding activity was directly proportional to immunoreactivity in both age groups.

The antiserum was prepared and characterized previously by Greene *et al.* (1977) and did not contain precipitating antibody. Therefore, the reaction between receptor and antibody was measured by the change in the receptor sedimentation coefficient from 4 S to 7–8 S on sucrose density gradients. At high ionic strength (0.3 M KCl) in the absence of antiserum, essentially all of the receptor was present in the 4 S form (at low ionic strength almost all receptor sedimented at 7–8 S). As the antiserum concentration was increased, however, more and more receptor was shifted to the 7–8 S form. No difference between the mature and senescent preparations in the proportion of receptor shifted to the 7–8 S form was observed at any concentration. At antiserum concentrations above 2.5 μl/250 μl cytosol, all receptor was shifted to the 7–8 S form (data not shown), as has also been reported by Greene *et al.*, 1977 for rat uterine estradiol receptors.

This type of titration was performed on a number of preparations from each age group, and the results are summarized in Figure 1. Again, no age difference in the amount of receptor reacting with any given concentration of antiserum could be detected.

The most important finding of this study was that comparable amounts of receptor-specific binding activity from mature and senescent uteri require comparable amounts of antiserum to shift 50% of the molecules from 4 S to 7–8 S. This suggests that the presence of immunologically cross-reacting or nonfunctional receptors in aged uteri does not account for the apparent reduction in concentration. Of course, it is still possible that the antiserum used in the study was directed against antigenic determinants that may not be accessible on nonfunctional receptors. This particular antiserum was prepared against calf uterine

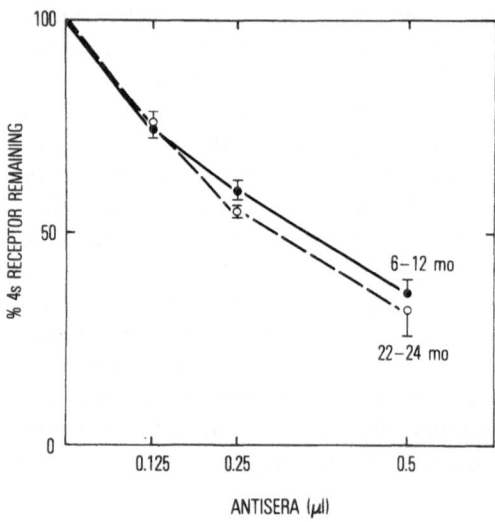

Figure 1. Effect of age on immunoreactive uterine estrogen receptor concentrations. Analyses were carried out as described previously (Gesell and Roth, 1981). The percentage of 4 S receptor converted to the 7–8 S form was determined at various antisera concentrations. Values represent the mean ± S.E. for six individual experiments for each age group.

estrogen receptors and shown to cross-react almost equally well with receptors from rat, mouse, and guinea pig uterus, as well as human breast cancer. Unfortunately, the availability of such antisera is still extremely limited. Although Jensen's laboratory has produced some newer batches, these are monoclonal and even more restricted in terms of immunogenic determinants, since cross-reactivity between species is relatively limited.

If nonfunctional receptors do exist in aged uteri, it may be particularly difficult to purify them for antisera preparation, since conditions required for their purification may differ from those necessary to isolate functional molecules. Probably, the most practical approach to this problem will be to test more batches of specific antisera when they become available. At present, however, at least in the case of apparent steroid receptor loss during aging a more likely explanation appears to be an altered control of biosynthetic rate rather than an alteration in molecular structure.

Effects of Aging on Estrogen Stimulation of Uterine RNA Polymerase II *in Vivo*

Despite the numerous studies reporting estrogen receptor loss cited above, it was still not completely clear to what extent decreased estrogen receptor levels relate to reduced uterine responses to estrogen in aged female rats, since little was known about aging effects on postreceptor mechanisms of uterine estrogen action. Subsequent to estrogen binding to receptor and nuclear translocation, the stimulation of messenger RNA synthesis is a primary prerequisite step for translation of estrogen-dependent proteins (Glasser *et al.*, 1972; Knowler and Smellie, 1973; Borthwick and Smellie, 1975; Ariz and Knowler, 1980), and polymerase II is responsible for messenger RNA synthesis (Lindell, 1980; Benz *et al.*, 1977). If loss of receptors is causally related to reductions in estrogenic responsiveness during aging, then stimulation of RNA polymerase II should be impaired. On the other hand, if receptor loss is not related to reduced response, then it is necessary to examine RNA polymerase II in order to determine whether independent changes in estrogen regulation of this enzyme may account for impairments in subsequent biological responses. Our next study attempted, therefore, to further elucidate the mechanisms of aging changes in uterine response to estrogen by examining stimulation of RNA polymerase II activity in isolated uterine nuclei of mature and senescent female rats.

Uterine nuclei from rats ovariectomized for 3 weeks were isolated by a modification of the method of Hewish and Burgoyne (1973) by Woll *et al.* (1981). All procedure were performed at $4\,^\circ\mathrm{C}$. Each uterus was minced and homogenized in 25 vol of 0.25 sucrose, 2 mM Na_2 EDTA, 0.5 mM Na_2EGTA, 60 mM KCl, 15 mM NaCl, 0.15 mM spermine (Sigma), 0.5 mM spermidine (Sigma), 1 mM phenylmethylsulfonylfluride (PSMF, Sigma), 0.5 mM dithiothreitol, and 15 mM Tris HCl (pH 7.4), with three 30-sec bursts of a polyton homogenizer at a power setting of 4. The homogenates were filtered through nylon mesh, mixed with 2 vol of 2.1 M sucrose, 0.1 mM Na_3EDTA, 0.1 mM Na_2EGTA, 60 mM KCl, 15 mM NaCl, 0.15 mM spermine, 0.5 Mm spermidine, 1 mM PSMF, 0.5 mM dithiothreitol and 15 mM Tris HCl (pH 7.4), and underlayed with 10 ml of the same 2.1 M sucrose solution in polycarbonate centrifuge tubes. The tubes were then centrifuged at $110,000g$ for 60 min. The resulting nuclear pellet was resuspended in 0.34 M sucrose and 1 mM $MgCl_2$. Nuclei from various tissues of rats and mice prepared in the presence of polyamines, dithiothorintol, etc., exhibit no RNAase activity at any age (Castle *et al.*, 1978).

Nuclear RNA polymerase II activities were measured by a modification of the method of Roeder and Rutter (1970) and Glasser *et al.* (1972). Isolated nuclei (approximately 50 μg DNA) were incubated at 27°C for 10 min in the reaction mixture (100 μl final volume), which contained 0.09 μmole each of ATP, GTP, and CTP, 0.004 μmole of $[^3H]$-UTP (2μCi), 10 μmoles of Tris HCl (pH 7.9), 0.3 μmole of $MnCl_2$, 0.05 μmole of dithiothreitol, 0.5 μmole of phosphoenolpyruvate, 5 μg of pyruvate-kinase, 0.005 μmole of S-adenosyl-1-methionine, 20 μmoles of $(NH_4)_2SO_4$, and 0.6 μmole of NaF in the absence or presence of 0.08 μg of α-amanitin (Sigma). Under these conditions, $[^3H]$-UTP incorporation into nuclear acid-insoluble material increased linearly for at least 10 min. The reactions were stopped after 10 min by adding 2 ml of 10% TCA and 1% sodium pyrophosphate. The precipitatable material was then collected on a nitrocellulose filter (Millipore 0.45 μm), washed, dried, and assayed for radioactivity in 4 ml of ACS (Amersham) in a Packard liquid-scintillation spectrometer. The counting efficiency was 33%. RNA polymerase II activity was calculated by subtracting the incorporated $[^3H]$-UTP in the presence of α-amanitin from that in the absence of α-amanitin and expressed as incorporated $[^3H]$-UTP pmole/mg DNA. The concentration of α-amanitin employed completely inhibits RNA polymerase II activity in rat liver and uterine nuclei without affecting polymerases I and III (Lindell, 1980; Webster and Hamilton, 1976).

Figure 2 shows the time course of uterine nuclear RNA polymerase II activity after estradiol (E_2) (3 μg/100 g body weight) injection in mature and senescent ovariectomized rats. The control ovariectomized rats in both age groups showed similar levels of this enzyme (mature; 267.6 ± 8.0,old; 280.2 ± 7.5 pmoles/mg DNA, $p > 0.1$). In mature ovariectomized rats, following intraperitoneal E_2 injection, RNA polymerase II activity did not change appreciably during the first 4 hr (data not shown) but increased significantly by 6 hr and reached a peak level as 12 hr (494.9 ± 12.0 pmoles/mg DNA, 185.0% of control value). On the other hand, in old ovariectomized rats, the stimulation of this enzyme activity was not observed until 12 hr after E_2 administration, after which a maximum level was reached at 18 hr (457.6 ± 14.2 pmoles/mg

Figure 2. Time course of.17β-estradiol effect on nuclear RNA polymerase II activity in uteri of mature and senescent ovariectomized rats. Ovariectomized mature (●——●) and senescent (○– – –○) rats were injected intraperitoneally with 17β-estradiol (3 μm/100 g body weight) and killed at 6, 12, and 18 hr after injection. Control rats were injected with vehicle alone. Uterine nuclei were isolated and nuclear RNA polymerase II activities were measured as described previously (Haji and Roth, 1984). Values at each time point are the means ± S.E. of 4–10 rats, as cited in parenthesis. The values for mature and senescent control rats (100%) were 267.6 ± 8.0 and 280.2 ± 7.5 pmoles/mg DNA, respectively.

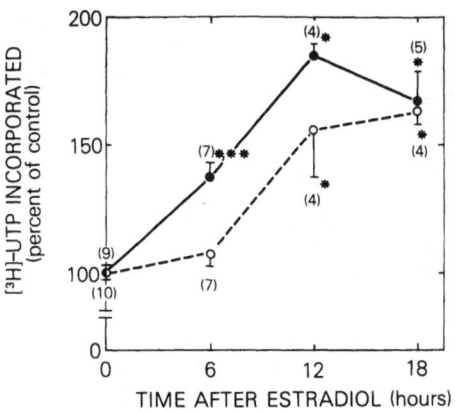

Asterisks indicate significant differences: (*) $p < 0.001$ compared with the corresponding control; (**) $p < 0.005$ compared with the corresponding senescent group.

DNA, 163.3% of control value). There was a small but significant difference ($p < 0.05$) in the maximum increments in RNA polymerase II activities between mature and senescent ovariectomized rats (mature maximum = 12 hr, senescent maximum = 18 hr). However, values were not significantly different between age groups at 18 hr ($p < 0.01$).

The effect of various E_2 doses on uterine nuclear RNA polymerase II activity was determined 6 hr after E_2 injection in mature and old ovariectomized rats (Figure 3). In mature ovariectomized rats, 1 μg of E_2/100 g body weight resulted in a significant increase relative to the control value, and a peak response was observed at dose above 3 μg of E_2/100 g body weight (an increase of 101.0 ± 14.3 pmoles/mg DNA or 37.7 ± 5.4% above the control value). In old ovariectomized rats, however, RNA polymerase II activity did not change at doses below 3 μg/100 g body weight of E_2, but 10 μg of E_2/100 g body weight stimulated the enzyme to essentially the same levels (an increase of 86.3 ± 26.4 pmoles/mg DNA or 30.8 ± 9.4% above the control value) as mature counterparts.

The main purpose of this study was to examine the effects of aging on nuclear RNA polymerase II activity in rat uterus. Although basal levels of nuclear RNA polymerase II activity in uteri of mature and old ovariectomized rats were similar, the response of this enzyme to E_2 was delayed and its sensitivity to E_2 was decreased in senescent ovariectomized rats, as compared with mature ovariectomized counterparts. Similar age-related reductions in RNA synthesis were observed in other tissues and other species. The basal RNA polymerase "I + II" and "I" activities and the initiation rate of glucocorticoid–receptor complex-induced total RNA synthesis in liver nuclei of male rats decreased with

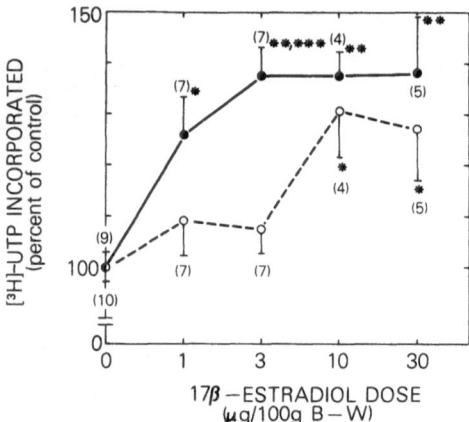

Figure 3. Dose effect of E_2 on nuclear RNA polymerase II activity in uteri of mature and senescent ovariectomized rats. Ovariectomized nature (●——●) and senescent (○-- -○) rats were injected intraperitoneally with 1–30 μg of 17β-estradiol/100 g body weight and killed 6 hr after injection. Control rats were injected with vehicle alone. Uterine nuclei were isolated and nuclear RNA polymerase II activities were measured as described previously (Haji and Roth, 1984). The value for each point is the mean ± S.E. of 4–10 rats, as cited in parenthesis. The values for mature and senescent control rats (100%) were 267.6 ± 8.0 and 280.2 ± 7.5 pmoles/mg DNA, respectively. Asterisks indicate significant differences: (*) $p < 0.005$, (**) $p < 0.001$, compared with the corresponding control; (***) $p < 0.005$ compared with the corresponding senescent group.

increasing age (Castle *et al.*, 1978; Bolla and Denckla, 1979; Miller and Bolla, 1981). The response of RNA synthesis to estrogen *in vivo* also declined in the uterus of aging ovariectomized mice (Soriero and Talbert, 1975; Soriero, 1980).

The delayed onset of E_2-stimulated RNA polymerase II activity was also reminiscent of the early studies of Adelman, Roth, and colleagues (Adelman, 1971; Adelman *et al.*, 1972; Roth *et al.*, 1974), in which the specific induction of various hepatic metabolic enzymes as well as DNA synthesis and cell division in salivary glands were delayed as age increased. Such age-related delays appear to be frequently associated with responses to hormones and similar agents administered *in vivo*.

The delayed and less E_2-sensitive responses of uterine nuclear RNA polymerase II activity in old female rats, as observed in this study, are consistent with decreased estrogen receptor concentration and reduced translocation of estrogen-receptor complexes from cytosol to nucleus in aged female rats as cited above. Furthermore, such changes may be a factor in the diminished estrogen-induced carbohydrate metabolism (Singhal *et al.*, 1969), total RNA synthesis (Soriero, 1980), and lack of decidual cell response (Saiduddin and Zassenahus, 1979) in senescent animals.

Despite age changes in time course and sensitivity, the increment in RNA polymerase II activity following E_2 administration in uterine nuclei of ovariectomized old rats was nearly the same as in mature counterparts. This result is somewhat different from previous correlations between receptor levels and responsiveness to steroid hormones in which absolute magnitude of response was diminished with age (Roth, 1975; Roth and Livingston, 1976). However, as stated above, such differences may be related to whether hormone is administered *in vivo* or *in vitro*.

Effects of Aging on Estrogen Stimulation of Uterine RNA Polymerase II *in Vitro*

In order to resolve the above discrepancies between *in vivo* and *in vitro* stimulation, it became necessary to establish a cell-free system consisting of isolated uterine nuclei and cytoplasmic receptors. Such a system could allow precise adjustment of nuclear and receptor concentrations as well as heterologous mixing of these components from both mature and senescent uteri. Some previous reports of analogous cell-free systems had appeared, but these has been restricted to immature animals (Mohla *et al.*, 1972; Arnaud *et al.*, 1971). Our next study, therefore, established and utilized a cell-free system from adult animals, and examined the effects of aging on the ability of isolated uterine nuclei and cytosol, respectively, to allow estrogenic stimulation of RNA polymerase II.

The preparation of cytosol receptor–E_2 complexes (R–E_2) and isolation and treatment of nuclei were carried out as previously described (Haji *et al.*, 1985). To allow receptor binding, 0.3 ml of nuclear solution was incubated with 0.3 ml of R–E_2 (1 nM final concentration) for 30 min at 4°C. After incubation, the nuclei were collected by centrifugation at 2000g for 10 min. The nuclear pellets were washed three times with the same nuclear buffer and resuspended in 0.34 M sucrose and 1 mM $MgCl_2$. For determination of basal RNA polymerase II activity, aliquots of nuclear solution were incubated with cytosol in the absence of E_2 and assayed exactly as described above. RNA polymerase II activity was then determined by the same procedure employed in the previous *in vivo* study (Haji and Roth, 1984).

In order to determine whether the ability of uterine nuclei or cytoplasmic receptor preparations to support stimulation of RNA polymerase II was altered during aging, various homologous and heterologous combinations of nuclei and cytosols were examined. In all cases an R–E_2 concentration of 1 nM was employed, and basal levels of polymerase activity were not significantly different between groups. Results of these experiments are summarized in Table II. First, it is obvious that mature nuclei allow three times more stimulation than senescent nuclei when mature cytosols are used. Second, mature cytosols are five times more effective than senescent cytosols when mature nuclei are used. In fact, stimulation by senescent cytosols is equally poor in mature and old nuclei. Finally, equimolar mixtures of mature and senescent cytoplasmic receptor preparations incubated with mature nuclei yield a value intermediate between those of mature and senescent cytosols separately. The fact that this value is intermediate between mature (~ 155%) and senescent (~ 30%) cytosols used alone, suggests that poor stimulation by senescent cytosol is not the result of a freely diffusable inhibitor. However, changes in cytoplasmic factors other than receptors can by no means be excluded as causes of reduced estrogenic stimulation of polymerase II by senescent cytosols.

Table II

Effect of Age on Stimulation of RNA Polymerase II by Isolated Uterine Nuclei and Cytosol[a]

Age group			RNA polymerase II activity (pmoles [³H]-UTP incorporated per mg DNA)		
Cytosol	Nuclei	N	Without E_2	With E_2	Percent increase
M	M	11	106.6 ± 5.4	267.2 ± 10.9	155.0 ± 13.0[b]
M	S	11	102.9 ± 5.9	157.3 ± 9.0	57.7 ± 11.3
S	M	10	95.5 ± 5.9	123.5 ± 6.1	31.5 ± 7.0[c]
S	S	10	98.7 ± 6.9	127.8 ± 9.8	31.2 ± 9.3
M + S	M	10	98.8 ± 3.8	170.6 ± 12.1	77.3 ± 10.8[d]

[a] Respective cytosols and nuclei from mature (M) and senescent (S) uteri were incubated together, and RNA polymerase II activity determined as described previously (Haji and Roth, 1984). Values are the means ± S.E. for the indicated numbers of experiments (N).
[b] Significantly different from all other values ($p < 0.001$).
[c,d] Significantly different from each other ($p < 0.005$).

Impaired responsiveness of senescent nuclei in the presence of mature receptor preparations indicates that aging also affects cellular components and/or processes in the scheme of estrogen action distal to cytoplasmic receptor activation. These might include penetration of the nuclear envelop, binding to chromatin, or subsequent steps in the actual stimulation of RNA polymerase II. In relation to these possibilities, only changes at the chromatin level have been examined thus far during aging, and alterations in DNA strand integrity, secondary structure, and transcription activity have been reported (for a review see Cutler, 1978), however, several groups have also reported reduced concentrations of nuclear estrogen receptors (Belisle and Lehoux, 1982; Belisle and Beaudry, 1982; Jiang and Peng, 1981), but it is not clear whether this reflects a defect in nuclear translocation or is simply the consequence of reduction in total cellular receptor content.

Effects of Aging on Nuclear Translocation of Uterine Estrogen Receptors

In light of the age-related deficits in cytoplasmic receptors and nuclei for estrogenic stimulation of RNA polymerase II and reports of reduced concentrations of nuclear estrogen receptors described above, it became necessary to determine whether defects in either aged receptors or nuclei truly result in impaired translocation in aged uteri. This question was addressed by the use of a cell-free system of isolated nuclei and cytoplasmic receptors according to the method of Kon and Spelsberg (1982) as previously modified by us (Chuknyiska et al., 1984). This is, to our knowledge, the first in vitro cell-free system for measurement of nuclear translocation that is sensitive to steroid injected in vivo.

For isolation and purification of cytosol receptor-[^{3}H]estradiol complex, rat uteri were excised, minced and homogenized in 3 vol of 0.3 M KCl, 100 mM Tris, 10 mM EDTA, 12 mM monthioglycerol, 5% glycerol, pH 7.5. The 105,000g supernatants were incubated with 50 nM [^{3}H]estradiol at 4°C overnight. Then the procedure of Kon and Spelsberg (1982) was followed with slight modification (Chuyknyiska *et al.*, 1984). The labeled crude cytosol was saturated with 35% ammonium sulfate. After centrifugation at 11,000g, the pellets were redissolved in buffer A (20 mM Tris, 1 mM EDTA, 12 mM monothioglycerol) and were mixed by rotation with CNBr-activated heparin–agarose for 90 min in the cold room. The nonbound [^{3}H]estradiol was subsequently washed out with buffer A and [^{3}H]estradiol–receptor complex was eluted with 2 bed volumes of buffer A plus 0.3 M NaCl. After dialysis of 50% ammonium sulfate-saturated eluant, the 11,000g supernatant was used further in the experiments.

Uterine nuclei were isolated by a modification of the method of Hewish and Burgoyne (1972) as performed by Woll *et al.* (1981). For binding of estradiol–receptor complex to rat uterine nuclei *in vitro*, incubation of purified [^{3}H]estradiol–receptor and nuclei was performed according to Kon and Spelsberg (1972). The total volume of 300 μl (200 μl of cytosol plus 100 μl of nuclei containing about 45–100 μg of DNA) was maintained at room temperature (20°C) for 2 hr with periodic vortexing. Each tube contained 0.3–6.0 fmole of [^{3}H]estradiol–receptor complex per microgram of DNA. Control samples, containing equal concentrations of receptor complex, were incubated with a 100-fold excess of nonlabeled 17β-estradiol for 30 min at 37°C and then used under the same conditions as the experimental tubes. Protein concentration was measured by the method of Lowry *et al.* (1951), and DNA was estimated according to the methods of Burton (1956) or Labarca and Paigen (1980).

It was initially determined that the time course of translocation was essentially identical (maximal after 90 min incubation of receptor–estradiol complexes with nuclei) for components derived from mature or senescent uteri (Chuknyiska *et al.*, 1985). Four nuclear-cytoplasmic mixtures were subsequently examined: mature cytosol–mature nuclei; senescent cytosol–mature nuclei; mature cytosol–senescent nuclei; and senescent cytosol–senescent nuclei.

Data from a number of individual analyses for the four nuclear-cytoplasmic mixtures described above are presented in Table III. Two facts are immediately obvious. First, the acceptor site concentrations of the mature cytosol–mature nuclei mixture is greater than any other mixture (one way analysis of variance, $p < 0.01$). Second, no differences between any groups are observed for Ka's or Kd's.

These findings indicate that both nuclei and cytoplasmic receptors from uteri or senescent rats are inferior to those obtained from mature animals in ability to support translocation. No age differences were observed in association and dissociation constants for the translocation reaction. The dependency of translocation on the concentration of receptor–estradiol complexes closely par-

Table III

Effect of Aging on K_a, K_d, and B_{max} for [3H]estradiol–Receptor Complex Translocation into Nuclei

	Mature cytosol/ mature nuclei	Senescent cytosol/ senescent nuclei	Mature cytosol/ senescent nuclei	Senescent cytosol/ mature nuclei
K_a	$3.6 \pm 0.3 \times 10^9$ M^{-1}	$3.2 \pm 0.3 \times 10^9$ M^{-1}	$3.1 \pm 0.4 \times 10^9$ M^{-1}	$3.2 \pm 0.3 \times 10^{-9}$ M^{-1}
K_d	$2.8 \pm 0.3 \times 10^{-10}$ M	$2.9 \pm 0.3 \times 10^{-10}$ M	$3.5 \pm 0.5 \times 10^{-10}$ M	$3.0 \pm 0.3 \times 10^{-10}$ M
B_{max} [b]	$2.0 \pm 0.2 \times 10^{-12}$	$1.3 \pm 0.2 \times 10^{-12}$	$1.3 \pm 0.2 \times 10^{-12}$	$1.5 \pm 0.1 \times 10^{-12}$

[a] Experiments were performed as described above and values obtained from six to seven individual Scatchard analyses for each group. Statistical analyses were carried out using a one-way analysis of variance.
[b] Values are fmoles/μg DNA.

allels that for *in vitro* cell free induction of RNA polymerase II as described above. Thus, it is possible that age-associated nuclear and cytoplasmic deficits in estrogenic stimulation of polymerase II may be related to impairments in translocation. However, correspondence between translocation and polymerase activation is not exact for all four mixtures of nuclei and cytoplasmic receptors. Thus, a rigorous analysis of both translocation and RNA polymerase II stimulation must be carried out in the same preparations for the various age mixtures. In so doing, various receptor–estradiol complex concentrations must be employed in order to determine the stoichiometric relationship between translocation and polymerase activation under all four conditions.

Summary and Conclusions

In summary, it appears that age-associated defects occur at various levels of the estrogenic signal transconduction apparatus in rat uterus. These range from quantitative and possibly qualitative deficits in the receptors themselves to changes in nuclei and the activation of RNA polymerase II. Our future studies will attempt to further elucidate the molecular nature of such age related alterations and to find ways to halt, prevent or reverse functional impairments by appropriate neuroendocrine, biochemical, or pharmacological manipulations.

References

Adelman, R. C., 1971, Age-dependent effects on enzyme induction—a biochemical expression of aging, *Exp. Gerontol.* **6**:49.

Adelman, R. C., and Roth, G. S. (eds.), 1982, *Endocrine and Neuroendocrine Mechanisms of Aging*, CRC Press, Boca Raton, Florida.

Adelman, R. C., Stein, G., Roth, G. S., and Englander, D., 1972, Age-dependent regulation of mammalian DNA synthesis and cell proliferation *in vivo*, *Mech. Ageing Dev.* **1**:49.

Ariz, B., and Knowler, J. T., 1980, The direction of messenger ribonucleic acid sequences in heterogeneous nuclear ribonucleic acid fractions of the estrogen-stimulated rat uterus, *Biochem. J.* **187**:265.

Arnaud, M., Beziat, Y., Guilleux, J. C., Hough, A., Hough, D., and Mousseron-Canet, M., 1971, Les recepteurs de l'oestradiol dans l'uterus de genisse stimulation de la biosynthese de RNA *in vitro*, *Biochem. Biophys. Acta* **232**:117.

Belisle, S., and Beaudry, C., 1982, Endocrine aging in CBA mice. Characterization of uterine cytosolic and nuclear sex steriod receptors, *Exp. Gerontol.* **17**:417.

Belisle, S., and Lehoux, J-G., 1982, Endocrine aging in C57BL mice—II. Dynamics of estrogen receptors in the hypothalamic-pituitary axis, *J. Steroid Biochem.* **18**:737.

Benz, E. W., Getz, M. J., Wells, D. J., and Moses, H. J., 1977, Nuclear RNA accumulation in cultured AKR mouse embryo cells stimulated to proliferate, *Exp. Cell Res.* **108**:157.

Bolla, R., and Denckla, W. D., 1979, Effect of hypothesectomy on liver nuclear ribonucleic acic synthesis in aging rats, *Biochem. J.* **184**:669.

Borthwick, N. M., and Smellie, R. M. S. 1975, The effects of oestradiol-17β on ribonucleic acid polymerases of immature rabbit uterus, *Biochem. J.* **147**:91.

Burton, K., 1956, A study of the conditions and mechanism of the diphenylamine reaction for the colorimetric estimation of DNA, *Biochem. J.* **62:**315.

Castle, T., Katz, A., and Richardson, A., 1978. Comparison of RNA synthesis by liver nuclei from rats of various ages, *Mech. Ageing Dev.* **9:**383.

Chang, W-C., Hoopes, M. T., and Roth, G. S., 1981, Biosynthetic rats of proteins having the characteristics of glucocorticoid receptors in adipocytes of mature and senescent rats, *J. Gerontol* **36:**386.

Chuknyiska, R. S., Haji, M., Foote, R. H., and Roth, G. S., 1984, Effect of *in vivo* estradiol administration on availability of rat uterine nuclear acceptor sites measured *in vitro*, *Endocrinology* **115:**836.

Chuknyiska, R. S., Haji, M., Foote, R. H., and Roth, G. S., 1985, Age-associated changes in nuclear binding of rat uterine estradiol receptor complexes, *Endocrinology* **116:**547.

Cohen, M. L., and Berkowitz, B. A., 1976, Vascular contraction. Effect of age and extracellular calcium, *Blood Vessels* **67:**139.

Cuatrecases, P., 1974, Membrane receptors, *Annu. Rev. Biochem.* **43:**169.

Cutler, R. G., 1978, Alterations with age in the informational storage and flow systems of the mammalian cell, in: *Genetic Effects on Aging* (D. Bergsma and D. H. Harrison, eds.), Alan R. Liss, New York, pp. 463–498.

Gee, M. V., Baum, B. J., and Roth, G. S., 1983, Stimulation of parotid cell glucose oxidation; Role of alpha-adrenergic receptors and calcium mobilization, *Biochem. Pharmacol.* **32:**335.

Gesell, M. S., and Roth, G. S., 1981, Decrease in rate uterine estrogen receptors during aging: Physio- and immuno-chemical properties, *Endocrinology* **109:**1502.

Glaser, S. R., Chytil, F., and Spelsberg, T. C., 1972, Early effects of oestradiol-17β on the chromatin and activity of deoxyribonucleic acid-dependent ribonucleic acic polymerases (I and II) of the rat uterus, *Biochem. J.* **130:**947.

Greene, G. L., Close, L. E., Fleming, H., DeSombre, E. R., and Jensen, E. V., 1977, Antibodies to estrogen receptor: immunochemical similarity of estrophilin from various mammalian species, *Proc. Natl. Acad. Sci. U.S.A.* **74:**3681.

Gregerman, R. I., and Bierman, E. L., 1981, Aging and hormones, in: *Textbook of Endocrinology* (R. H. Williams, ed.), Saunders, Philadelphia, pp. 1192–1212.

Guarnieri, T., Filburn, C. R., Zitnik, G., Roth, G. S., and Lakatta, E. G., 1980, Contractile and biochemical correlates of β-adrenergic stimulation of the aged heart, *Am. J. Physiol.* **239:**H501.

Haji, M., and Roth, G. S., 1984, Impaired estrogen stimulation of RNA polymerase II activity in uterine nuclei of senescent rats, *Mech. Ageing Dev.* **25:**141.

Haji, M., Chuknyiska, R. S., and Roth, G. S., 1985, Isolated uterine nuclei and cytosol receptors of aged rats exhibit impaired estrogenic stimulation of RNA polymerase II, *Proc. Natl. Acad. Sci. U.S.A.* **81:**7481.

Henry, J. M., and Roth, G. S., 1985, Effect of aging on recovery of striatal dopamine receptors following N-ethoxycarbonyl-2-ethoxy-1, 2-dihydroquinoline (EEDQ) blockade, *Life Sci.* **35:**899.

Hess, G. D., and Roth, G. S. (eds.), 1984, Receptors and Aging, in: *Aging and Cell Functions* (J. E. Johnson, ed.), Plenum Press, New York, pp. 149–185.

Hewish, D. R., and Burgoyne, L. A., 1973, The calcium dependent endonuclease activity of isolated nuclear preparations, relationships between its occurrence and the occurrence of other classes of enzymes found in nuclear preparations, *Biochem. Biophys. Res. Comm.* **53:**475.

Holinka, C. F., Nelson, J. F., and Finch, C. E., 1975, Effect of estrogen treatment on estradiol binding capacity in uteri of aged rats, *Gerontologist* **15:**30.

Hseueh, A. J. W., Erickson, G. F., and Lu, K. H., 1979, Changes in uterine estrogen receptor and morphology in aging female rats, *Biol. Reprod.* **21:**793.

Huang, H. H., Steger, R. W., Bruni, J., and Meites, J., 1978, Changes in patterns of sex steroid and gonadotropin secretion in female rats, *Endocrinology* **103:**1855.

Ito, H., Baum, B. J., Uchida, T., Hoopes, M. T., Bodner, L., and Roth, G. S., 1982, Diminished alpha adrenergic responsiveness in rat parotid acinar cells with normal receptor characteristics, *J. Biol. Chem.* **246**:9532.

Jensen, E. V., 1980, Monoclonal antibodies as probes for estrogen receptor structure, *3rd Intl. Colloq. on Physiol and Chemical Information Transfer in the Regulation of Reproduction and Aging*, Varna, Bulgaria, p. 47.

Jiang, M. J., and Peng, M. T., 1981, Cytoplasmic and nuclear binding of estradiol in the brain and pituitary of old female rats, *Gerontology* **27**:51.

Joseph, J. A., Filburn, C. R., and Roth, G. S., 1981, Development of dopamine receptor denervation supersensitiviey in the neostriatum of the senescent rat, *Life Sci.* **29**:575.

King, R. J. B., and Mainwaring, W. I. P. (eds.), 1974, *Steroid–Cell Interactions*, University Park Press, Baltimore.

Knowler, J. T., and Smellie, R. M. S., 1973, The oestrogen-stimulated synthesis of heterogeneous nuclear ribonucleic acid in the uterus of immature rats, *Biochem. J.* **131**:689.

Kon, O. L., and Spelsberg, T. C., 1982, Nuclear binding of estrogen–receptor complex: Receptor-specific nuclear acceptor sites, *Endocrinology* **111**:1925.

Labarca, C., and Paigen, K., 1980, A simple, rapid and sensitive DNA assay procedure, *Anal. Biochem.* **102**:344.

Larson, L. L., Spilman, C. H., and Foote, R. H., 1972, Uterine uptake of progesterone and estradiol in young and aged rabbits, *Proc. Soc. Exp. Biol. Med.* **141**:463.

Lindell, T. J., 1980, Inhibition of mammalian RNA polymerase, in: *Inhibitors of DNA and RNA Polymerases* (P. S. Savin and R. C. Gallo, eds.), Pergamon Press, New York, pp. 111–137.

Lowry, O. H., Rosebrough, N. J., Farr, A. L., and Randall, R. J., 1951, Protein measurement with the Folin phenol reagent, *J. Biol Chem.* **193**:265.

Masoro, E. J. (ed.), 1981, *Handbook of Physiology in Aging*, CRC Press, Boca Raton, Florida.

Miller, J. K., and Bolla, R., 1981, Influence of steroid-hormone-receptor–protein complexes on initiation of ribonucleic acid synthesis in liver nuclei isolated from rats of various ages, *Biochem. J.* **196**:373.

Mohla, S., DeSombre, E. R., and Jensen, E. V., 1972, Tissue-specific stimulation of RNA synthesis by transformed estradiol–receptor complex, *Biochem. Biophys. Res. Comm.* **46**:661.

Nelson, J. F., Holinka, C. F., and Finch, C. E., 1976, Age related changes in estradiol binding capacity of mouse uterine cytosol, *Abstrs. 29th Ann. Mtg. of the Geront. Soc.*, p. 86.

O'Malley, B. W., and Means, A. R. (eds.), 1978, *Receptors for Reproductive Hormones*, Plenum Press, New York.

Orida, N., and Feldman, J. D., 1982, Age related deficiency in calcium uptake by mast cells, *Fed. Proc.* **41**:822.

Pavlik, E. J., and Coulson, P. B., 1976, Modulation of estrogen receptors in four different target tissues: Different effects of estrogen vs. progesterone, *J. Steroid Biochem.* **7**:1.

Peng, M. T., and Peng, Y. M., 1973, Changes in the uptake of tritiated estradiol in the hypothalamus and adenohypophesis of old female rats, Fertil. Steril. **24**:534.

Peterson, C., and Gibson, G. E., 1983, Amelioration of age-related neurochemical and behavioral deficits by 3, 4-diaminopyridine, *Neurobiol. Aging* **4**:25.

Pitha, J., Hughes, B. A., Kusiak, J. W., Dax, E. M., and Baker, S. P., 1982, Regeneration of beta-adrenergic receptors in senescent rats; study using an irreversible antagonist, *Proc. Natl. Acad. Sci. U.S.A.* **26**:4424.

Pradham, S. N., 1980, Central neurotransmitters and aging, *Life Sci.* **26**:1843.

Roeder, R. G., and Rutter, W. J., 1970, Multiple ribonucleic acid polymerases and ribonucleic acic synthesis during sea urchin development, *Biochemistry* **9**:2543.

Rosner, B. A. and Cristofalo, V. J., 1981, Charges in specific dexamethasone binding during aging in WI-38 cells, *Endocrinology* **108**:1965.

Roth, G. S., 1975, Reduced glucocorticoid responsiveness and receptor concentration in splenic leukocytes of senescent rats, *Biochem. Biophys. Acta* **399**:145.

Roth, G. S., 1979, Horome action during aging, alterations and mechanisms, *Mech. Ageing Dev.* **9**:497.

Roth, G. S., and Hess, G. D., 1982, Changes in the mechanisms of hormone and neurotransmitter action during aging: Current status of the role of receptor and post-receptor alterations, *Mech. Ageing Dev.* **20**:175.

Roth, G. S., and Livingston, J. N., 1976, Reductions in glucocorticoid inhibition of glucose oxidation and presumptive glucocorticoid receptor content in rat adiporytes during aging, *Endocrinology* **99**:831.

Roth, G. S., Karoly, K., Britton, V. J., and Adelman., R. C., 1974, Age dependent regulation of isoproterenol-stimulated DNA synthesis in rat salivary gland *in vivo*, *Exp. Gerontol.* **9**:1.

Saiduddin, S., and Zassenhaus, H. P., 1979, Estrus cycles, decidualization response and uterine estrogen and progesterone receptor in Fisher 344 virgin aging rats, *Proc. Soc. Exp. Biol. Med.* **161**:119.

Shock, N. W., 1962, The physiology of aging, *Sci. Am.* **29**:100.

Singhal, R. L., Valadares, H. R. E., and Ling, G. M., 1969, Estrogenic regulation of uterine carbohydrate metabolism during senescence, *Am. J. Physiol.* **217**:793.

Soriero, A. A., 1980, Autorodiagraphic study of the effect of estrogen on *in vivo* incorporation of ^{3}H-uridine into uterine smooth muscle and stromal RNA in the aging ovariectomized mouse, *J. Gerontol.* **35**:167.

Soriero, A. A., and Talbert, G. B., 1975, The effect of estrogen on protein and RNA concentration and on *de novo* synthesis of RNA in the uterus of aging overiectomized mice, *J. Gerontol.* **30**:264.

Steger, R. W., Huang, H. H., and Meites, J., 1979, Relation of aging to hypothalamic LHRH content and serum gonadal steroids in female rats, *Proc. Soc. Exp. Biol. Med.* **161**:251.

Webster, R. A., and Hamilton, T. H., 1976, Comparative effects of estradiol 17-β and estriol on uterine RNA polymerases I, II, and III *in vivo*, *Biochem. Biophys. Res. Comm.* **69**:737.

Woll, W. W., Duffy, J. J., Giese, N. A., and Lindell, T. J., 1981, Nuclear isolation by a modified method of Hewish and Burgoyne: Implication for the study of nuclear enxymology, *Life Sci.* **29**:2709.

Zava, D. T., Harrington, N. Y., and McGuire, W. L., 1976, Nuclear estradiol receptor in the adult rat uterus: A new exchange essay, *Biochemistry* **15**:4292.

III

AGING AND THE REPRODUCTIVE YEARS

8

PATERNAL AGE AND EFFECTS ON CHROMOSOMAL AND SPECIFIC LOCUS MUTATIONS AND ON OTHER GENETIC OUTCOMES IN OFFSPRING

ERNEST B. HOOK

Introduction

Investigation of effects of parental aging upon genetic effects in offspring is of significance for several reasons: (1) accurate description of these effects will provide appropriate data for genetic counseling and enable informed decisions regarding reproduction; (2) the identification of high-risk groups by age will facilitate efficient delivery of specific secondary "preventive" techniques, (e.g. amniocentesis and prenatal diagnosis) to appropriate target groups in the population—at least to those to whom such techniques are ethically acceptable; (3) identification of the biological processes responsible for the association of aging with adverse genetic outcomes may eventually lead to primary prevention and new methods of secondary prevention.

In such an analysis one must distinguish "simple" genetic disorders from diseases whose genetic contribution is more complex but less certain quantitatively. The "simple" genetic disorders include: (1) monogenic or so called specific locus disorders—the classical Mendelian recessive and dominants; (2) germinal structural chromosomal rearrangements such as rings, translocations, deletions, etc.; and (3) numerical chromosome abnormalities such as $47, +21$

ERNEST B. HOOK • Bureau of Maternal and Child Health, New York State Department of Health, Albany, New York 12237; and Department of Pediatrics, Albany Medical College, Albany, New York 12208.

resulting in Down syndrome. (One could also add to this list disorders of the genes and "chromosomes" of mitochondria and other cytoplasmic inclusion bodies, but we have very little knowledge as yet of these.) The simple genetic disorders are a small proportion of all diseases. The much greater number of disorders in the remaining "complex" category have varying genetic determination involving polygenes of small effect and in some instances major genes of significant effect. In addition these outcomes tend to be more subject than the "simple" disorders to environmental influence. Among adverse reproductive outcomes, such events as fetal deaths and most birth defects may be regarded as examples of genetically "complex conditions. A strict sorting out of genetic contribution to these outcomes is often difficult. One may attempt to define some of the simple genetic components of such "complex" events by doing for example, chromosome studies on all fetal deaths, or on all children with birth defects, or by enumerating the detectable monogenic disorders among those with birth defects. But this will reveal only a porportion of those affected. Among the remaining cases the genetic contribution may be important, but difficult to determine qualitatively or quantitatively.

The paternal contribution to *simple* genetic disorders in offspring may be influenced by three "primary" genetic processes: *mutation* (in the broad sense, including nondisjunction), *recombination*, and *segregation*. We know most about age effects (or lack of them) on mutation. Little is known about recombination. Indeed, in humans the effects of recombination may sometimes be difficult to separate formally from those of mutation. Hardly anything is known of paternal factors that influence segregation. Association of paternal age with simple genetic disorders may reflect effects upon all three of these processes. In addition; *expression* of a simple genetic disorder transmitted to an offspring may be affected by paternal factors. (For example, a paternal mutation at one locus may effect expression in the offspring of an allele at another locus.) Moreover, expression of simple genetic mutants may of course be influenced by nongenetic processes associated with paternal factors. (See the discussion below on Huntington's disease.)

Disorders with *complex and uncertain genetic contribution* may be influenced by three "primary" genetic factors just noted as well as nongenetic mechanisms. The father may affect the developing fetus indirectly. He may transmit possible teratogenic contaminants during gestation for instance. His effect may be even more indirect, through influence upon socioeconomic, nutritional, psychological, and medical factors affecting the fetus or infant. Thus the association of changes of paternal age with complex categories of disorders in offspring such as fetal death or some birth defects may, in principle, be mediated both by both genetic and nongenetic factors. But I believe at the present state of our knowledge concerning birth defects of "complex" etiology, the presence of an association with advanced paternal age (independent of maternal age) is suggestive (but not definitive) evidence for a genetic effect, one mediated by specific locus mutations. The evidence for this is discussed below.

Methodological Issues

Parental ages and the differences between parental ages may vary considerably with space, time, and population. Maternal and paternal ages are very closely correlated so that the effects of one may be confounded with that of the other. (Birth order effects are another possible confounding factor. These however appear to be of significance only for some of the "complex" genetic outcomes, not the "simple" genetic disorders.) A simple method of analysis is to compare the paternal age-maternal age differences between cases and controls. This is often done crudely because frequently controls are not matched precisely on maternal age or selected from the same jurisdiction or temporal period as the cases.

Because paternal age–maternal age differences may vary with maternal age, one method of controlling for this is to *match* each case with a control or, preferably, group of (appropriate) controls of the same maternal age and analyze the differences in paternal ages or equivalently the maternal age–paternal age differences. Ideally, matches should be by 1-year maternal age interval and of course the other population variables mentioned should also be controlled (see Hook *et al.*, 1981a).

A disadvantages of this method is that while an effect may be demonstrated, the results are not directly useful for clinical prediction and genetic counseling. The mean paternal ages of cases may be, say, 3 years greater than that of controls, but this provides no obvious guide as to counseling a 50-year-old man about this relative risk of having an affected child. A regression approach in which paternal age and maternal ages are entered into the equation will yield equations predicting age-specific rates (or risk ratios). Both of these methods however, have a disadvantage in that if applied crudely, a positive paternal age effect at the upper extreme of age may be obscured if not masked completely by an effect at the lower extreme of age. A comparison of variances, however, might suggest such a pattern in the delta analysis. Complex regression analyses involving several powers of age or, preferably, application of simple regression analyses to restricted ranges of ages would also reveal such variation in effects. In addition, nonparametric tests comparing expected and observed numbers above certain paternal age boundaries might also reveal such variation (Stene *et al.*, 1981). The latter method also must be strictly controlled by 1-year maternal age interval because some types of comparison of observed and "expected" values can lead to pitfalls (Lamson *et al.*, 1980), particularly if there is a very strong independent maternal age effect as in 47, +21 (Hook and Cross, 1982).

An illustration of the necessity for matching on maternal age in case-control analyses is provided by the data in Table I.

These give paternal age–maternal age differences in New York. The data show variation in time, race, and maternal age.

Numerical Chromosome Disorders

The bulk of the data on this category of outcomes are on Down syndrome. Most studies (on live births) are on the phenotype only, but as about 95% of such instances are attributable to 47,+21, the inferences derived from such phenotypic studies will apply primarily to 47,+21.

Hook and Cross (1982) summarized some of the history of these investigations as follows. Down's syndrome was known from early studies to be markedly associated with elevated parental age. Jenkins (1933) and Penrose (1933) showed independently that maternal age was primarily responsible for the association, but could not exclude a modest paternal age effect. Penrose in many of his later writings appeared to assume that paternal age was not of importance and that only maternal age was significant (e.g., Penrose and Smith, 1966). But some observers pointed out a modest paternal age effect could not be excluded by the analyses done up to that time (Mantel and Stark, 1966; see also Lilienfeld, 1969). The issue was reopened with the discovery that in about 20% of cases of 47,+21, the extra chromosome is of paternal origin (e.g., Hansson and Mikkelsen, 1978). Despite the fact there is no known positive paternal age effect for 47,XYY, in which the extra chromosome is known to be of paternal origin (see below), the discovery of instances of two patroclinous 21 chromosomes led some observers to the apparent inference that paternal age was also a significant risk factor for Down's syndrome. (Indeed, the article by Holmes [1978a] implying this was misinterpreted by some of the lay press to mean that it was *exclusively* paternal and not maternal age that was associated with Down's syndrome. [Hook, 1978; Holmes 1978b].)

Numerous recent investigations of a possible paternal age effect have been done. One of the limitations of such investigations is that maternal age and paternal age are so highly correlated that it is very difficult to demonstrate a modest effect of one variable in the face of a strong effect of the other. In addition, maternal age-specific rates of Down's syndrome rise rapidly with maternal age while fertility plunges even more quickly, (see, e.g., Hook, 1981) leading to possible statistical artifacts. As an example, one of the first recent studies of paternal age since the discovery of patroclinous extra 21st chromosomes was that of Stene *et al.* (1977), who claimed "statistically significant" evidence for a twofold paternal age effect for cases born to men 55 and over. While no evidence was reported for an effect in men younger than this age, they also inferred a strong paternal age effect at ages below 55. Their analysis, however, was by crude age intervals. Erickson (1978) showed that he could construct an artifactual paternal age effect using the methodology of Stene et al. (1977) applied to U.S. data. This disappeared when he used a more appropriate method. If he analyzed data by 1-year interval he found no evidence for a paternal age effect. Similarly, Matsunaga et al. (1978) reported statistically significant evidence for a paternal age effect for men 55 and over in data analyzed by 5-year interval but Lamson et al. (1980) showed that the methodology used here could produce artifacts which disappeared when analysis was by 1-year interval. Subsequent reanalysis of some of these data by Cross and Hook (unpublished) by 1-year interval revealed that a trend originally reported to an increase in men 55 and over could still be found in the data of Matsunaga et al, but was weaker and no longer significant at the 0.05 level.

Investigations of live births by other investigators that have controlled by 1-year maternal age interval have yielded differing trends. Some showed no evidence for a paternal age effect, (Erickson, 1978; Regal et al., 1980, Hook et al., 1981a); others found a suggestive but not a significant effect, (Erickson, 1979); and others found significant positive effects, albeit not of strong magnitude (Hook et al., 1981a; Erickson and Bjerkedal, 1981). An interesting recent report from Ohio found statistically significant evidence for a *negative* paternal age effect in some data (Roecker and Huether, 1983). Before the latter study was published, it was suggested that a weak effect, of the order of about 1% increase in risk with each year of paternal age, was consistent with the negative and positive reports to date (Hook et al., 1981a). Tables have been constructed projecting what these expected rates should be at various combinations of maternal and paternal ages (Hook and Cross, 1982). But in view of the Ohio report even these may project too strong an effect. No studies in *live births* to date done by 1-year interval revealed positive effects as large as those claimed by Stene et al. (1977).

With regard to prenatal studies, Stene et al. (1981) have published data on 60 cases of 47, +21 diagnosed at amniocentesis and claimed a very strong paternal age effect similar in magnitude to that claimed earlier by Stene et al. (1977) in analysis of livebirths by 5-year interval. They projected from these observations putative risks for women of *any* age married to men of 41 or older sufficient to justify amniocentesis (Stene et al., 1981). But no evidence for such a paternal age effect was found in other, larger studies in New York State (Hook and Cross, 1982), in a European collaborative study (Ferguson-Smith and Yates, 1984) or in France (Roth et al., 1983). Moreover, there is no evidence that paternal age is increased in 47, +21 livebirth cases in which the extra chromosome is of paternal origin (Hook and Regal, 1984).

It is uncertain why analyses by Stene and other Danish workers got strong positive results (or at least results interpreted by them this way) while others got negative, indeterminate, or weaker positive results. Stene has implied that

Table I

Paternal Age–Maternal Age Differences by Maternal Age, Birth Year, Race, and Region in New York State

Maternal Age	1960			1970			1980		
	N (1000s)	Mean	S.D.	N (1000s)	Mean	S.D.	N (1000s)	Mean	S.D.
Upstate New York whites									
15–19	13	3.8	3.2	13	3.2	3.2	7	3.7	3.7
20–24	52	3.4	3.4	51	3.2	3.2	31	3.7	3.7
25–29	51	3.0	3.7	46	2.8	3.8	40	2.3	3.6
30–34	36	2.8	4.1	20	2.9	4.3	22	1.8	3.9
35–39	18	2.4	4.5	9	2.5	4.5	5	1.7	5.1
40–44	4.4	1.9	4.9	2.4	2.1	4.7	0.8	1.7	5.7
45–49	0.2	2.1	4.5	0.1	1.9	4.3	0.03	1.5	6.2
Upstate New York blacks									
15–19	1.1	4.2	3.3	1.3	3.4	3.4	1.0	3.7	4.0
20–24	2.9	3.8	4.3	2.8	3.3	4.0	2.3	3.9	4.7
25–29	2.1	3.7	4.9	2.0	3.3	4.9	1.9	3.1	4.8
30–34	1.3	3.6	5.4	1.0	3.5	5.3	1.2	2.4	5.0
35–39	0.6	3.4	5.6	0.5	3.0	6.2	0.4	1.9	5.5
40–44	0.1	3.3	6.2	0.1	3.1	5.9	0.05	2.5	7.1
45–49	0.005	1.8	2.3	0.01	−0.5	5.1	<0.005	—	—

New York City whites[a]

15–19	7.3	4.3	3.8	4.3	4.8	4.7
20–24	31	3.5	3.9	14	4.0	4.6
25–29	28	3.2	4.3	17	3.0	4.4
30–34	13	2.9	4.9	11	2.3	4.9
35–39	5.0	2.3	5.3	3.6	1.8	5.9
40–44	1.2	1.8	5.4	0.5	1.6	6.8
45–49	0.05	0.4	6.5	0.02	0.2	8.0

New York City blacks[a]

15–19	3.0	3.4	3.3	2.2	4.0	4.3
20–24	8.5	3.4	3.9	6.1	4.1	4.8
25–29	6.9	3.2	4.5	6.1	3.5	5.2
30–34	3.8	2.8	5.1	4.1	2.7	5.5
35–39	1.6	2.6	5.8	1.6	2.0	6.2
40–44	0.4	1.9	5.4	0.3	1.1	6.7
45–49	0.015	0.2	5.2	0.015	3.5	6.5

[a]1960 data unavailable for New York City.

methodological errors, incomplete sampling, and other such factors are responsible for at least some of the failure by others to confirm his reports. Some suggested that temporal and geographical variation in effects may be responsible (Hook and Cross, 1982) although this does not explain why there should be such variation. Others have speculated Stene's results were the results of chance effects (Ferguson-Smith and Yates, 1984). But the probability of observing results by chance alone in the second study of Stene *et al.* (1981) appears very small.

At present, I believe there appear no grounds to regard paternal age as a (positive) risk factor for 47, +21 in most if not all jurisdictions. Admittedly at the uppermost extremes of paternal age, e.g., 55 and over, where there are relatively few observations, it is still difficult to exclude an effect. But certainly none has been established.

Studies of paternal age effects upon other chromosome abnormalities are much less extensive. Data on 47, +13 and 47, +18 live births from the New York State Chromosome Registry appear in Table II. Maternal age and year of birth is controlled by 1-year interval. (The "control" age for any case was the mean paternal age for all live births born to mothers of the same age as the

Table II

Paternal Age in Selected Chromosome Disorders (Live Births): Data from New York State Chromosome Registry[a,b]

Trisomies	Cases	Mean	S.D.	Maternal age-matched controls Mean	S.D.	Difference Mean	S.E.M.
Trisomies							
47, +21	1637	32.8	8.9	32.9	6.9	−0.1	0.3
47, +13	72	28.7	5.7	29.0	4.8	−0.3	0.9
47, +18	142	32.3	8.5	32.5	6.8	−0.3	0.9
Unbalanced *de novo* Robertsonian translocations (mutants)							
D/21 translocation	32	28.7	7.0	28.3	4.9	+0.4	1.5
G/21 translocation	41	30.8	8.0	30.0	5.1	+0.8	1.5
D/13	12	30.1	7.3	29.9	5.6	+0.2	2.7
Unbalanced inherited Robertsonian translocations							
D/21	11	27.5	4.1	28.1	4.6	−0.6	1.9
G/21	2	28.5	12.0	28.7	9.4	−1.2	10.8
D/13	5	31.2	8.2	29.8	5.0	+1.4	4.3

[a]From Hook (1984).
[b]All cases matched exactly with controls on maternal age. (Control data is from vital statistics.)

case's mother, and in the same year.) As may be seen there is even less evidence for a paternal age effect here than for $47, +21$. (It should be noted that in these and related studies, when paternal age is controlled there is a major effect of maternal age.)

In studies of the XXY genotype in livebirths, Carothers *et al.* (1978) using regression methods, found evidence for a positive (independent) maternal age effect but no paternal age effect. (It is of interest that in about 75% of cases with XXY genotype the "extra" X, when it can be identified, is found to be of maternal origin.)* For the XXX genotype there was also a rather strong positive parental age effect, but it could not be decided whether it was paternal or maternal age that mediated the effect. For the XYY genotype, there was a small but significant *negative* relationship between parental age and rate of abnormality but again it could not be decided if it was maternal and paternal.

With regard to the 45,X, the predominant chromosome pattern associated with Turner syndrome, the situation is similar to that for 47,XYY. There is a slightly negative association with parental age but it is not possible to determine if this is attributable to a paternal age or maternal age effect. The relative risk for a livebirth at age 40 (maternal or paternal) is about 0.6 of the risk at age 30 (Carothers *et al.*, 1980). Similar observations have been made in abortuses (Warburton *et al.* 1980). Of interest the parental origin of the X chromosome in 45,X individuals was determined in 35 cases by Xga typing, (15% of the total studied) and found to be of maternal origin in 34 cases. This means that in these 45,X cases it was primarily the *paternal* sex chromosome that was lost, either through paternal nondisjunction, anaphase loss, or later during mitoses of the conceptus. In the sole case in which the X was found to be of paternal origin—implying loss of the maternal X—the maternal age was 36 and paternal age 33 (compared to values of 27.2 ± 0.90, and 29.9 ± 1.46 for the others).

Hatch (1983, 1984) reported on paternal age effects on chromosomal abnormalities in embryonic and fetal deaths. She found for monosomy X, (45,X) about a 1-year increase in paternal age over controls, significant at the 0.07 level. This is of interest since maternal age of 45,X individuals has been found to be younger than controls (see above).

Hatch also observed a significant positive effect on triploidy ($\beta = 1.34$) but cites work by others (presumably the Hawaii group) showing a negative association, so that in balance no firm conclusion is possible.

With regard to trisomies in general, in phase 1 of her study there was a significant positive paternal age effect that varied with age of the mother (Hatch, 1983). The paternal effect was slightly raised for older mothers ($+1.5$ years for fathers married to mothers ≥ 40) and even higher for younger mothers ($+3$

---

*Ferguson-Smith and Yates (1984) however, did observe a positive paternal age effect for 23 cases of 47, XXY diagnosed prenatally.

years for fathers married to women age 20). In phase 2 of her study, if anything the trends were reversed, leaving no statistically significant evidence for any paternal age effect in the entire sample (Hatch, 1984). This is consistent with the results in most studies on livebirths and at amniocentesis regarding 47, +21, 47, +18, and 47, +13 (see above). Hatch did not report results on specific trisomies.

Structural Chromosome Abnormalities

Structural chromosome rearrangements are likely to be similar in etiology to at least some specific locus mutations. Both involve some alteration in the structure of the genetic material. Indeed, formally it may be hard to distinguish very small structural chromosome changes from so-called specific locus mutations. At some level (above alteration in a single base in a codon) a strict separation may be impossible.

There are two categories of data on structural chromosome rearrangements, those on mutants and those on inherited rearrangements. Most of the results are on the former.

Chamberlin and Magenis (1980) in a review of parental origin of *de novo* chromosome rearrangements found that for 20 Robertsonian translocations 13 were of maternal origin and seven paternal. On the other hand for 17 presumptive mutant rearrangements that were *not* Robertsonian, 13 were of paternal origin and four of maternal origin. If paternal age effects for mutant structural chromosome abnormalities are similar to those for specific locus mutations (see below); then the results cited above would predict a higher paternal age effect upon non-Robertsonian rearrangements than Robertsonian rearrangements. But the sparse data available are contrary to this expectation.

Butler and Palmer (1983) studied the Prader–Willi syndrome, which may be produced by a interstitial deletion of chromosome 15 (q11 to q13). They found such a detectable lesion in 21 of 39 individuals studied. In 13 of these 21 cases, all of which were known mutants they could identify the parent in which the mutation occurred. In all 13 the mutation had occurred in the father, consistent with and even stronger than the trend observed by Chamberlin and Magenis. The mean parental ages for these 13 cases were 28.8 years for the father and 26.2 years for the mother (a difference of 2.6 years). The mean ages for the cases with apparently normal chromosomes were 30.3 years for the father and 25.9 for the mother, a difference of 4.4. These data are strong evidence against a (positive) paternal age effect for mutant 15q deletions. (They even suggest a negative effect. See below.) The data are all the more significant with regard to lack of a positive effect because the structural chromosome mutations have been shown to occur in the father.

The above results are on mutations in live births. Data on fetuses diagnosed at midtrimester amniocentesis are also available.

Structural mutants diagnosed "incidentally" at prenatal amniocentesis were compared with 13,040 normal cases (Hook *et al.*, 1983). There was no significant association with paternal age. While extensive data on maternal age were given, no specific results were presented for paternal age. Further unpublished data from that study are presented in Table III.

Niebuhr (1978) in a study of the cri du chat syndrome accumulated from cases around the world born in many different years used 1962 data from New York City births for controls. The paternal mean age associated with sporadic cases of 5p − (simple deletions or other causes that were the result of presumptive mutation in the most recent generation) was 30.8 years (± 5.7) and for maternal ages was 26.0 years (± 5.4). The difference of 3.9 years appears somewhat higher than would be expected, but there is no way to determine if it is statistically significant or what category of structural rearrangements is involved.

For structural rearrangements resulting in X *iso*-chromosomes, in livebirths there was no association with parental ages (Carothers *et al.*, 1980) in a sib comparison. But mean paternal age–maternal age difference was greater for

Table III

Mutant Structural Chromosomal Rearrangements Detected Incidentally at Amniocentesis and Reported to the New York State Chromosome Registry[a]

	N	Maternal age		Paternal age		Paternal–maternal age difference	
		Mean	S.D.	Mean	S.D.	Mean	S.D.
Unbalanced							
Robertsonian	1	38	—	41	—	+ 3	—
Markers, fragments	13	38.5	2.7	38.3	4.9	−0.2	3.6
Deletions	1	37	—	37	—	0	—
Other	1	37	—	35	—	−2.0	—
All	17	38.3	2.4	38.2	4.6	0.1	4.4
Balanced							
Robertsonian	4	36.0	1.8	38.0	7.5	+2.0	6.9
Inversions	1	36	—	55	—		—
Other	6	36.5	3.1	35.7	3.9	−0.8	4.3
Reciprocals	11	36.3	2.5	38.3	7.5	+2.0	7.6
All Robertsonian	5	36.4	1.8	38.6	6.7	2.2	6.0
All abnormalities except markers and fragments	15	36.5	2.2	38.1	6.7	1.6	6.7
All abnormalities	28	37.4	2.6	35.2	5.8	0.8	5.4
All normals	13,040	35.8	±4.0	37.6	±6.0	+1.8	5.7

[a]Analysis of data in Hook *et al.* (1983; unpublished).

this abnormality associated with Turner syndrome than for the 45,X pattern in all four series examined. In 14 of 80 cases (17.5%) the father was 10 or more years older than the mother, whereas for 45,XO this proportion was 11/226 = 4.9%. Carothers *et al.* suggest the weight of evidence from this study as well as some others (see their paper for references) on live births favors a "straightforward" increase in risk of iso-chromosome X with paternal age. They attribute the discrepancies, specifically the failure to detect a paternal age effect in their sibship analysis, to statistical fluctuation. (Of interest is a report by Mayo *et al.* (1976) on single gene disorders, which suggests that a sib analysis results in a *bias* to *under*estimation of parental age effects.) Nevertheless, at least one well-controlled study (Breg *et al.*, 1980) failed to find evidence for a paternal age effect for iso-chromosome X.

Data on mutant unbalanced Robertsonian translocations appear in Table II. A mutant Robertsonian translocation may result in an offspring with either unbalanced or balanced abnormality. Mutant balanced abnormalities are difficult to detect because they are not associated with abnormal phenotype. In live births almost all data are on the unbalanced Robertsonian translocations. Any factor associated with *de novo* unbalanced Robertsonian rearrangements could thus result from either an effect upon mutation producing a translocation, or segregation. That is, such a factor could increase the rate of formation of such mutations or, without any effect on mutation rate, could increase the rate of de novo *unbalanced* zygotes involving such translocations. In the data there is a very slight but nonsignificant positive paternal age effect associated with the three *de novo* unbalanced Robertsonian translocation in livebirths.

With regard to structural chromosome abnormalities known to be *inherited*, Carothers *et al.* (1978) found no evidence for a significant parental age effect, either paternal or maternal, upon 113 heterogeneous abnormalities in live births. (These were controls for the sex numerical chromosome abnormalities discussed above.) Of course trends in one direction for one category of structural chromosome abnormality could have been obscured by trends in the other direction for another.

Other data indicate no trend for inherited Robertsonian abnormalities (Table II) although inferences may be biased by ascertainment of cases.

With regard to inherited abnormalities diagnosed at amniocentesis, summary data appear in Table IV. It is emphasized that the presence of structural abnormalities were discovered "incidentally" (Hook *et al.*, 1984). They were not suspected prior to the study (i.e., cytogenetic investigation) so that age effects are not biased by ascertainment through birth of a previous affected child, as might have been the case for the results of Carothers *et al.* (1978). Of interest, there is evidence in the data for a paternal age effect, but not a maternal age effect, upon the inherited reciprocal balanced rearrangements. Such an effect might be mediated by effects upon segregation.

Table IV

Parental Ages and Age Differences For Inherited Rearrangements Detected
Incidentally at Prenatal Diagnosis[a]

	Origin	N	Maternal age		Paternal age		Paternal–Maternal age difference	
			Mean	S.D.	Mean	S.D.	Mean	S.D.
Unbalanced								
Markers	Maternal	5	35.8	2.6	34.8	4.4	−1.0	5.3
	Paternal	1	36	—	28	—	−8.0	—
	Both	6	35.8	2.3	33.7	4.8	−2.2	5.6
Derived abnormality	Maternal	1	40	—	37	—	−3.0	—
	Paternal	1	39	—	36	—	−3.0	—
	Both	2	39.5	0.7	36.5	0.7	−3.0	0.0
All unbalanced	Maternal	6	36.5	2.9	35.2	4.0	−1.3	4.8
	Paternal	2	37.5	2.1	32.0	5.7	−5.5	3.5
	Both	8	36.8	2.6	34.4	4.3	−2.4	4.7
Balanced								
Robertsonian	Maternal	2	37.5	5.0	42.0	8.5	+4.5	3.5
	Paternal	3	35.7	1.2	33.3	7.8	−2.3	8.6
	Both	5	36.4	2.8	36.8	8.4	+0.4	7.4
Inversions	Maternal	5	35.8	1.9	36.8	6.5	+1.0	4.9
	Paternal	7	36.1	3.4	38.6	4.4	+2.4	3.8
	Both	12	36.0	3.0	41.5	7.8	+5.1	6.4
Other (reciprocal)	Maternal	8	35.4	3.2	37.4	5.7	+2.0	4.1
	Paternal	11	37.2	2.7	44.6	8.0	+7.4	6.9
	Both	19	36.4	3.0	41.5	7.8	+5.1	6.4**
All Balanced	Maternal	15	35.8	2.9	37.8	6.1	+2.0	4.2
	Paternal	21	36.6	2.8	41.0	7.9	+4.3	7.0
	Both	36	36.3	2.8	39.6	7.2	+3.4	6.0
Balanced and Unbalanced		44	36.4	2.8	38.7	7.1	+2.3	6.2
Normal karyotypes		12,035	36.3	3.5	38.0	6.4	+1.8	5.8

[a]Analysis of data in Hook et al. (1984).
**$P < 0.05$.

Specific Locus Mutations

There is little if any maternal age effect upon specific locus mutations. But statistically significant paternal age effects have been demonstrated for several autosomal dominant disorders (see below). This sex difference in effect of aging upon specific locus mutations may be attributable, at least in part to sex differences in gametogenesis. Vogel and Motulsky (1979) summarize the rationale for this as follows.

Germ cells in the ovary enter meiosis before birth of the individual. Between the 3rd month and 7th month of gestation in the female fetus they progress through zygotene up to diplotene of the first meiotic (reduction division). By the 9th month of gestation they have entered an "arrested" stage called dictyotene. If it does not subsequently degenerate, the germ cell will persist in dictyotene until ovulated after puberty. It then completes meiosis-I and enters meiosis-II, the reduction division. But the latter is completed only after fertilization. Between birth of a female child and fertilization of an ovum in adulthood, the egg will apparently undergo no further replications and only two meiotic divisions. Before birth of a female there have been, it is estimated, 20 cell divisions between the time of formation of the primordial germ cell and development of the ova. Thus, in adulthood at the time of fertilization of the ova, there have been about 22 cell divisions from the primordial germ cell to the fertilized egg in the next generation. The total number of germ cells in the female embryo and fetus increase it is estimated from 6×10^5 at 2 months of gestation to 6.8×10^6 at 5 months gestational age, but then diminishes to 2×10^6 at birth. These numbers may be contrasted with the numbers of germ cells in the male.

Spermatogonia are formed in the testes through progressive continuous division of germ cells. By puberty there are an estimated 4.3×10^8 to 6.4×10^8 spermatogonia. These have resulted from 29 to 30 cell divisions per testes from the time of male primordial germ cells. At the onset of sperm formation, at puberty, the undifferentiated spermatogonia (Ad or dark cells) continue to divide. The Ad cells have it is estimated a 16-day division cycle to the time of production of two Ad daughter cells. But some divisions of Ad cells now result in different daughter cells, Ap (pale) spermatogonia, in which final sperm development sequence begins. After two further divisions the latter form spermatocytes-1 which then undergo the first (reductional) meiotic division to become spermatocytes-2. They then undergo the second (equational) meiotic division to form spermatids. These then mature into spermatozoa.

After puberty, there are about 23 divisions per year of the undifferentiated spermatogonia. It is estimated sperm produced at age 28 have resulted from about 380 cell divisions and at age 35 to 540 divisions. This is in contrast to the estimated 22 cell divisions for ova at both of these ages at the time of fertilization.

[For references and further discussion concerning the above see Vogel and Motulsky (1979).]

At least some specific locus mutations are known to be dependent upon DNA replication. Each mitotic cell division is associated with one previous doubling of DNA. As there are a fixed number of cell divisions prior to meiosis in the female, irrespective of age, but an increasing number of cell divisions in the males with increasing age, a paternal age effect of greater strength than a maternal age effect (if any exists) may be expected. In addition, at any age of reproduction, there have been more previous cell divisions resulting in sperm than ova. Thus there might well be higher average specific locus mutation rates in males than in females. There are data consistent with both of these theoretical expectations although agreement is not unanimous.

At present, sex differences in mutation for autosomal dominant traits are not easily studied. For X-linked recessive lethal disorders, however, indirect evidence is available. The ratio of mutant cases to inherited cases for X-linked lethal recessives is expected to be 1 : 2 if mutations rates are equal in the sexes but will be lower if the mutation rate is higher in males (and conversely). In fact, ratios much lower than 1 : 2 have been reported for several X-linked disorders including Lesch–Nyhan disease, hemophilia, and muscular dystrophy (for references and discussion see Francke *et al.*, 1981; Lubs, 1981; Bucher *et al.*, 1980). But agreement is by no means unanimous (see, e.g., Thompson *et al.*, 1981; Yasuda and Kondo, 1982). I believe the best evidence is on Lesch–Nyhan disease (Francke *et al.*, 1981), where despite some methodological problems, the paucity of inherited cases—which implies a higher mutation rate in males—appears convincing.

With regard to an effect for paternal aging upon specific locus mutation, there are numerous studies, but few have controlled strictly for maternal age. One of the difficulties is that cases of mutant disorders are usually collected from a variety of clinical sources and heterogeneous reports, over a period of some years, and in a number of different jurisdictions. But reference data on controls are usually derived from a single jurisdiction and time period, which is not necessarily appropriate for comparison. To cite an extreme example, Connor and Evans (1982) in a study of fibrodysplasia ossificans progressiva from all of England and Wales had to use reference data on paternal age reported from the city of London and from Australia.

I will take the data in the literature on paternal age effects for specific locus mutation at face value but I have serious reservations about how some of the comparisons were made and the validity of the differences claimed. Almost all of the analyses have been crude case-''control'' comparisons. In most instances the reference data are not clearly appropriate for comparison with all cases included.

In his review of the data, Vogel (1983) has cited achondroplasia, acrocephalocephalosyndactyly (Apert's syndrome), and Marfan's syndrome among others as showing strong evidence of a paternal age effect (see also Table V). On the other hand, bilateral retinoblastoma, neurofibromatosis, and tuberous sclerosis, which are also autosomal dominants, show little if any paternal age effect. (Neurofibromatosis is of interest in that its mutation rate, about 10^{-4} per

Table V

Paternal–Maternal Age Differences for Mutations Resulting in Various
Defined Dominant Disorders[a]

	Cases	Controls	Difference ("excess" in cases of paternal age)
Achondropolasia	6.6	4.7	+1.9
	5.6	3.3	+2.3
	3.4	2.3	+1.1
	7.2	2.6	+4.6
Apert's syndrome	5.1	3.0	+2.1
	4.2	3.2	+1.0
Marfan's syndrome	8.0	2.5	+5.5
	7.3	3.3	+4.0
Bilateral retinoblastoma	4.1	3.4	+0.7
	5.5	3.8	+1.7
	1.8	3.1	−1.3
Neurofibromatosis	4.8	4.7	+0.1
Osteogenesis imperfecta	3.4	4.7	−1.3
Tuberous sclerosis	4.6	4.7	−0.1
	2.8	2.3	+0.5
Aniridia	3.2	4.6	−1.4

[a]Calculated from data in Vogel and Rathenberg (1975).

gamete per generation, is the highest known for any specific locus mutation.)
It may be coincidental that the latter group all involve tumors. But osteogenesis
imperfecta also has no evidence for effect.

Riccardi *et al.*, (1984) analyzed data on mutations resulting in neurofibromatosis. They controlled for year of birth but not for geographical region. They
reported an increase in both maternal age (+1.4 years) and paternal age (+3.15
years) and concluded the maternal age association was secondary to the paternal
age effect. They used the method of Matsunaga (1978) to control on maternal
age in evaluation of paternal age. This method, however, has been shown to
result in artifacts leading to overestimates of paternal age effects (Lamson *et
al.*, 1980). But while this objection may hold for Down's syndrome—the subject of Matsunaga's study—which has a strong maternal age effect, it may not
hold for disorders such as neurofibromatosis, which do not have strong maternal
age effects. It is not clear in the Riccardi study why nationwide U.S. data were
used for controls, whereas the cases affected were those seen at Baylor University in Houston and were, presumably, primarily from Texas. This geographic
difference between cases and controls may have introduced a distortion in apparent parental age effects.

One of the difficulties in study of achondroplasia is the relatively recently recognized problem of heterogeneity. Other entities may have been included with achondroplasia in early mutation studies. Oberklaid *et al.* (1979) in a study from Victoria, Australia, probably ascertained every case of achondroplasia and carefully distinguished them from other dwarfs. They report that their analysis of "expected" numbers affected using statistics from Victorian births "in the appropriate years" demonstrated by a "sophisticated analysis" a paternal age effect, but no maternal age effect. While the results are not quantitated, they suggest that heterogeneity of cases is not confounding paternal age effects.

With regard to retinoblastoma specifically, some of data were reviewed earlier by Tunte (1972). There is a great deal of diversity in the results on

Table VI

Paternal–Maternal Age Differences for Various Sporadic Cases of Dominant Disorders and Disorders of Unknown Etiology[a]

	N	Observed difference	Excess difference correct for population value	
			Mean	S.E.M.
Defined dominants				
Basal-cell syndrome	12	+5.2	+2.5	2.0
Waadenburg syndrom	22	+4.3	+1.6	1.1
Crouzon disease	38	+5.3	+2.9	6.9
Cleidocranial dysostosis	32	3.3	+1.1	0.9
Oculodentodysplasia	11	4.2	+1.7	2.0
Treacher–Collins syndrom	98	4.2	+1.0	0.6
Multiple exostosis	14	4.5	+1.7	0.6
Pfeiffer syndrome	10	+1.8	−0.5	0.7
Mean for all 8 disorders			+1.5	0.4
Sporadic disorders of unknown etiology				
Progeria	18	+4.4	+2.6	1.3
Acrodysostosis	11	+4.9	+2.4	0.9
Hallerman–Streiff syndrome	37	+3.4	+1.0	0.7
Cerebral gigantism	50	.9	+0.6	0.7
Goldenhar syndrome	52	+3.2	+0.4	0.4
Cornelia de Lange syndrome	59	+3.5	+0.9	0.6
Russell–Silver syndrome	39	+3.7	+1.2	0.8
Rubinstein–Taybi syndrome	153	+2.8	+0.2	0.3
Prader–Willi syndrome[b]	50	+1.3	−1.2	0.4
William's syndrome	29	+2.3	+0.4	0.6
Mean for all 10 disorders			+0.85	0.3

[a]Calculated from data of Jones *et al.* (1975).
[b]Chromosome disorder associated with 15q deletion.

sporadic retinoblastoma. But if there is any paternal age effect, it is for bilateral, not unilateral, tumors (as might be expected since it is only the former that are "classical" dominants).

Some data are available on dominants in which the biochemical lesion is known. Stamatoyannoupoulos *et al.* (1981) have a repository of data on abnormal hemoglobins involving hemoglobin M and unstable hemoglobins. Both these produce phenotypes inherited as dominants. They analyzed parental age effects within these data but, unfortunately, did not have appropriate control data for comparison with all cases. Suggestive trends to a paternal, but not a maternal, age effect were present but were not formally significant.

An interesting study of paternal age effect upon sporadic disorders, some of which were known to be dominant was done by Jones *et al.* (1975). A retabulation of some of these results appears in Table VI.

The first part of the table is for known dominants not previously studied. The "excess" in paternal age–maternal age difference over the reference control value used varied from -0.5 for Pfeiffer syndrome to $+2.9$ for Crouzon disease. Unfortunately, the authors did not investigate the well-studied dominant disorders (i.e., achondroplasia and the others cited above) so a strict comparison with the results on the latter conditions is not possible. In addition the authors used control data from U.S. live births in 1955 or 1966 only, depending upon which was the closest to the *mean* birth year of the patients affected. Thus, if half the cases with a disorder were born in 1971 and half in 1941, the 1955 reference value would have been used. This is preferable to the method used in many other studies, but it is not at all clear that this approach would adjust completely for temporal variation in appropriate control values. The results must be regarded as suggestive but not definitive.

The values for Crouzon disease and multiple exostosis are different from zero and statistically significant. But the variation among the disorders may well be attributable to chance. All the confidence intervals overlap.

For the ten sporadic disorders of unknown etiology there is also considerable variation. Only for acrodysostosis is the paternal age effect significantly greater than zero. (For Prader–Willi syndrome the difference is significantly less than zero.) It is of interest that for nine out of ten of these disorders the parental age difference is positive, i.e., greater than in controls. This trend is highly significant ($P = 0.01$). This may mean that a significant fraction of these disorders of unknown etiology are either "simple" dominants or else are disorders of complex etiology but with a strong influence by dominant genes. On the other hand, the trend may also reflect simply biases resulting from the method in which the control data were chosen (see above). Nevertheless, several things are striking about the list:

1. The only disorder on either list with a significant *negative* association with paternal age—Prader–Willi syndrome—has been shown since this study to be associated with, if not uniformly caused by, a chromosome deletion of 15q

or by rearrangements resulting in such a deletion (see above). While many sporadic cases may result from mutation, others may result from inherited rearrangements, but there is no reason to believe Prader–Willi syndrome results from a classical specific locus mutation.

2. Williams syndrome has one of the lowest paternal age effects on the list. There is suggestive evidence that environmental factors may contribute significantly to the etiology of the disorder.

3. Among the 10 cases of unknown etiology the mean "excess" paternal age is $+0.85$ (S.E.M. $= 0.3$). If Prader–Willi syndrome is excluded, the mean is $+1.1$ (S.E.M. $= 0.3$). But for the eight known dominant disorders the mean excess paternal age is $+1.5$ (S.E.M. $= 0.4$). As the dominant disorders were identified as such independent of knowledge of any paternal age effect, the difference between the two groups, while not significant, is at least in the expected direction.

The above is consistent with the inference that observation of a paternal age effect for a sporadic disorder of unknown etiology suggests that there is a strong specific locus mutagenic component to the disease.

In addition to effects upon mutation, parental age may modify *expression* of *inherited* traits in ways that are not entirely understood but may result from effects on gene interaction. Those with early onset of Huntington's chorea, for example, appear to have inherited the disorder more often from their father than the mother. (This was first reported by Barbeau, 1970.) Brackenridge (1974) reported that the proportion of those with Huntington's disease who are affected with *rigidity* decreases linearly with increasing paternal age (but increases with maternal age). Later multivariate analysis (Brackenridge, 1980) revealed that paternal age at conception is less important in affecting the proportion with rigidity than is age of onset of the disease in the parent, presence of rigidity in the affected parent, and sex of the affected parent. The precise reasons for the association of rigidity with paternal age at conception remain unexplained but may be mediated by social processes indirectly as well as by biological effects (see also the discussion of "complex" traits below).

Because of the great difficulties in studying mutations for Huntington's disease there, are to my knowledge no data on parental age effects upon mutations for this condition. The proportion of apparent sporadic cases (which may result from a mutation, "phenocopy," illegitimacy, or misdiagnosis) may vary from 2% to 20% in various series. But documenting a true mutation is extremely difficult (Shaw and Caro, 1982; Baraitser *et al.*, 1983). Thus there are no data on parental age effects upon mutation for this condition.

What are the practical implications of the observations of a paternal age effect upon dominant mutations?

1. Clinicians often infer that if paternal age is significantly elevated for an observed sporadic cases of a known dominant disorder, then it is likely the case is result of mutation and not inherited from a nonpenetrant

carrier (see, e.g., Hernandez *et. al*, 1979; Zilber *et al.*, 1984). At the present time, however, the likelihood that such a conclusion is correct is impossible to determine. (Precise estimates would depend upon knowledge of the mutation rate, age effect, and penetrance, knowledge that is lacking for most dominant conditions.)

2. If a paternal age, but not a maternal age, effect can be demonstrated for sporadic cases of a disorder of unknown etiology, then this suggests (but does not prove) that there are at least some genetic determinants to this disorder, if it is not indeed the consequence of a simple specific-locus dominant disorder. For example, as a result of the findings of Jones *et al.* (1975) on acrodysostosis, this sporadic disorder is regarded as most likely dominant. Again other factors associated with paternal age must be excluded.

3. Attempts may be made for genetic counseling regarding the risk of older men to have children with a mutant dominant genetic disorder.

With regard to the latter, Friedman (1981) estimated at five paternal age categories the "minimum absolute frequency" of individuals with all diseases attributable to new dominant mutations in sperm. I believe these estimates are in error for various reasons discussed in detail below, but it is worth discussing this approach because this is the only such attempt to my knowledge and is widely cited. Friedman estimated rates of 0.11 per 1000 at ages ≤ 29, 1.1 per 1000 at age 30–34, 1.3 per 1000 at age 35–39, 4.5 per 1000 at age 40–44, and 3.7 per 1000 at ages ≥ 45. These are, presumably, livebirth prevalence rates (or rather what I prefer to term livebirth "potential prevalence" rates in that newborns with an autosomal dominant gene may not manifest the disease coded by the gene until later in life). It is not usually realized that these estimates do *not* include a component attributable to mutations in ova. Under the same assumptions Friedman used, I calculate a rate of 0.5 per 1000 at all ages from such a source. Under these assumptions this leads to an estimate of a total rate of livebirths (with a mutant dominant disorder) per 1000 of about 0.4, 1.4, 1.6, 4.8, and 4.0 at the five successive paternal age categories. These rates for all dominant disorders are independent presumably of maternal ages.

There are at six specified assumptions made by Friedman in deriving his values, and several unspecified ones. I believe at least some of these assumptions are incorrect. The most critical erroneous assumption is that the paternal age effect in achondroplasia is typical of all autosomal dominant diseases. But, as discussed above, this is far from correct. The assumption will tend to result in gross *overestimates* of paternal age effects. In his specific analysis of achondroplasia, moreover, he uses reference data for 1940 only, although probands were born between 1897 and 1967. This will have an uncertain effect upon his estimates. He also assumes an average gene mutation rate of 10^{-6} per haploid male genotype, i.e., per sperm, and (implicitly) of 0.5×10^{-6} for ova. Thus the "average" mutation rate per gamete assumed is 0.75×10^{-6}. But, many geneticists believe that the mutation rates derived for frequent, easily studied dominant traits are not typical of all dominant disorders. Friedman argues this makes no difference because he is concerned with the recognized mutations that will be reflected in the more common dominant disorders. But it is probable that there are many dominant disorders with mutation rates of 10^{-7} or lower which have not been studied. Ten such disorders will result in the same number of affected individuals as one disorder with a mutation rate of 10^{-6}.*

Thus it cannot be assumed that the 1000 dominant disorders Friedman assumed all have *either* the mutation rates *or* the paternal age effects of achondroplasia.

An implicit assumption he made is that all mutations have 100% penetrance. As they do not (for achondroplasia penetrance is perhaps 90%), this assumption will also tend to result in an overestimate of rates of the prevalence of diseases associated with mutation.

Admittedly, the assumption that there are only 1000 dominants will result in a bias to underestimate. It is probably true that there are more, perhaps considerably more, than 1000 dominants. But this underestimate is, I suspect, more than offset by what I believe to be the smaller *average* mutation rate for each condition than that he assumed.†

For these reasons, I believe that the figures cited by Friedman or derived from his estimates are unlikely to be "minimum absolute frequencies" but are better regarded as likely upper boundaries of rates with age. It should be noted nevertheless that some of the assumptions made (that the male mutation rate is on the average twice the female mutation rate or that the distributions of the reference series for paternal age is an appropriate control group) could result in errors in either direction.

Recombination

With regard to recombination, Hamerton (1971) cited data on chiasma frequency in human males. Between ages 15 to 79 there was a mean of 54.4 chiasm per primary spermatocyte, or of 2.34 chaisma per bivalent. There was no evidence from the data reviewed of any effect of age on crossover frequency in males.

It is also worth noting that there is a sex difference in crossing over in humans and mice, it being higher in females than in males (for references see Vogel and Motulsky, 1979, p. 115). A paper by Weitkamp (1972) cited by

---

Addendum: Recent observations on protein variants, however, are consistent with an average mutation rate of about 0.5×10^{-5} or even greater (Neel, 1985).

†*Addendum*: But note Neel's (1985) estimates are consistent with this higher rate.

Vogel and Motulsky (1979, p. 115) reported increased incidence of recombination with birth order for eight pairs of loci. This is consistent with a parental age effect. Weitkamp (personal communication) has not followed up this work further.

Mayo (1974) suggested a curvilinear association between paternal age and chiasma frequency, but according to Conneally and Rivas (1980) the "numbers were not high enough for convincing statistical tests." Lange et al. (1975) in review of the data in the literature found a small nonsignificant negative regression of male chaisma frequency with age.

Elston et al. (1975) in investigation of Gm : Pi linkage concluded that while a parental age effect on recombination could not be excluded or even presumed negligible, it could at least be ignored in estimation of recombination frequency "from any comparable set of data." Subsequently, Elston et al. (1976) reported a small nonsignificant *negative* regression on paternal age for recombination between ABO and nail patella loci.

There are no strong grounds at present for inferring any paternal age effect on recombination in humans.

Reproductive Outcomes of Complex Etiology and Undefined Genetic Contribution

There have been relatively few studies of these outcomes compared to disorders of known simple genetic determination. A finding of a paternal age effect for such outcomes (independent of maternal age, birth order, or other possible confounding agents) would establish a prima facie case that at least some contribution to the outcomes is mediated by fresh dominant mutations at specific loci. Such an observation would also imply that (for nonlethal outcomes) inherited alleles at specific loci also contribute to population variation. Again, it is emphasized that paternal age association can only be regarded as presumptive evidence of such genetic determinants, not strict proof. Social factors operating through indirect mechanisms could also exert effects, and it may be difficult to identify and measure the effects of such factors. Moreover, in view of the evidence cited above, the *absence* of a paternal age effect may not be interpreted as strong evidence against a genetic contribution, although it would imply that many fresh dominant mutations at different loci make little contribution.

Fetal Death

In a study of fetal deaths at age 20 weeks of gestation or older, Selvin and Garfinkel (1976) used multiple logistic analysis to analyze parental ages and birth order. In upstate New York births for 1959 to 1967 they estimated a 2.7% increase per year of "risk" associated with paternal age, independent of maternal age and birth order. Of interest, the increase in the independent "risk" for maternal age was 3.2% per year, which is not much higher than the paternal age effect. (A separate birth order effect was described although it was not characterized in detail.) At least some maternal age effect would be expected because of the known associations of trisomies with fetal death and of trisomies with increasing maternal age. (There is also evidence for a parental age effect [maternal and/or paternal] for fetal deaths with normal karyotypes.) The data cited suggest that specific locus mutations may account for a significant proportion of fetal deaths. This is because the paternal age effect upon specific locus mutation is much weaker than that upon trisomies, yet the observed ratio of the paternal to maternal age effect (2.7/3.2) is relatively high. Further data and sophisticated analyses of the data are clearly needed. Selvin and Garfinkel, incidentally, cite two earlier studies that noted paternal age effects independent of maternal age upon fetal death but that did not adjust for birth order.

With regard to paternal age effects upon embryonic and fetal death (EFD), in Hatch's study (1984), no statistics were reported for *all* EFD, but among those with normal karyotypes there was no significant relationship. The regression coefficient, however, was positive ($\beta = 0.08$). This is contrary to the strong trend reported by Selvin and Garfinkel in vital record studies.

Ressequie (1976) in analysis of Wisconsin "stillbirths" (fetal deaths at 20 weeks or older), interpreted his data as showing that rates do not increase with father's age independent of maternal variables. He implies that previous studies that found positive effects did so because of failure to control such variables as education and marital status. Nevertheless, he did observe a tendency to increased stillbirths in older fathers (≥ 35 years) married to "younger women" (20–24), which he dismisses as not biologically significant because it was not found in the 25–29 maternal age range. (The tables and text of this paper imply that analysis was restricted to offspring of women ages 20–29, a curious limitation, although nowhere can I find this stated explicitly.)

Jayasekara and Street (1978) analyzed dyslexia and reported both paternal and maternal age effects, but did not control for either variable in analyses of the other. Of interest, no birth order effect was observed in a separate analysis that was uncontrolled for parental age. The latter negative finding is puzzling because if there is a parental age effect, at least a maternal age effect, than an uncontrolled study of birth order should reveal an association with this variable also. Thus, there are grounds for skepticism.

Low Birth Weight

With regard to low birth weight, Selvin and Garfinkel (1972) studied factors affecting the proportion of infants under 2500 g. They found a rough "U-shaped curve" at all maternal ages, although the relative height of each arm of the U varied somewhat by maternal age. As sexes differ in the proportion under 2500 g (it is greater for females), this could have been confounded somewhat by sex ratio effects (see below). A genetic explanation for the increased rate of those with lower birth weight at the lower paternal ages is not clear.

Birth Defects

There are many single genes in humans that are known to produce birth defects, either isolated anomalies or patterns of multiple defects (McKusick, 1983). The genes for achondroplasia and Apert's syndrome are, as discussed above, known to have an associated paternal age effect. There has been, however, relatively little investigation of paternal age effect upon broad categories of congenital malformation. Milham and Gittlesohn (1965) reported no evidence for an independent paternal age effect upon ten broad types of malformations reported upon upstate New York vital certificates. Polednak (1976) analyzed the data from the same source for years 1968–1973. There was no consistent pattern to an increase with paternal ages at all maternal ages although there were some suggestive trends at the older ages. The strongest evidence was for syndactyly, which can result from numerous (different) dominant genes. Because of small numbers the results could be attributable to chance, but they are at least suggestive of an effect for some disorders. Previous reports cited of an association of paternal age and oral clefts were found in only one maternal age category.

Padma *et al.* (1979) report an interesting study on cataracts in India. Congenital cataracts are known to have many single gene causes. Sporadic cataracts, especially sporadic juvenile cataracts, may be attributable to dominate mutations or to recessive inheritance among other mechanisms. Nevertheless, there was little if any evidence for any paternal age effect in sporadic cases compared to familial cases for this disorder.

Lints and Parisi (1981) reported that within-pair differences in total-finger-ridge count (a dermatoglyphic variable) decreased with maternal age in monozygotic twins but increased with maternal age in dyzgotic twins. They commented that they could not exclude the possibility that paternal age was responsible for the association. Thus, the calculated values of "heritability"—the proportion of variation in the trait attributable to genetic factors—increased with parental age. The heritability values reported at maternal ages ≥ 27, 28–33, and ≥ 34 were respectively 0.83, 0.92, and 0.95 by one method and 0.87, 0.94, and 0.95 by another. It is not clear however, that the reported trend is statistically significant.

Sex Ratio

The sex of an offspring is, of course, not in itself a sign of pathology. But a gross deviation in the live-birth population sex ratio from unity may reflect significant pathology, particularly effects on fetal death. Sex ratio may be affected by both age effects upon sperm *development* at meiosis and *mutations* that affect the likelihood of fertilization or survival of the conceptus during gestation. The Y chromosome has relatively little genetic material. Mutations at any loci on the Y chromosome would appear likely to manifest fewer severe effects upon survival during gestation than X-linked mutations. Thus, if there are paternal age effects upon mutations at loci on sex chromosomes, one would expect XX fetuses (females) to be more vulnerable than XY fetuses (males) and the sex ratio to rise with father's age. For analogous reasons the sex ratio would fall with increase in mother's age, but, for the reasons discussed above, not as markedly. This is a theoretical expectation, however, and is subject to confounding by a number of factors.

The results of many studies of paternal age upon sex ratio have been summarized by Imaizumi and Murato (1979), who also present their own data from Japan. Despite theoretical predictions, most studies of sex ratio suggest a *negative* association of increased paternal age with sex ratio although complex interactions between paternal age, maternal age, and birth order prevent clear inferences. Indeed, the strongest effect is of a negative association of birth order with sex ratio, independent of parental age. The birth order effect, I believe, may be mediated in part by sex-linked or sex-limited immunological factors that affect fetal survival.

Conclusions

The weight of available evidence provides no grounds for inferring a paternal age effect for any numerical chromosome abnormality. Not all have been studied intensively, however.

For structural chromosome aberrations, there are conflicting data for different abnormalities. For all abnormalities considered together, if there is a positive effect, it appears likely to be quite weak. For rearrangements resulting in a 15 (q11 or q12) deletion associated with the Prader–Willi syndrome there in fact may be a negative effect, but this trend may not hold for all structural abnormalities.

Among "simple" genetic outcomes, only for autosomal dominant mutations have relatively strong paternal age effects been demonstrated, and these have only been found for some specific conditions, not all dominants. A reported six- to eightfold greater rate of mutations at paternal ages 40 and over then at ages under 30 is likely to be the maximum effect.

For outcomes whose etiology is regarded as "multifactorial" there is variable evidence for associations with advanced paternal age. The strongest evidence is on fetal deaths. Possibly a significant proportion may be associated with fresh dominant mutations.

Paternal age also may conceivably influence genetic outcomes through effects on segregation, gene expression, and recombination. There is at least slender evidence consistent with such effects on segregation and gene expression.

References

Baraitser, M., Burn, J., and Fazzone, T. A., 1983, Huntington's chorea arising as a fresh mutation, *J. Med. Genet.* **20**:459–475.

Barbeau, A., 1970, Parental ascent in the juvenile form of Huntington's chorea, *Lancet* **2**:937.

Brackenridge, C. J., 1974, Relation of parental age to rigidity in Huntington's Disease, *J. Med. Genet.* **11**:136–140.

Brackenridge, C. J., 1980, Parental factors associated with rigidity in Huntington's disease, *J. Med. Genet.* **17**:112–114.

Breg, W. R., Hook, E. B., Magenis, R. E., Palmer, C. G., Pasztor, L. M., and Summitt, R. L., 1980, Parental ages in Turner syndrome: Lack of evidence of advanced paternal age in i(Xq) mosaic cases. *Am. J. Hum. Genet.* **32**:64A (abstract).

Bucher, K., Ionasescu, V., and Hansen, J., 1980, Frequency of new mutants among boys with Duchenne Muscular Dystrophy, *Am. J. Med. Genet.* **7**:27–34.

Butler, M. G., and Palmer, C. G., 1983, Paternal origin of chromosomes 15 deletion in Prader–Willi syndrome, *Am. J. Hum. Genet.* **35**:128A.

Carothers, A. D., Collyer, S., deMey, R., and Frackiewicz, A., 1978, Parental age and birth order in the aetiology of some sex chromosome aneuploidies, *Ann. Hum. Genet.* **41**:227–287.

Carothers, A. D., Frackiewicz, A., deMey, R., Collyer, S., Polani, P. E., Osztovics, M., Horvath, K., Papp, Z., May, H. M., and Ferguson-Smith, M. S., 1980, A collaborative study of the aetiology of Turner's syndrome, *Ann. Hum. Genet.* **43**:355–368.

Chamberlin, J., and Magenis, R. E., 1980, Parental origin of de novo chromosome rearrangements, *Hum. Genet.* **53**:343–347.

Conneally, P. M., and Rivas, M. L., 1980, Linkage analysis in man, *Adv. Hum. Genet.* **10**:209–266.

Connor, J. M., and Evans, D. A. P., 1982, Genetics aspects of fibrodysplasia ossificans progressiva, *J. Med. Genet.* **19**:35–39.

Elston, R. C., Namboodiri, K. K., Lange, K., and Gedee-Dahl, T., 1975, Effect of age on the Gm:Pi linkage, *Birth Defects Original Article Series* **11**(3):298–301.

Elston, R. C., Lange, K., and Namboodiri, K. K., 1976, Age trends in human chiasma frequencies and recombination fractions. II. Method for analyzing recombination fractions and application to the ABO:nail patella linkage, *Am. J. Hum. Genet.* **28**:69–76.

Erickson, J. D., 1978, Down syndrome paternal age, maternal age and birth order, *Ann. Hum. Genet.* **41**:289–298.

Erickson, J. D., 1979, Paternal age and Down syndrome, *Am. J. Hum. Genet.* **31**:498–497.

Erickson, J. D., and Bjerkedal, T., 1981, Down syndrome associated with father's age in Norway, *J. Med. Genet.* **18**:22–28.

Ferguson-Smith, M. A., and Yates, J. R. W., 1984, Maternal age specific rates for chromosome aberrations and factors influencing them, *Prenatal Diagnosis.* **4**:5–43.

Francke, U., Winter, R. M., Lin, D., Bakay, B., Seegmiller, J. G., and Nyhan, W. L., 1981, Use of carrier detection tests to estimate male to female ratio of mutation rates in Lesch-Nyhan disease, in: *Population and Biological Aspects of Human Mutation*, Academic Press, New York, pp. 117–130.

Friedman, J. M., 1981, Genetic disease in the offspring of older fathers, *Obstet. Gynecol.* **57**:745–749.

Hamerton, J. L., 1971, *Human Cytogenetics*, Vol. 1, *General Cytogenetics*, Academic Press, New York, pp. 102–104.

Hansson, A., and Mikkelsen, M., 1978, The origin of the extra chromosome 21 in Down Syndrome, *Cytogenet. Cell Genet.* **20:**194–203.

Hatch, M. C., 1983, Paternal risk factors for spontaneous abortion. Doctoral thesis. Columbia University, New York.

Hatch, M. C., 1984, Male risk factors for spontaneous abortion. Address to the Society for Epidemiologic Research, Houston, Texas, June.

Hernandez, A., Aguire-Negrete, M. G., Ramirez-Soltero, S., Gonzalez-Mendoza, A., Martinez, Y., Martinez, R., Velaquez-Cabrera, A., and Cantu, J. M. 1979, A distinct variant of Ehlers-Danlos syndrome, *Clin. Genet.* **16:**335–339.

Holmes, L. B., 1978a, Genetic counseling for the older pregnant woman: New data questions, *New Engl. J. Med.* **298:**1419–1421.

Holmes, L. B., 1978b, Genetic counseling for older pregnant woman, *New Engl. Med.* **299:**836.

Hook, E. B., 1978, Genetic counseling for the older pregnant woman, *New Engl. J. Med.* **299:**835–836.

Hook, E. B., 1981, Down syndrome: Frequency in human populations and factors pertinent to variation in rates, (F. de la Cruz and P. S. Gerald, eds.), in: *Trisomy 21 (Down syndrome): Research perspectives,* University Park Press, Baltimore, pp. 3–67.

Hook, E. B., 1984, Parental age and unbalanced translocations associated with Down syndrome and Patau syndrome: Comparison with maternal and paternal age effects for 47, +21 and 47, +13, *Ann. Hum. Genet.* **48:**313–325.

Hook, E. B., and Cross, P. K., 1982, Paternal age and Down's syndrome genotypes diagnosed prenatally: No association in New York State data, *Hum. Genet.* **62:**167–174.

Hook, E. B., and Porter, I. H. (eds.), 1981, *Population and biological aspects,* Academic Press, New York, pp. 1–435.

Hook, E.B., and Regal, R. R., 1984, A search for a paternal age effect upon cases of 47, +21 in which the extra chromosome is of paternal origin, *Am. J. Hum. Genet.* **36:**413–421.

Hook, E. B., Cross, P. K., Lamson, S. H., Regal, R. R., Baird, P. A., and Uh, S. H., 1981a, Paternal age and Down syndrome in British Columbia, *Am. J. Hum. Genet.* **33:**123–128.

Hook, E. B., Cross P. K., and Schreinemachers, D. M., 1981b, The evolution of New York State Chromosome Registry, in (E. B. Hook and I. H. Porters eds.), *Population and Biological Aspects of Human Mutation,* Academic Press, New York, pp. 389–428.

Hook, E. B., Schreinemachers, D. M., Willey, A. M., and Cross, P. K., 1983, Rates of mutant structural chromosome rearrangements in human fetuses: Data from prenatal cytogentic studies and associations with maternal age and parental mutagen exposure, *Am. J. Hum. Genet.* **35:**96–103.

Hook, E. B., Schreinemachers, D. M., Willey, A. M., and Cross, P. K. 1984, Inherited structural cytogenetic abnormalities detected incidentally in fetuses diagnosed prenatally: comparison with rates of mutant abnormalities, parental age associations, sex ratio trends, *Am. J. Hum. Genet.* **36:**422–443.

Imaizumi, Y., and Murata, M. 1979, The secondary sex ratio, paternal age, and birth order in Japan, *Ann. Hum. Genet.* **42:**457–465.

Jayasekara, R., and Street, J., 1978, Parental age and parity in dyslexic boys, *J. Biosoc. Sci.* **10:**255–261.

Jenkins, R. L., 1933, Etiology of mongolism, *Am. J. Dis. Childh.* **44:**506.

Jones, K. L., Smith, D. W., Harvey M. A. S., Hall, B. D., and Quan, L., 1975, Older paternal age and fresh gene mutation: Data on additional disorders, *J. Pediatr.* **86:**84–88.

Lamson, S. H., Cross, P. K., Hook, E. B., and Regal, R. R., 1980, On the inadequacy of analyzing the paternal age effect on Down syndrome rates using quinquennial data, *Hum. Genet.* **55:**49–51.

Lange, L., Page, B. M., and Elston, R. C., 1975, Age trends in chiasma frequencies and recombination fractions. I. Chiasma frequencies, *Am. J. Hum. Genet.* **27:**41–48.

Lilienfeld, A., 1969, Epidemiology of Down's syndrome, Johns Hopkins Press, Baltimore, pp. 27–29.

Lints, F. A., and Parisi, P., 1981, The variation of heritability as a function of parental age, in: *Twin Research 3: Epidemiological and Clinical Studies* (L. Geddas *et al.*, eds.), Alan R. Liss, New York, pp. 225–230.

Lubs, M. L., 1981, Mutation rates for human autosomal recessives, in: *Population and biological aspects of human mutation* (E. B. Hook and I. H. Porter, eds.), Academic Press, New York, pp. 91–100.

Mantel, N. and Stark, E. R., 1966, Paternal age in Down's syndrome, *Am. J. Ment. Defic.* **71:**1025.

Matsunaga, E., Tonomura, A., Oishi, H., and Kikuchi, Y., 1978, Reexamination of paternal age effect in Down's syndrome, *Hum. Genet.* **40:**259–268.

Mayo, O., 1974, Effect of age on chiasma number in man, *Hum. Hered.* **24:**144–150.

Mayo, O., Murdoch, J. M., and Hancock, T. W., 1976, On the estimation of parental age effects on mutation, *Ann. Hum. Genet.* **39:**427–431.

McKusick, V. A., 1983, Mendelian inheritance in man, Johns Hopkins Press, Baltimore, 1378 pp.

Milham, S., and Gittlesohn, A. M., 1965, Parental age and malformation, *Hum. Biol.* **37:**13–22.

Neel, J. V., 1985, Spontaneous nucleotide mutation rates as seen through protein eyes. *Am. J. Hum. Genet.* **37:**A203.

Niebuhr, E., 1978, The cri du chat syndrome, *Hum. Genet.* **44:**274–278.

Oberklaid, F., Danks, D. M., Jensen, F. Stace, L., and Rosshandler, S., 1979, Achondroplasia and hypochondroplasia, *J. Med. Genet.* **16:**140–146.

Padma, T., Murty, J. S., and Siva Reddy, P., 1979, Parental ages, birth order and reproductive fitness in cataracts, *Ind. J. Ophtal Mol,* **2:**35–38.

Penrose, L. S., 1933, The relative effects of paternal and maternal age in mongolism, *J. Genet.* **27:**219–224.

Penrose, L. S., 1955, Paternal age and mutation, *Lancet* **2:**312.

Penrose, L. S., 1957, Parental age in achondroplasia and mongolism, *Am. J. Hum. Genet.* **9:**167–169.

Penrose, L. S., and Smith, G. F., 1966, *Down's Anomaly*, J. and A. Churchill, London pp. 167–171.

Polednak, A. P., 1976, Paternal age in relation to selected birth defects, *Hum. Biol.* **48:**727–739.

Regal, R. R., Cross, P. K., Lamson, S. H., and Hook, E. B., 1980, A search for evidence for a paternal age effect independent of a maternal age effect in birth certificate reports on Down's syndrome in New York State, *Am. J. Epidemiol.* **112:**650–655.

Resseguie, L. J., 1976, Paternal age, stillbirths and mutations. *Ann. Hum. Genet.* **40:**213–219.

Riccardi, V. M., Dobson, C. E., II, Chakraborty, R., and Bontke, C., 1984, The pathophysiology of neurofibromatosis: IX. Paternal age as a factor in the origin of new mutations, *Am. J. Med. Genet.* **18:**169–176.

Roecker, G. O., and Huether, C. A., 1983, An analysis for paternal-age effects in Ohio's Down syndrome births, 1970-1980, *Am. J. Hum. Genet* **35:**1297–1306.

Roth, M. P., Stoll, C., Taillemite, J. L., Girard, S., and Boue, A., 1983, Paternal age data, *Prenatal Diag.* **3:**327–335.

Schneider, E. L., and Kram, D., 1981, Animal models for studying parental age effects, in: *Trisomy 21 (Down syndrome): Research perspectives* (F. de la Cruz and P. S. Gerald eds.), Baltimore, University Park Press, pp. 275–280.

Selvin, S., and Garfinkel, J., 1972, The relationship between parental age and birth order with the percentage of low birth-weight infants, *Hum. Biol.* **44:**501–510.

Selvin, S., and Garfinkel, J., 1976, Paternal age, maternal age and birth order and the risk of a fetal loss, *Hum. Biol.* **48:**223–230.

Shaw, M., and Caro, A., 1982, The mutation rate of Huntington's chorea, *J. Med. Genet.* **19:**161–167.

Stamatoyannopoulos, G., Nute, P. E., and Miller, M., 1981, De novo mutations producing unstable hemoglobins or hemoglobins M—Establishment of a depository and use of data to test for an association of de novo mutation with advanced parental age, *Hum. Genet.* **58:**396–404.

Stene, J., Fisher, G., Stene, E., Mikkelsen, M., and Petersen, E., 1977, Parental age effect in Down's syndrome, *Ann. Hum. Genet.* **40:**229–306.

Stene, J., Stene, E., Stenge,-Rutkowski, S., Murken, J-D., 1981, Paternal age and Down's syndrome. Data from prenatal diagnoses (DFG), *Hum. Genet.* **59:**119–124.

Thompson, M. W., Percy, M. E., Hulton, E. M., and Williams, W. R., 1981, Mutation in the muscular dystrophies, in: *Population and Biological Aspects of Human Mutation* (E. B. Hook and I. H. Porter, eds.), Academic Press, New York, pp. 101–160.

Tunte, W., 1972, Human mutations and paternal age, *Humangenetik* **16:**77–82.

Vogel, F., and Rathenberg, R., 1975, Spontaneous mutations in man, *Adv. Hum. Genet.* **5:**223–317.

Vogel, F., and Motulsky, A. G., 1979, *Human Genetics: Problems and Approaches*, Springer-Verlag, Berlin, pp. 301–311.

Vogel, F., 1983, Mutation in man, in: Vol. 1 (A. E. H. Emery and D. Rimoin, eds.), *Principles and Practices of Medical Genetics*, Churchill and Livingston, London, pp. 20–46.

Warburton, D., Kline, J., Stein, Z., and Suzzer, M., 1980, Monosomy X: A chromosomal anomaly associated with young maternal age, *Lancet* **1:**167–169.

Yasuda, N., and Kondo, L., 1982, The effect of parental age on rate of mutation for Duchenne Muscular Dystrophy, *Am. J. Med. Genet.* **13:**91–99.

Warburton, D., 1983, Parental age and X-chromosome aneuploidy, in: *Cytogenetics of the Mammalian X Chromosome, Part B* (A. A. Sandberg, ed.), Alan R. Liss, New York, pp. 23–33.

Weitkamp, L. R., 1972, Human autosomal linkage groups, in: *Proc. 4th Int. Congr. Hum. Genet.* (Paris, 1971), Excerpta Medica, Amsterdam, pp. 445–460.

Zilber, N., Korczyn, A. D., Kahana, E., Fried, K., Alter, M., 1984, Inheritance of idiopathic torsion dystonia, *J. Med. Genet.* **21:**13–20.

9

FERTILITY RATES AND AGING

JANE MENKEN and ULLA LARSEN

How much does the capacity to reproduce decline with age? Until quite recently, it was believed that fecundity declined slightly from age 20 to the early thirties, but more sharply after 35. Then a French study (Federation CECOS, 1982) published in February, 1982, reported results from a group of women who had undergone artificial insemination (Table I). Approximately 74% of women who were not over 30 conceived within 12 menstrual cycles. The percentage fell to only 62 for women 31–35, and to 56 for women 36–40. An accompanying editorial and many later newspaper and magazine articles suggested that risks of infertility rose sharply starting as early as age 30. The purpose of this chapter is to review evidence of the level and age pattern of decline in fecundity for women and for men and then to focus on delayed childbearing.

Overall fertility rates from contemporary populations are so contaminated by the effects of voluntary control of reproduction that they are nearly useless for any study of the relationship between fecundity and age. They are nonetheless of considerable interest in that they reflect a combination of biological and social influences. Figure 1 shows age-specific fertility rates for 15- to 44-year old women in the U.S. from 1955 to 1981, the latest years for which figures are available. In all years, rates are highest for women in their twenties, and fall rapidly thereafter. Until the mid-1970s, rates in all age groups followed a downward trend, dropping most precipitously for women over 35. Since then, there has been a turnaround among 20- to 40-year-old women that is most marked for women in their thirties. The upturn is modest; for no age group in 1981 does the rate approach even half the corresponding 1955 baby boom level.

JANE MENKEN and ULLA LARSEN • Office of Population Research, Princeton University, Princeton, New Jersey 08544.

Table I
CECOS Estimates of Percent of Women
Who Conceive within 1 Year[a]

Age	Number	Percent conceiving in 12 cycles
<25	371	73
26–30	1079	74
31–35	599	62
36–40	128	56
Total	2177	69

[a]From Federation CECOS (1982).

The rates for women over 45 (not shown) have also dropped continuously, so that they are about one-twentieth of the already extremely low rates for 40- to 45-year-old women—on the order of 0.2 per 1000 in 1981. Women aged 45–49 never really participated in the baby boom; their rates have been falling consistently since the 1920s or even earlier.

The decline in fertility since the mid-1950s has accompanied major changes in marriage and the initiation of parenting. Taking a slightly longer time frame, 40 years, marriage and childbearing first moved to progressively younger ages

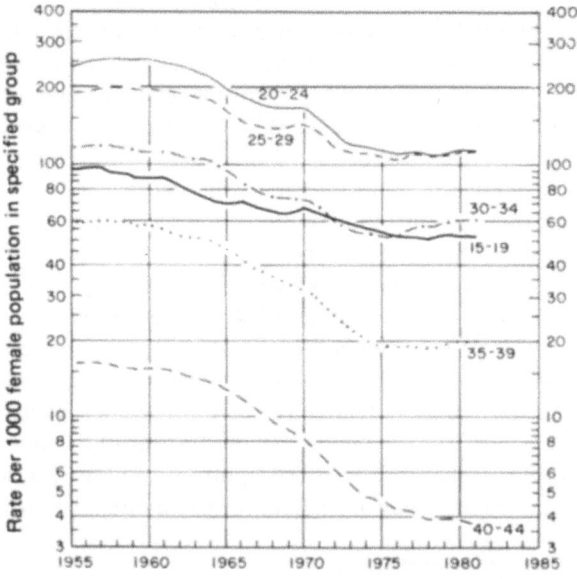

Figure 1. Age-specific fertility rates: U.S. 1955–1981. Source: National Center for Health Statistics (1983a).

and then reversed direction decisively. Figure 2, taken from the U.S. cohort fertility tables, shows the proportion childless, for women 20–40, every 5 years from 1941 to 1981. For all women and for white women, the swings have been substantial; yet the proportions childless are certainly no greater than in 1941 and for over 30s they are not as high. For nonwhite women, there is quite a different picture. Proportions childless over 30 have dropped since the early 1950s. Only among the 20- to 30-year-old groups have there been recent increases. Thus, the increase in childlessness is far greater for whites than nonwhites, for whom a decline in infertility has been suggested (cf. McFalls, 1979). For white women, it could be that the women who are childless at older ages are more likely to be planning to become mothers than childless women in previous decades.

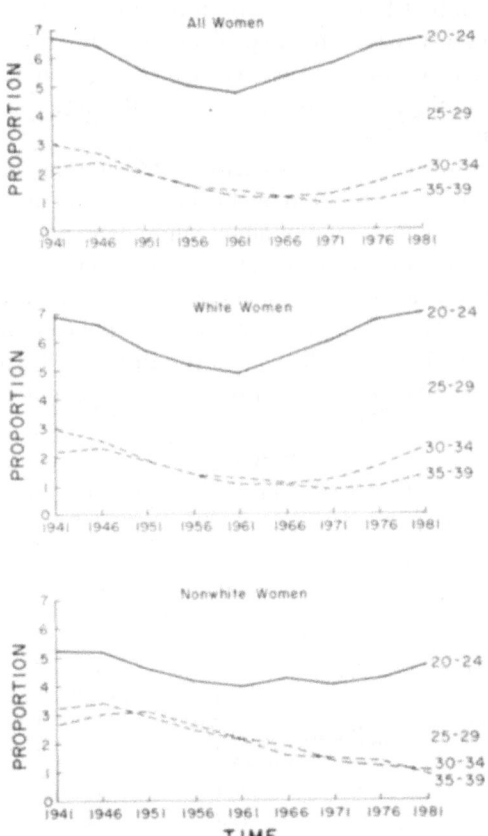

Figure 2. Proportions childless by age: U.S. 1941–1981. Sources: Heuser (1976); Vital Statistics of the United States (1976); National Center for Health Statistics (1983b).

This statement is not, however, born out by examination of the age-specific first birth rates and first birth probabilities shown in Table II for women 25–40. These figures differ in that the numerator is always first births but the denominator in the first case is the total number of women in the age group while in the second it is the number of *childless* women. For white women, both the rates and probabilities are generally lower in 1980 than in 1941. Since about 1975, the first birth rates have been increasing for all these age groups and the probabilities have risen for women in their thirties (Ventura, 1982). Even if these trends have continued, it is unlikely that the figures are higher than in 1955. Unless there has been a widespread increase in infertility, there is no evidence, over the long term, of a major rise in the desire of nulliparous women to have children.

Among nonwhite women, the pattern again is startlingly different, showing a steady rise both in first birth rates and probabilities at least since 1940. For white women, the most reasonable explanation of the recent levels and trends in fertility is related to individual choice regarding marriage and family formation; for others in the population, decreased infertility seems the only plausible interpretation.

None of these figures, however, help us to determine how fecundity and conception rates change due to biological aging, either among women who start childbearing while they are young or among those who postpone their families. For help, we have to turn to historical data on populations in which family limitation was not practiced. These data come from censuses and from family reconstitution studies or geneological records, usually taken from parish or civil registers. The are available for all women (or all married women), as well as women grouped by age at marriage. From these studies it can again be seen

Table II
Birth Rates and Birth Probabilities: U.S. 1941, 1955, and 1980[a]

Age range	First birth rates			First birth probabilities		
	1941	1955	1980	1941	1955	1980
White women						
25–29	44.9	35.0	39.6	105	144	94
30–34	17.7	12.0	12.8	59	83	61
35–39	5.2	4.7	2.5	24	31	20
Nonwhite women						
25–29	13.2	19.1	30.0	34	67	141
30–34	4.7	8.3	12.6	15	32	125
35–39	1.9	3.7	3.2	7	13	29

[a]Sources: For 1941 and 1955, Heuser (1976), Tables 3B, 3C, 9B, and 9C. For 1980, S. Ventura (unpublished data from the National Center for Health Statistics).

that, contrary to the impressions one would receive from the popular press, delayed childbearing is not a new phenomenon—or even a twentieth century phenomenon. In fact, it is a return to a fashion of behavior that characterized the U.S. before World War II and, to an even greater extent, much of Western Europe and parts of North America for perhaps several hundred years extending well into the twentieth century. In the past, delayed childbearing was accomplished not by contraception but primarily through postponement of marriage. Family reconstitution studies for Belgium, England, France, Germany, and Scandinavia between 1780 and 1820 show that only in England was the mean age at marriage under 27. In Scandinavia it was close to 30. Thus the proportions of women who delayed marriage beyond age 30 were sizable in all these countries (cf. Hajnal, 1965; Flinn, 1981).

In the remaining sections of this chapter, we will look at the historical and contemporary evidence of decline in fecundity and increase in infertility and sterility with age, first for all individuals and then for family delayers. We believe that these data permit an assessment of biological changes with age. Before concluding that these results transfer directly to contemporary women and men in developed countries, we must, however, consider how people today differ in sexual behavior from their counterparts in earlier times. In the past, women who delayed marriage and, thereby, childbearing, were, for the most part, also delaying participation in sexual intercourse. The link between marriage and intercourse and the link between marriage and pregnancy after a short time have both been weakened if not broken. Intercourse regularly precedes marriage for much of the population; the timing of parenthood within marriage has become more variable. The use of contraception or abortion to control both timing and number of children is the norm. Thus, in addition to looking at biological decline in fecundity due to aging, we must consider other effects on fertility that may be due to the revolution in sexual practices taking place concomitantly with delayed childbearing and lowered desired numbers of children.

In other words, we can ask whether there is increased infertility *within* an age group, as indeed several demographers suggest has occurred (Mosher and Pratt, 1982; Mosher, 1982). The factors implicated as major potential causes of increased infertility are contraceptive use, abortion, and pelvic inflammatory disease that leads to occlusion of the fallopian tubes. Here, our discussion is limited by lack of data to brief scraps of information and suggestions for needed research.

We turn first to an examination of age patterns of marital fertility in historic populations. Over 25 years ago, the French demographer, Louis Henry (1961), collected data from population groups in which there was evidence of little or no family limitation. We (following Coale and Trussell, 1974) have retained the ten for which age reporting is relatively accurate. Although the level of fertility varies, the age patterns of decline, shown in Figure 3, are quite similar

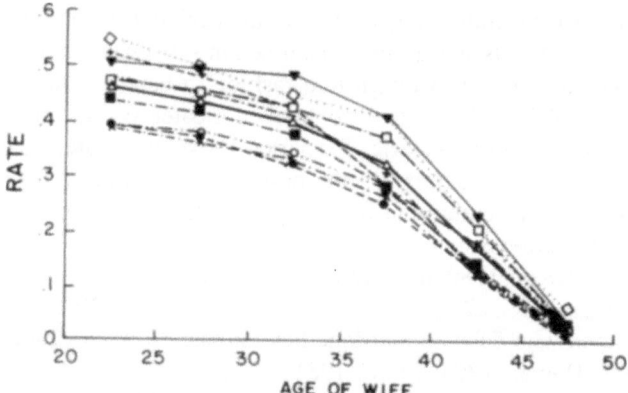

Figure 3. Marital age-specific fertility rates: Ten historical populations and common age pattern (solid line). (See Table III.)

and, in general, not large until after age 35. To quantify a common age pattern, a model (predicted observation = overall mean + age effect + population effect) was fit to the observations by using a mean polish (Tukey, 1977). The correspondence between the predicted and observed data is quite good indeed. As a very simple measure of goodness of fit, we can take

$$1 - \frac{\Sigma\ (\text{observed} - \text{predicted value})^2}{\Sigma\ (\text{observed value} - \text{mean})^2}$$

a measure comparable to the usual multiple correlation coefficient, R^2. The fit of the additive model, which estimated age and population effects, was excellent. The value of the goodness of fit measure was 0.98. The fertility rates and the population effects estimates are shown in Table III, along with the typical age pattern (mean + age effect). Taking the rates for women aged 20–24 as a base, the rates drop by 6, 14, and 31% in successive 5-year age groups and far more sharply after 40 (as shown in the last column of Table III). The typical age pattern may be interpreted as evidence of the level of fecundity and the final column as indicating the decline due to aging. In both cases, these figures represent some unknown combination of decreasing ability to reproduce and sterility. They also, of course, represent *couple* fecundity, not the fecundity of the woman alone.

Sterility can be estimated separately using a variety of demographic methods applied to data on the proportions of women who have a birth after particular ages. Bongaarts (1982) reported proportions sterile calculated from historical data for seventeenth and eighteenth century France and for the Hutterites in the early part of the twentieth century (Table IV). In these populations it was

Table III
Marital Age-Specific Fertility Rates in Populations with No Evidence of Fertility Limitation[a]

Age	Population[b]										Typical age pattern	Percent decline
	1	2	3	4	5	6	7	8	9	10		
20–24	550	509	475	525	468	480	440	396	395	389	463	0
25–29	502	495	451	485	430	450	420	380	370	362	435	6
30–34	447	484	425	429	402	410	375	341	325	327	397	14
35–39	406	410	374	287	324	315	280	289	255	275	322	31
40–44	222	231	205	141	190	125	140	180	130	123	169	64
45–49	61	30	29	16	13	10	10	41	20	19	25	95
Column mean	362	360	327	314	304	298	278	271	249	249	301	
Population effect	63	58	25	12	3	–3	–24	–30	–52	–52		

[a]Source: Henry (1961).
[b]The populations included are: (1) Hutterites, marriages, 1921–1930. (2) Canada, marriages, 1700–1730. (3) Hutterites, marriages, before 1921. (4) Bourgeoisie of Geneva, husbands born 1600–1649. (5) Tunis, marriages of Europeans 1840–1959. (6) Normandy, marriages and births from 1760–1790. (7) Normandy marriages, 1674–1742. (8) Norway marriages, 1874–1876. (9) Iran (villages) marriages, 1940–1950. (10) Bourgeoisie of Geneva husbands born before 1600.

154

JANE MENKEN and ULLA LARSEN

Table IV
Estimates of Percent Sterile by Age[a]

Age	Percent sterile		
	France	Hutterites	Average
20–24	5	3	4
25–29	6	5	5
30–34	10	9	9
35–39	18	22	20

[a] Source: Bongaarts (1982).

estimated that proportions sterile increased from 4% of 20- to 24-year-olds to 5% for 25–29 to 9% for 30–34, and then sharply to 20% for 35- to 39-year-olds.

    The marital fertility rates discussed earlier have frequently been interpreted as representing biological capacity to reproduce at each age. There is, however, a complicating factor. Several studies have reported that fertility rates decline (and sterility or infertility increases) with marriage duration as well as with age (Page, 1977; Mineau and Trussell, 1982), quite possibly due to a combination of decreased coital frequency and reproductive impairment associated with parity. Thus the modest increase in infertility and sterility in the 30- to 34-year-old groups as compared to 25- to 29-year-olds of much less than 10% and the sharper increases at older ages may be considered upper bounds for the true increases if, as we suspect, reproductive impairment related to childbearing itself has declined because women are bearing fewer children and receiving better medical care.

    The only study we found that considered age of both spouses is Mineau and Trussell's (1982) analysis of Mormon geneological data for cohorts of women born in the mid-nineteenth century. There are two groups: women born prior to 1860 (who are considered unlikely to have limited their families) and those born after that date (for whom voluntary control of fertility was suspected). In their analysis, the marital age- and duration-specific fertility rate was treated as the product of a "typical value," a female age effect, a male age effect, and a duration effect. Estimates of these effects are shown in Table V. They take as the base the estimated value when both husband and wife are 20–24 and in the first 5 years of marriage. The results are clearcut and remarkable. In the early group, fertility remains at approximately the same level for women aged 20–35. Adolescent subfecundity may be present, but it has a quite small effect—perhaps because married women in this age group tended to be closer to 20 than 15. Even for women in their late thirties, fertility remained high; it was 90% of the level for women who were over 20 years younger. It should be noted that these figures are not inconsistent with the results for marital fertility

Table V

Effects on Marital Age-Specific Rates of Age of Wife and Husband
and Marital Duration: Mormon Geneological Data[a,b]

	Wives born 1840–1859	Wives born 1860–1879
Base value[c]	471	499
Effect of wife's age		
15–19	0.96	0.95
20–24	1.00	1.00
25–29	1.03	0.97
30–34	0.99	0.87
35–39	0.90	0.77
40–44	0.62	0.45
45–49	0.14	0.07
Effect of husband's age		
15–19	0.90	0.83
20–24	1.00	1.00
25–29	0.99	1.01
30–34	1.04	0.99
35–39	0.97	0.94
40–44	0.83	0.95
45–49	0.82	0.83
50–54	0.73	0.81
55–59	0.48	0.69
60–64	—	0.56
Effect of duration of marriage		
0–4	1.00	1.00
5–9	0.89	0.84
10–14	0.81	0.77
15–19	0.79	0.79
R^2	0.92	0.94

[a] Source: Mineau and Trussell (1982).
[b] The model is observed value = base value (wife's age effect) (husband's age effect) (marriage duration effect).
[c] Base value is taken as the value when husband and wife are both 20–24 and in the first 5 years of marriage.

given earlier in Table III. There the common pattern rate for women aged 35–39 is 69% of that for women aged 20–24. The value given here is based on the assumptions that these women have husbands who are 20–24 and were married less than 5 years. In fact, if we assume their husbands were 40–45 and their marriages had taken place 10–14 years previously, the expected marital fertility rate would be only 61% ([0.90][0.83][0.81]) of the baseline level and quite close to the value in Table III.

Fertility declines far more slowly with age for men as compared to women;

men aged 50–54 have fertility rates 73% as high as youngsters who are in their early twenties. At that age, women who are capable of childbearing are so rare that they appear as single case studies and not in demographic rates.

The decline with increased marital duration is evident in the final set of estimates in Table V.

It is also worthwhile to compare the results for the earlier and later cohorts. The declines with age for men are comparable, as are the marital duration effects. Fertility drops far more sharply for women in the second group, giving vivid indication that deliberate control of fertility was being exerted.

These data and results must be interpreted with some caution. The Mormon geneologies were collected from surviving kin. They may be biased toward high fertility because the greater the number of children a couple had, the greater the likelihood of one or more descendants being alive and able to contribute their family history.

Data on conception rates by age also exist. Bongaarts (1982) responded to the CECOS study by describing the crisis as a "false alarm." He made the telling point that artificial insemination appears to be less likely to result in conception than natural insemination and that *all* conception rates reported by CECOS were very low. His comparison group was women in an English study (Vessey *et al.*, 1978) of contraceptive users who stopped using methods other than the pill because they wanted to conceive (Table VI). There, women who had no previous children and whose average age was 29, roughly comparable to the French group, had over 80% of their number become pregnant within a year; for parous women, the figure was closer to 90%. By comparison, overall, only 69% of the French women conceived within the same period. (In fact, the British study considered only pregnancies leading to a live birth, whereas the French counted all recognized conceptions, so the discrepancy is even greater.)

His conclusions that female fertility decline with age was modest until the late thirties, but real, were criticized by several demographers (Hendershot *et al.*, 1982) on the grounds that in the British study there was selection for higher social class among women attending a family planning clinic and that women

Table VI
Percent Conceiving within 1 Year after
Stopping Use of Contraceptive Methods
Other than the Pill[a]

Parity	Number	Percent conceiving in 12 cycles
0	779	~ 80
1 +	1343	~ 89

[a]Source: Vessey *et al.* (1978).

were excluded if they had evidenced any conditions related to infertility. The data on fertility rates presented here are, however, consistent with Bongaarts's position. The same demographers also suggested that the results for historical populations, in which most of the older women had already had children, might not hold for nulliparous women, those who delayed childbearing, in populations of either the past or the present. (They also speculated that the Hutterites may be genetically selected for very high reproductive capacity.)

It seems reasonable, therefore, to examine biological aging in reproductive capacity among women who delay childbearing. We sought information on fertility by age at marriage in population groups that met three criteria: (1) late marriage was common; (2) there was evidence of little or no deliberate fertility control; (3) parity distributions by age at marriage were available or could be calculated. Seven groups meeting these standards were located (Menken and Larsen, 1984). For the first four, data came from censuses while, for the last three, they derived from family reconstitution studies. The percentages married over 30 range from 11 to 35. It must be noted that for some of these women, the marriage is not the first. Remarriage is, however, uncommon in all of these populations.

Table VII gives, in Panel A, the proportions of these women who had at least one child and, in Panel B, the proportions who had at least two children. They also are given graphically in Figure 4. The pattern of decline is quite similar in all these population groups and, quite clearly, is not large until over 35. (It should be noted that when women pregnant at marriage were excluded from the reconstitution data sets and first birth probabilities recalculated, little difference was found. Therefore, we chose to work with the data for all women since they are more comparable to census information.)

The common age pattern of proportion fertile, overall mean + age effect, is given in Column 8 and the risk of childlessness and the percent having fewer than two children in Column 9. The figures in Panel A, in these noncontracepting populations, are most likely involuntary childlessness. The risk of childlessness jumps from 6 to 9% from 20–24 to 25–29, and more sharply thereafter. For women who married in their early thirties these risks are about 15%. The increase in the risk of having fewer than two children goes up more rapidly, in part because women are older when they could have that second child. The last column shows the proportions of those who had a *first* child, who did not have a second one. Again, only for those married after age 35 does this risk rise sharply.

These risks are slightly higher for each age group than Bongaarts's estimates, as is expected. His figures refer to women who are married throughout an entire 5-year period, say from 30 to 34 and who are, on average 32.5. Ours come from women who *marry* within an age group and are thus less likely to be exposed to the risk of childbearing at the youngest, most fertile ages within that period.

Table VII

Numbers, per 1000 Marriages, Having at Least One and at Least Two Children, by Age at Marriage and Results of Mean Polish; Seven Populations[a,b]

	1	2	3	4	5	6	7	8	9	10
	Ireland	Scotland	Quebec		England		Germany			
			Rural							
Age at marriage	1911	1911	<1876	1876-85	Nineteen parishes	Quakers	Fourteen villages	Row mean	Percent childless	
				Panel A: Number having at least one child/1000 marriages						
20–24	938	948	937	935	948	941	957	943	6	
25–29	909	901	898	903	911	889	939	907	9	
30–34	852	824	838	795	840	873	893	845	15	
35–39	725	656	687	654	747	701	759	704	30	
40–44	483	310	311	316	443	317	371	364	64	
Column mean	781	728	734	721	778	744	784	753[c]		
Column effect	28	-25	-19	-32	25	-9	31			

Panel B: Number having at least two children/1000 marriages

									Percent having 0-1 children	Percent not having second child[d]
20-24	900	897	902	903	902	894	855	893	11	5
25-29	859	816	844	842	855	841	827	841	16	7
30-34	778	700	743	697	720	794	762	742	26	12
35-39	601	456	512	473	507	472	494	502	50	29
40-44	334	141	143	145	173	146	105	170	83	53
Column mean	694	602	629	612	631	629	6.09	630[c]		
Column effect	64	-28	-1	-18	1	-1	-21			

[a] Sources: (1) Ireland (1913); (2) Scotland (1914); (3 and 4) Eighth Census of Canada (1946); (5) Schofield (personal communication); (6) J. Knodel (personal communication).
[b] The mean polish expresses the observation as: observation = overall mean + row effect + column effect, where column effect = column mean − overall mean and row effect = row mean − overall mean.
[c] Here only the row means are given (Column 8) because they represent the common age patterns.
[c] Overall mean.
[d] Percent of those having at least one child who do not have at least two children.

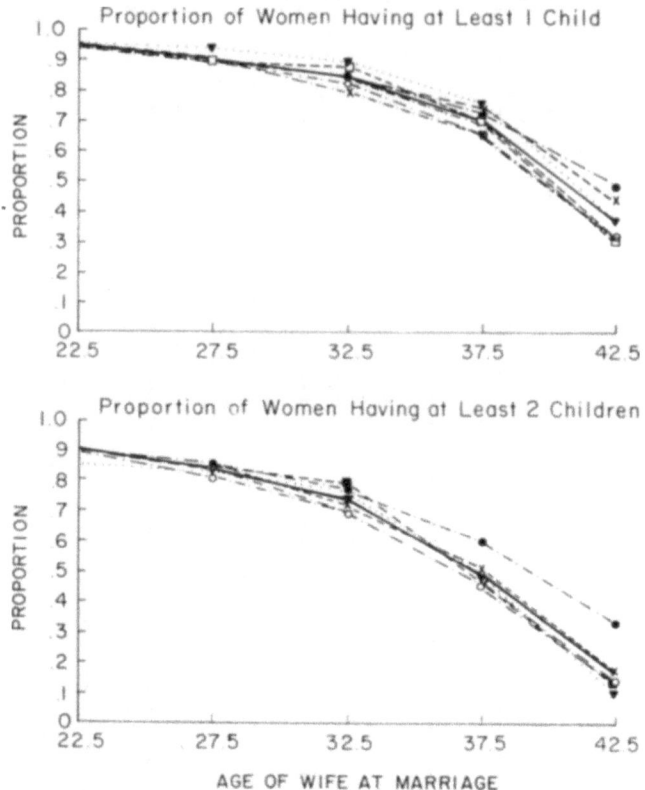

Figure 4. Proportions having at least one and two children, by age at marriage: Seven historical populations and common age pattern (solid line). (See Table VII.)

These figures may be viewed as an upper bound to the risk of involuntary childlessness due to biological aging. The *actual* risks today for women who want children are likely to be lower because some of these couples may have chosen to remain childless and because with modern medical advances in the treatment of infertility, some couples whose conditions are now treatable could have borne children.

Therefore, the evidence that infertility rises, but only moderately, with age is to our minds persuasive. How late a woman waits to have children, whether a first or later birth, is a matter of balancing her work and family life plans, of assessing, as in almost every other aspect of life, the risks, the costs, and the benefits that go along with alternative choices. Reassuringly for many, the biological risks of involuntary childlessness or infertility do not rise at nearly so dramatic a pace as the rash of articles would have had us believe. For men, the decline in fecundity appears to be rather small until at least middle age.

If not due to biological aging, why, then, has there been so much attention to infertility recently in the popular press? In part, this attention is in response to the availability of new methods for treating infertility and new problems in managing infertility—test-tube babies, new surgical procedures for both men and women, artificial insemination, surrogate mothers—and to the decline in availability of infants for adoption; all call attention to infertility without giving a clue as to whether or not it has increased in magnitude. Certainly the evidence goes in the opposite direction for black women over the last 30 years—there has been a continuous decline in childlessness. However, data from fertility surveys taken between 1965 and 1973 have been analyzed and interpreted by some demographers (Mosher, 1982; Mosher and Pratt, 1982) as indicating increased infertility among married women, particularly young black women. They tabulated the proportion of married women who either had a known nonsurgical condition causing sterility or had not used contraception for at least twelve months and had not conceived since they stopped using. This proportion rose for 20- 24-year-olds from 3% in 1965 to 9% in 1973 and then from 9 to 15% in the next 3 years—between 1973 and 1976. The rise in proportion with long intervals without conception is statistically significant; yet, as an indicator of increased infertility, it must be viewed with some skepticism. Because of the need to find good ways to estimate infertility and sterility in entire populations, it seems worthwhile to comment in some detail on why we believe this statement is true.

In each age group, the sample is quite small—there were 203 married black women 20–24 in the 1965 sample, 410 in 1976. The calculation was not, but should have been, based on women continuously married for at least 12 months at the time of the survey—they are the only ones who could have long intervals. We recalculated the proportion infertile in 1976 in this way. The sample size, when women whose fertility status was undefined were excluded, dropped to 247 and the percent infertile increased to 18. That means that fewer than 50 women were counted as infertile under the long-interval definition. In response to direct questions as to whether they had any reason to believe they had trouble conceiving, 75% of these women had responded in the *negative*. Obviously, this finding may only indicate that the subjective measure of fecundity is inadequate. There are, however, additional problems related to the operational definition of infertility.

Even if all conceptions are reported, any estimate derived from this kind of survey has to be an *underestimate* of the true level of infertility. All women who are using contraception are automatically considered fecund. Some of them are infertile, but we (and they) don't know it. It may be, on the other hand, an *overestimate* of infertility for at least two reasons: twelve months is too short a time for detecting true infertility. In many populations, time to conception is over ten months *on average*. For 20- to 24-year-old black women, for example, the percent not conceiving decreased from 18 to 14 to 10 as the definition of long interval increased from 12 to 18 to 36 months (source: Tabulations from 1976 National Surveys of Family Growth (NSFG) Public Use Tape). In addition, unreported pregnancies terminated through either spontaneous or induced abortions would raise the estimate of infertility. In fact, of women who report a pregnancy termination in the year prior to survey, some 21% of blacks and 16% of whites do report a nonlive birth as the outcome. There are, however, few reports of induced abortions—about 1% of currently married women, both for blacks and for whites, report an abortion. Data from The Alan Guttmacher Institute (S. Henshaw, personal communication) show that about 3% of black married women have abortions each year (actually 1979) and about 1% of whites. Clearly, there is underreporting of both pregnancies and abortions and the error is greater for blacks.

Finally, any attempt to assess changes over time is severely affected by changes in contraceptive use (since the proportions *eligible* to be counted as infertile are altered) and by changes in marriage duration.

These first attempts to measure infertility in a population are both necessary and laudable. They do cover national samples, at least of married women, and they are characterized by major efforts to consider the definitional problems. It is to be hoped that the new attempts made in the 1982 cycle of the NSFG to develop the kinds of questions that will enable us to obtain adequate estimates of infertility from representative surveys will yield better results.

Clearly, defining infertility by failure to conceive within relatively short periods is sure to lead to overestimates. Historical data from England from the mid-sixteenth to mid-nineteenth century (Table VIII) shows that about half or more of women who married by age 30 and who bore no children in the first 2 years of marriage had a child in the next 3 years. For women who married in their thirties, the figure was over 30%. Obviously defining infertility by failure to have a child within 2 years would lead to gross overestimates. Even when infertility is medically diagnosed today, rather high proportions of couples may go on to conceive a child, even if their infertility problems are not treated. In fact, 41% of treated and 35% of untreated couples conceived within a 2- to 7-year followup period in the recent study by Collins *et al.* (1983).

There is little evidence that either contraceptive use or legal abortions themselves lead to later difficulty in conceiving and delivering a full-term infant although questions have been raised about repeated abortions (cf. Hogue *et al.*, 1982, 1983). It appears, however, that both IUD use and abortion are connected to infertility through pelvic inflammatory disease (PID).

It is pelvic inflammatory disease that may pose the greatest danger to fertility. In particular, when PID spreads upward from the lower genital tract and involves the fallopian tubes, infertility may follow because of tubal occlusion or because of increased risk of ectopic pregnancy. According to the best evi-

Table VIII

Proportion of Women with a Live Birth by Specified Durations of Marriage and by Age at Marriage: England, 1550–1849[a]

	Age at Marriage						
	15–19	20–24	25–29	30–34	35–39	40–44	45–49
Duration of Marriage (months)							
24	73	77	72	68	59	25	6
60	88	90	86	80	72	36	7
Percent of women with no birth by 24 months who bear a child by 60 months	56	59	48	37	32	14	1

[a]Source: Trussell and Wilson (1984).

dence currently available, women who use IUDs have considerably higher risks of contracting PID than women who use no contraceptives. The relative risk is in the neighborhood of 1.5–4.0 times as high for IUD users. By contrast, the risk for pill users is only 0.3–0.9 as high as for noncontraceptors (Senanayake and Kramer, 1980). It is suspected that the tail attached to the IUD permits organisms to climb more easily into the uterus. The inhibitory effect of the pill, it has been suggested, works through a "cervical resistance factor" that disappears during ovulation and menstruation, when PID appears more frequently. The pill reduces the frequency of ovulation and, therefore, may increase this cervical resistance factor (Weström, 1980).

The overall enormous increase in PID provides indirect evidence that infertility is likely to be rising. Between 1965 and 1975, the reported cases of gonorrhea tripled and then remained at the level of about 1 million per year for the next 5 years (Curran, 1980). Other agents also produce PID and their incidence has increased as well. Despite a number of efforts there are no good estimates either of the proportion of women who have ever had PID and, more importantly, salpingitis, or of the proportion who have consequently become infertile. The best information on infertility comes from studies in Sweden by Weström (e.g., 1980) and his colleagues. They followed women with well-diagnosed salpingitis for nearly a decade and found some 15% were subsequently infertile. Infertility from a single episode increased with age and with the severity of infection. It also increased sharply with the number of infections. They also found a seven- to tenfold increase in the risk of subsequent ectopic pregnancy.

The main problem in transfering these figures to infertility estimates is lack of good data on *rates* of PID. Apparently these rates are highest for women under 20 and decrease sharply with age, although they have increased for all ages. Most of this evidence comes from the Weström study and another small one in Atlanta. No reporting system in the U.S. appears to provide good enough numerator and denominator information to calculate rates. If, however, the entire increase in infertility estimated by Mosher as going from 3% to 15% is due to PID in women, and we use Weström's estimate that 25% of women with salpingitis become sterile, then over 80% of married black women aged 20–24 have had PID severe enough to involve the fallopian tubes. Thus, it appears unlikely that infertility in women is as extensive a problem as Mosher estimates. Because we were able to find even less data on changing infertility among men, it is impossible to assess the possibility that these figures reflect infertility problems of the couple. Perhaps the best quantitative evidence of changing infertility is the reported increase in etopic pregnancies, from under 14,000 in 1967 to over 41,000 a decade later (Curran, 1980).

Suffice it to say, there is an enormous need to document infertility and its causes in the U.S. It seems possible that *disease* may be leading to decreased capacity to bear children and that the cumulative effect may become evident in greater reproductive difficulty in the latter part of the usual childbearing years.

ACKNOWLEDGMENTS

Roger Schofield furnished the necessary tabulations for nineteen English parishes from the sixteenth through the nineteenth centuries and for English Quakers of the same period. These data were collected under the auspices of the Cambridge Group for the History of Population and Social Structure. John Knodel provided similar results from the data he has collected on German villages. Barbara Vaughan prepared the tabulations from the public use tapes of the National Surveys of Family Growth. Carol Ryner supervised all manuscript preparation with care and skill.

References

Bongaarts, J., 1982, Infertility after age 30: A false alarm, *Fam. Plann. Perspect.* **14**(2):75–78.

Coale, A., and Trussell, J., 1974, Model fertility schedules: Variation in the age structure of childbearing in human populations, *Popul. Index* **40**:185–201.

Collins, J. A., Wrixon, W., Janes, L. B., and Wilson, E. H., 1983, Treatment-independent pregnancy among infertile couples, *New Engl. J. Med.* **309**:1201–1206.

Curran, J. W., 1980, Economic consequences of pelvic inflammatory disease in the United States, *Am. J. Obstet. Gynecol.* **138** (7, Part 2):848–851.

Eighth Census of Canada, 1946, Vol. 3, Ages of the population, Ottawa.

Federation CECOS, Schwartz, D., and Mayaux, H. J., 1982, Female fecundity as a function of age, *New Engl. J. Med.* **307**:404–406.

Flinn, M., 1981, *The European Demographic System, 1500–1820*, The Johns Hopkins University Press, Baltimore.

Hajnal, J., 1965, European marriage patterns in perspective, in: *Population in History* (D.V. Glass and D.E.C. Eversley, eds.), Edward Arnold, London.

Hendershot, G. E., Mosher, W. D., and Pratt, W. F., 1982, Infertiltiy and age: An unresolved issue, *Fam. Plann. Perspect.* **14**(5):287–290.

Henry, L., 1961, Some data on natural fertility, *Eugenics Q.* **8**:81–91.

Heuser, R. L., 1976, Fertility tables for birth cohorts by color, DHEW publication No. (HRA) 76-1152, U.S. Government Printing Office, Washington, D.C.

Hogue, C. J. R., Cates, W., and Tietze, C., 1982, The effects of induced abortion on subsequent reproduction, *Epidemiol. Rev.* **4**:66–71.

Hogue, C. J. R., Cates, W. and Tietze, C., 1983, Impact of vacuum aspiration abortion on future childbearing: A review, *Fam. Plann. Perspect.* **15**:119–126.

[Ireland], Great Britain Parliament House of Commons, 1913, Parliamentary papers, accounts and papers, Vol. 118, 1912-1913. "Census of Ireland, 1911. General report, with tables and appendix." H.M. Stationary Office, London.

McFalls, J., 1979, *Frustrated Fertility: A Population Paradox, Popul. Bull.* 34, No. 2.

Menken, J., and Larsen, U., 1984, Age and fertility: How late can you wait? Paper presented at the annual meeting of the Population Association of America; 1984 May, 3–5; Minneapolis, Minnesota.

Mineau, G., and J. Trussell, 1982, A specification of marital fertility by parents' age, age at marriage, and marital duration, *Demography* **19**(3):335–350.

Mosher, W. D., 1982, Infertility trends among U.S. couples: 1965-1976, *Fam. Plann. Perspect.* **14**(1):22-27.

Mosher, W. D., and Pratt, W. F., 1982, *Reproductive impairments among married couples, United States*, Vital and Health Statistics. Series 23, No. 11. National Center for Health Statistics. U.S. Government Printing Office, Washington.

National Center for Health Statistics, December, 1983a, Advance report of final natality statistics, 1981, *Monthly Vital Statistics Report*, Vol. 32, No. 9.

National Center for Health Statistics, 1983b, Unpublished data.

Page, H., 1977, Patterns underlying fertility schedules: A decomposition by both age and marriage duration, *Popul. Stud.* **31**:85-106.

[Scotland], Great Britain, Parliament House of Commons, 1914, Parliamentary papers, accounts and papers, Vol. 44, 1914. Census of Scotland, 1911. Report on the Twelfth Decennial Census of Scotland. H.M. Stationery Office: London.

Senanayake, P., and Kramer, D. G., 1980, Contraception and the etiology of pelvic inflammatory disease: New perspectives, *Am. J. Obstet. Gynecol.* **138** (7, Part 2):852-860.

Trussell, J., and Wilson, C., 1985, Sterility in a population with natural fertility, *Popul. Stud.* **39**: 269-286.

Tukey, J. W., 1977, *Exploratory Data Analysis*, Addison-Wesley, Reading.

Ventura, S., 1982, Trends in first births to older mothers, 1970-79, National Center for Health Statistics, *Monthly Vital Statistics Report*, Vol. 31, No. 2 Supp. (2).

Vessey, M. P., Wright, N. H., McPherson, K. and Wiggens, P., 1978, Fertility after stopping different methods of contraception, *Br. Med. J.* **1**(4 Feb.):265-267.

Vital Statistics of the United States, 1976, Vol. 1, Natality, Hyattsville, Maryland, NCHS, 1980.

Weström, L., 1980, Incidence, prevalence and trends of acute pelvic inflammatory disease and its consequences in industrialized countries. *Am. J. Obstet. Gynecol.* **138**(7, Part 2):880-892.

10

AGE-RELATED CHANGES IN GESTATION AND PREGNANCY OUTCOME

ROBERT RESNIK

Introduction

Various segments of the medical community, as well as women contemplating pregnancy, have long been concerned about what risks the older gravida and her infant may be subject to during gestation. Although somewhat difficult to quantitate, it is clear from numerous studies that this population represents a special group, requiring medical attention to unique problems that may develop with respect to advanced reproductive age.

This issue has emerged as one of greater public interest as a result of population statistics in the United States that reveal a tendency to delay childbearing. Whether this reflects a change in personal family planning choices or the fact that within many family units both partners opt for careers prior to raising children, the trend is quite clear. Considerable data is available that demonstrates these trends. For example, the number of women in their early thirties having their first child has increased 66% between 1970 and 1979, from 7.3/1000 to 12.1/1000 live births (National Center for Health Statistics, 1982). More recently, the National Center for Health Statistics (1983) has reported that, since 1972, the first-birth rate has increased by 4% for women aged 30–34; 8% for women aged 35–39; and 33% for women aged 40–44. Current projections for the 1980s, based upon data accumulated by the United States Bureau of the Census, suggest that the percentage of births to women 35 years and over will increase 37% by 1990 (Adams, 1982). Although the lay popu-

ROBERT RESNIK • Department of Reproductive Medicine, School of Medicine, University of California, San Diego, San Diego, California 92103.

lation may not ordinarily read documents such as these, it is of particular interest that the data was recently summarized in the right-hand lead on the first page of a national newspaper publication (*USA Today*, 1984).

Given this data base reflecting changes in family planning in the United States population, it is no surprise that members of the medical community are being queried with greater frequency about risks, real or otherwise, to which the reproductively older woman may be exposed. The purpose of this discussion is to enumerate and delineate the extent of risks to the mother and offspring in order that a reasonable counseling format may be employed.

Fecundity as a Function of Maternal Age

Although it has been general knowledge that the prognosis for conception and pregnancy is inversely related to advancing maternal age, recent data obtained from studies on artificial insemination with donor semen (AID) provide documentation of this phenomenon. In a study of 2193 women with azospermic husbands treated by AID, cumulative success rates reveal a decrease in conception rate per cycle that is slight but significant after 30 years of age, and marked after 35 years (Federation CECOS, 1982). Specifically, the probability of success of AID for 12 cycles was 73% for women under the age of 30, and dropped to 61% for those aged 31–35 years, and 54% for those over 35 years of age. More recently, Virro and Shewchuk (1984) have confirmed these findings. Among 30 couples treated with AID, conception rates decreased linearly, and the number of AID cycles required to achieve conception increased as a function of maternal age.

Chromosomal and Other Fetal Abnormalities

The incidence rates of certain types of birth defects is increased with advancing maternal age. This is most notable in defects due to chromosomal abnormalities. The incidence of Down's syndrome alone has been observed to increase as follows: 1/885 live births at maternal age 30 years; 1/365 at age 35 years; 1/109 at age 40 years; and 1/32 at age 45 years (Hook and Lindsje, 1978). If one considers chromosomal abnormalities other than Down's syndrome, the incidence is even higher.

In an interesting analysis on the potential for reducing birth defect risk by amniotic fluid chromosome analysis in pregnancy termination, Goldberg *et al.* (1979) reemphasized that the increase in birth defect risk with advancing maternal age is largely due to chromosomal rather than other types of abnormalities (Figure 1). Based on their data, the authors point out that if all women beyond the age of 35 years utilized available diagnostic techniques, the estimated risk reduction was approximately 46% for those aged 35–39 years, and 68% for

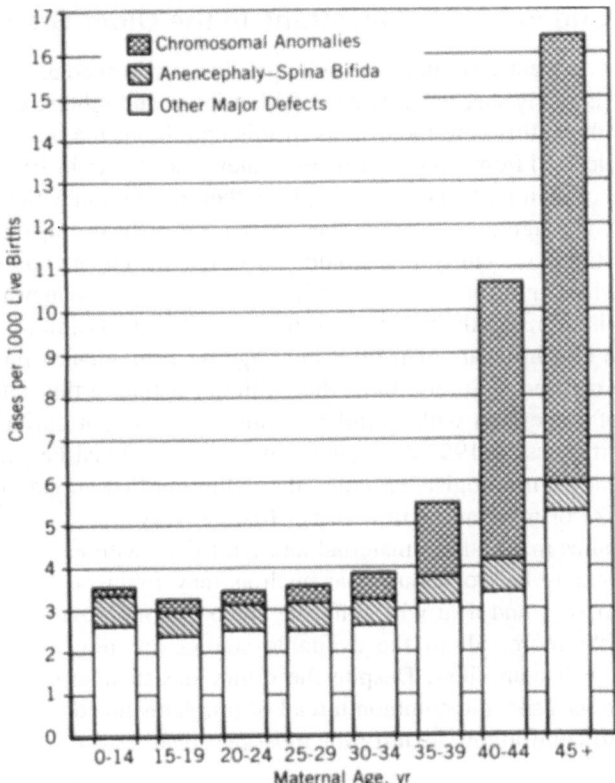

Figure 1. Demonstration of the increasing incidence of selected major birth defects by maternal age. Note that the increase is largely due to chromosomal abnormalities. From Goldberg *et al.* (1979).

those aged 40–44 years. Although the routine use of diagnostic ultrasound in pregnant women is highly controversial, and the authors did not address this issue in their study, it seems reasonable to assume that appropriate utilization of ultrasound would allow for early diagnosis of other types of fetal abnormalities as well.

The cost effectiveness of genetic surveillance programs, as in other areas of medicine, is under constant scrutiny. In a recent cost–benefit analysis of prenatal detection of Down's syndrome and neural tube defects in British Columbia, Sadovnick and Baird (1981) concluded that prenatal screening becomes cost beneficial at maternal age 34 years or more at the time of conception. They emphasize the complexities and imponderables in any such analysis and note that the outcome of any similar study may differ depending upon factors such as the system of medical care, demographic features, and differing costs.

Maternal and Fetal Complications in the Older Gravida

Current convention defines the older gravida as one aged 35 years or more. Although it is likely that the writings of Mauriceau and others, dating back to the eighteenth century, are the origins of this definition, the modern definition is probably derived from a publication by Waters and Wager in 1950. Following presentation of their data, the authors opened their discussion with the following comment: "The age at which one becomes an elderly primipara necessarily must be arbitrary. . . . Thirty-five would seem to be the choice optional year for indicating when aging becomes directly a concern in pregnancy management and prognosis. It marks the half-time of life as well as the last third of menstrual life, and the closing years of fruitful marriage or of pregnancy possibilities in the recently married." Surely, the rather arbitrary nature of this definition must be a bone of contention with countless numbers of women during the last 35 years! Nevertheless, in 1958 the Council of the International Federation of Obstetricians and Gynecologists accepted the definition that an elderly primipara is one aged 35 or more at the time of her first delivery.

The studies that address maternal and fetal risks with advancing maternal age are extremely heterogenous, inasmuch as they frequently do not separate patients by parity, and deal with differing socioeconomic, national, and racial groups. Furthermore, all of the available studies are retrospective and frequently not well controlled. Despite the difficulties in interpretation of these reports, there is a specific common thread of problems noted so frequently that one is able to comfortably identify the most pertinent issues.

Pregnancy in Women over 35 Years of Age

In a 5-year review (1968–1972) of 127 primigravid women over age 35 years, compared to those under age 35, Morrison (1975) observed an 11% incidence of small-for-gestational age (SGA) infants and a corrected perinatal mortality rate (PMR) of 47/1000 live births. Although neonatal morbidity, defined as a low Apgar score and requirement for neonatal intensive care, was three times higher in the older maternal age group, 75% of this morbidity could be attributed to prolonged gestation and prolonged labor. The cesarean section rate in this Canadian study was 31%, compared to 3% in the control group, and other maternal complications were not reported.

A somewhat different population base was examined by Grimes and Gross (1981) at Grady Memorial Hospital in Atlanta. Among 788 black women older than age 35, of whom only 3.6% were primigravidas, the PMR was 47 vs. 28/1000 live births. However, when patients with hypertension were excluded, the differences in PMR compared to controls disappeared. Hypertension, particularly associated with primiparity, constituted the major neonatal risk factor. The incidence of primary cesarean section was also significantly increased. The authors emphasize what appears to be a consistent and highly pertinent finding, specifically, the fact that age alone apparently is not an important risk factor in the healthy women over 35 years of age as it relates to the outcome of her pregnancy.

Messinis *et al.* (1982) have reported outcomes of 1179 primigravid women over the age of 35 years in Athens. Hypertension and uterine fibromyomas were significantly increased in frequency compared to controls. Neonatal problems included prematurity, SGA, and a two fold increase in the incidence of congenital malformations, although the latter were not described in detail. The PMR was 71/1000 live births in the study group versus 30/1000 live births among controls.

Pregnancy in Women over 40 Years of Age

Two separate studies from Finland demonstrate that the spectrum of problems continues, as might be expected, beyond the age of 40 years. Among 558 women in the study group, Kajanoja and Widholm (1978) report a 31% incidence of hypertension compared to 7.3% in controls. Overall cesarean section rates for women over 40 years of age was 31%, but of the 130 in their first pregnancy, the rate increased to 63%. At Oulu University Central Hospital, Kujansuu *et al.* (1981) observed similar findings. Hypertension was, again, the most serious and frequent maternal complication (25.8%) and the cesarean section rate for the primigravid patient was 47%. A similar high incidence of hypertension, cesarean section, as well as an elevated perinatal mortality rate has been reported by Horger and Smythe (1977) and Biggs (1973).

Caspi and Lifshitz (1979) studied outcomes in 494 women over age 40 years and again noted hypertension and higher rate of cesarean section and congenital malformations. The important distinction in this report is the observation of a decrease in perinatal mortality rate between the years 1968 and 1974. During the period 1968–1970, the perinatal mortality rate was 72/1000 live births; this decreased sharply to 14/1000 live births in the time period 1971–1974.

The salient observations from the studies of pregnancy outcome in women over age 35 years and over 40 years are summarized in Tables I and II, respectively.

A variety of other observations pertaining to the offspring of the reproductively older woman are worthy of mentioning. Gillberg *et al.* (1982) have studied 5- and 6-year old children born to women with a mean age of 39.4 years, compared to matched contols born to mothers with a mean age of 27.9 years. The frequency of fine motor problems was five times greater among children of women in the older age group; visual-perceptual dysfunction and attentional deficit signs were also more common. In a more positive vein, Stanford–Binet

Table I

Pregnancy Outcomes in Women Aged 35 Years and Over

Institution	Study population/controls	Maternal complications	PMR	Comments
Grady Memorial (Grimes and Gross, 1981)	788 parity mixed	Hypertension 22.5% C-section 17 vs. 10%	47 vs. 28/1000	No difference when excluded. Primiparas at greater risk
Kaplan Hospital, Hebrew University (Kessler et al., 1980)	98 primipara/100 multipara >35, 100 young primipara	↑Incidence toxemia ↑Myomata ↑C-section	No difference	↑Incidence of prematurity
Health Sciences Center, Winnipeg (Morrison, 1975)	127 primipara/randomly selected primipara <35	C-section 31 vs. 3%	47/1000 controls not stated	34 > 40 years SGA 11 vs. 2% Low Apgar and NICU 18 vs. 6%
Athens University (Messinis et al., 1982)	1179 primipara/66,819 (1966–1976)	Toxemia (×2) Myoma (×6)	73 vs. 30/1000 (uncorrected)	↑Prematurity ↑SGA (×3)
Margaret Hague Maternity Hospital (Waters and Wager, 1950)	649 primipara/649 primipara <25	Toxemia (×1.8) ↑C-section (16 vs. 3%)		

Table II

Pregnancy Outcomes in Women Aged 40 Years and Over

Institution	Study population/controls	Maternal complications	PMR	Comments
Tel Aviv University (Caspi and Lifshitz, 1979)	484 (grand multipara 60%, primipara 2.6%/298 <40 years	Hypertension 16 vs. 7% ↑Postpartum hemorrhage ↑Uterine inertia	44 vs. 18/1000	PMR dropped from 104 in 1965 to 16 in 1975
Oulu University Hospital Finland (Kujansuu et al., 1981)	174 (primipara 11%/ remainder of deliveries 1975–1977	Preeclampsia 25.9% ↑C-section rate (16 vs. 8%)	No difference	No ↑ risk of hypertension if primiparas excluded
Royal Women's Hospital Brisbane, Australia (Biggs, 1973)	231 (primipara 2%)/13,841 deliveries 1968–1970	↑C-section rate (17 vs. 5%)	48 vs. 28/1000	
Medical University of South Carolina (Horger and Smythe, 1977)	345 (primipara 2%)/1517	Hypertension 34% ↑Abruptio placentae	101 vs. 31/1000	22 of 31 fetal deaths in hypertensive women. If excluded, not significant. 3.4% congenital malformations including 2 trisomies
Helsinki University Hospital (Kajanoja and Widholm, 1978)	558 (primipara 29%)/41,409	Hypertension 31% C-section 31 vs. 12%	28 vs. 19/1000	↑SGA 13 with trisomy 21

I.Q. tests have been reported to increase with advancing maternal age (Lobl *et al.*, 1971).

A more recent and provocative report by Wagener *et al.* (1983) demonstrates a relationship between Type I diabetes mellitus and advancing maternal age. The authors studied 1006 families of children admitted to the Children's Hospital of Pittsburgh with a diagnosis of insulin-dependent diabetes. They noted a greater overall cumulative incidence in this diagnosis among children born to women over age 35 years; also, the rate of increase in cumulative incidence was more rapid in this group of children.

Conclusions

It is clear from the evaluation of numerous disparate studies that advanced maternal age in pregnancy is associated with a variety of special features. Most notable and consistently observed are an increased frequency in chromosomal abnormalities, maternal hypertension, and an elevated perinatal mortality rate. However, the risk of chromosomal defects can be reduced and for all practical purposes eliminated with widespread genetic counseling programs. Although the frequency of maternal hypertension may not be reducible, the associated problems of premature delivery and varying degrees of fetal growth deficiency are manageable with modern obstetrical and neonatal care. This is perhaps evidenced by the optimistic finding of a decreasing perinatal mortality rate among the offspring of these women over the last several years. Thus, although faced with an increasing population of reproductively older women who are pregnant, it seems reasonable that one might provide accurate, optimistic and enthusiastic counseling.

References

Adams, M. M., Oakley, G. P., Jr., and Marks, J. S., 1979, Maternal age and births in the 1980s, *JAMA* **247**:493–494.

Biggs, J. S. G., 1973, Pregnancy at 40 years and over, *Med. J. Aust.* **1**:542–544.

Caspi, E., and Lifshitz, Y., 1979, Delivery at 40 years of age and over, *Israel J. Med. Sci.* **15**:418–421.

Federation CECOS, Schwartz, D., and Mayaux, M. J., 1982, Female fecundity as a function of age, *N. Engl. J. Med.* **306**:404–406.

Gillberg, C., Rasmussen, P., and Wahlstrom, J., 1982, Minor neurodevelopmental disorders in children born to older mothers, *Dev. Med. Child. Neurol.* **24**:437–447.

Goldberg, M. F., L. D., and Oakley, G. P., 1979, Reducing birth defect risk in advanced maternal age, *JAMA* **242**:2292–2294.

Grimes, D. A., and Gross, G. K., 1981, Pregnancy outcomes in black women aged 35 and older, *Obstet. Gynecol.* **58**:614–619.

Hook, E. J., and Lindsje, A., 1978, Down's Syndrome in live births by single-year maternal age interval in a Swedish study: Comparison with results from a New York State study, 1978, *Am. J. Hum. Genet.* **30**:19–27.

Horger, E. O., and Smythe, A. R., II, 1977, Pregnancy in women over forty, *Obstet. Gynecol.* **49:**257-161.

Kajanoja, P., and Widholm, O., 1978, Pregnancy and delivery in women aged 40 and over, *Obstet. Gynecol.* **51:**47-51.

Kessler, I., Lancet, M., Borenstein, R., Steinmetz, A., 1980, The problem of the older primipara, *Obstet. Gynecol.* **56:**165-169.

Kujansuu, A., Kivinen, S., and Tuimala, R., 1981, Pregnancy and delivery at the age of forty and over, *Int. J. Gynaecol. Obstet.* **19:**341-345.

Lobl, M., Welcher, D. W., and Mellits, E. D., 1971, Maternal age and intellectual functioning of offspring, *Hopkins Med. J.* **128:**347-357.

Messinis, I., Malamitsi-Puchner A., Hadjigeorgiou, E., Lolis, D., and Nicolopoulos, D., 1982, Preinatal complications in elderly primigravidas, *Padiatrie und Padologie* **17:**597-602.

Morrison, I., 1975, The elderly primigravida, *Am. J. Obstet. Gynecol.* **121:**465-470.

National Center for Health Statistics, 1982, *Monthly Vital Statistics Report*, Vol. 31, No. 2, Supp. 2, May 27, 1982.

National Center for Health Statistics, 1982, *Monthly Vital Statistics Report*, Vol. 32, No. 9, Supp., December 29, 1983.

Sadovnick, A. D., and Baird, P. A., 1981, A cost-benefit analysis of prenatal detection of Down Syndrome and neural tube defects in older mothers, *Am. J. Med. Genet.* **10:**367-378.

USA Today, 1984, Gannett Co., Inc., Washington, D.C., April 26, 1984, p. 1.

Virro, M. R., and Shewchuk, A. B., 1984, Pregnancy outcome in 242 conceptions after artificial insemination with donor sperm and effects of maternal age on the prognosis for successful pregnancy, *Am. J. Obstet. Gynecol.* **148:**518-524.

Wagener, D. K., LaPorte, R. E., Orchard, T. J., Cavender, D., Kuller, L. H., and Drash, A. L., 1983, The Pittsburgh Diabetes Mellitus Study. 3: An increased prevalence with older maternal age, *Diabetologia* **25:**82-85.

Waters, E. G., and Wagner, H. P., 1950, Pregnancy and labor experiences of elderly primipara, *Am. J. Obstet. Gynecol.* **59:**296-304.

IV

THE CLIMACTERIC

11

OVARIAN CHANGES DURING THE CLIMACTERIC

SANTO V. NICOSIA

Introduction

The main known functions of the human ovary are the cyclic release of oocytes and the elaboration of steroid hormones. These two functions are closely interrelated during a reproductive life span that may last up to four decades. During this period, cyclic and repetitive changes in estrogen and progesterone secretion are integrated with follicle maturation, ovulation, corpus luteum formation and regression. Cessation of reproductive life is characterized by a decline in gametogenic and steroidogenic functions. This decline is gradual and thus ovarian aging represents the conclusion of a chain of events that begins in childhood and terminates with death. This chapter will consider these events from a structural point of view. Greater emphasis will be placed on those morphologic changes that characterize the human ovary between cessation of reproduction and senescence.

Ovarian aging represents a classic example of physiologic aging, and it has always aroused interest in gerontological research (Timiras and Meisami, 1972). This interest is particularly justified today since symptomatic women in the climacteric age range may approximate 35 million in the United States alone (Hammond and Maxson, 1982). Conceivably, a better understanding of ovarian aging and its consequences may improve the quality of their life, currently projected to last 28 years after menopause (Hammond and Maxson, 1982).

SANTO V. NICOSIA • Department of Obstetrics and Gynecology and Department of Pathology, University of Pennsylvania, School of Medicine, Philadelphia, Pennsylvania 19104. Present address: Department of Pathology, University of South Florida, College of Medicine, Tampa, Florida 33612.

Ovarian Morphology from Intrauterine Life to Menopause: Basic Facts

The human ovary originates from two somatic mesodermal derivatives, the coelomic epithelium and the mesonephric mesenchyme, and from an extragonadal derivative, the primordial germ cell (Nicosia, 1983). The primitive ovary is formed at approximately 22 weeks after fertilization when primordial germ cells become enclosed by prefollicular cells and when fibrovascular and stromal tissues proliferate in a ventrodorsal direction to form a rudimentary cortex and medulla. Continuation of these processes will produce, at birth, an organ that possesses most of the basic structures necessary for gametogenic and steroidogenic function. During childhood, the human ovary is not a static organ as it undergoes a 30-fold increase in weight between birth and menarche (Nicosia, 1983). To a great extent, this volumetric increase is due to continuous waves of abortive follicle maturation and incorporation of atretic follicle (theca) cells into an expanding stroma (Peters *et al.*, 1976; Hughesdon, 1978; Nicosia, 1983; Reeves, 1980).

The main morphologic features of the human ovary become established after a peripubertal period of anovulatory and atretogenic cycles. When observed in a midsagittal plane, the adult ovary consists of an outer and an inner zone. The former zone or cortex is covered by a mesothelium (surface epithelium) and contains the outer connective tissue or tunica albuginea, follicular-luteal complexes at various stages of development and regression, microvessels, and a heterogeneous cellular stroma (Reeves and Jacobs, 1980). The latter zone or medulla contains a variably dense connective tissue that enmeshes stromal cells, larger blood vessels, lymphatics, and nerves. The most salient feature of the human ovary is ovulation. To prepare for this central event in reproduction, the ovary develops during each menstrual cycle a cohort of follicles. Of these, only one will ovulate and form a corpus luteum while the remaining others will regress and become atretic. Unless fertilization takes place, the corpus luteum will also regress after a finite life span and its cells, as those of atretic follicles (Figure 1A, B), will become part of the ovarian stroma. These morphogenetic changes will continue throughout the entire reproductive life. However, their occurrence will become less regular as the climacteric approaches when anovulatory and atretogenic cycles once again will dominate, as during menarche, the ovarian morphology.

Ovarian Morphology during the Climacteric

The ultimate consequence of ovarian changes before and during the reproductive years is the transformation of the ovary from a follicle-rich organ into an organ whose predominant tissue is the corticomedullary stroma. As already alluded, the ovarian changes mainly responsible for this phenomenon are the

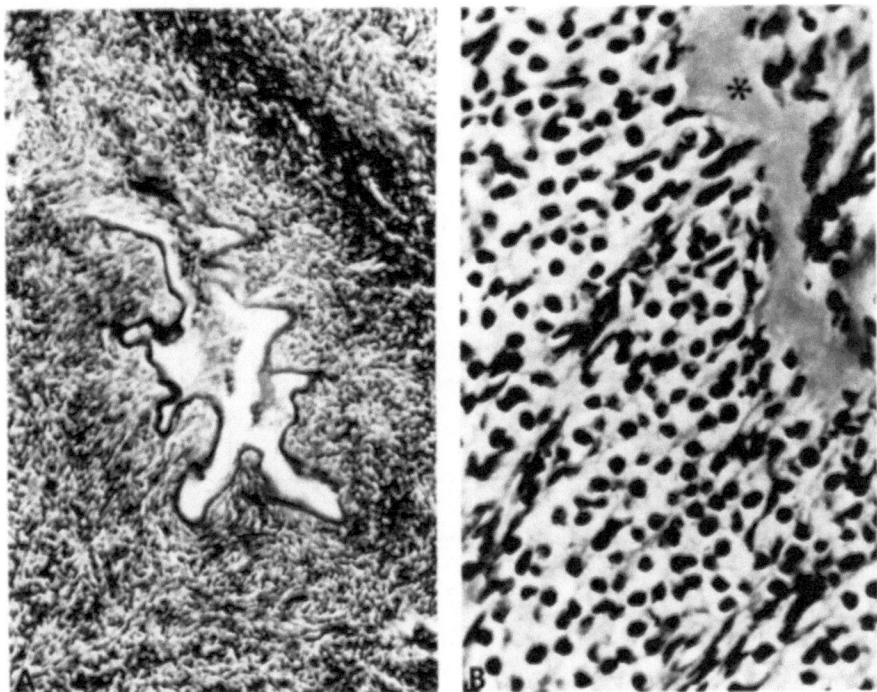

Figure 1. Ovary of a 30-year-old woman. (A) Atretic follicle; ×110. (B) Well-preserved theca cells at edge of atretic follice (asterisk); ×450. (Reproduced at 80%.)

continuous attrition of the oocyte's capital (Figure 2), through ovulation and follicular atresia (Hertig, 1944), and the relentless incorporation of atretic theca cells into the ovarian stroma (Hughesdon, 1978; Nicosia, 1983; Reeves, 1980). This tissue remodeling affects profoundly the physical characteristics of the ovary. For instance, the adult ovary weighs approximately 7 g while the weight of the postmenopausal ovary is only 5 g or less (Figure 3) (Nicosia, 1983). This volumetric decrease, as well as involutional vascular changes (Figure 4) (Lang and Aponte, 1967), deposition of "wear and tear" pigments (Thung, 1961), cortical fibrosis, and declining steroidogenesis, lends support to the common belief that the climacteric is accompanied by ovarian involution. However, there is considerable evidence suggesting that the climacterial ovary is, in many instances, an active organ. This evidence will be reviewed now. As defined by the First International Congress on Menopause, the climacteric is that phase of life which marks the transition from the reproductive to the nonreproductive years (Utian, 1976). Arbitrarily, the climacteric can be divided into the three phases of premenopause, menopause, and postmenopause. By definition, the menopause is the date of the last menstrual period, which in North American

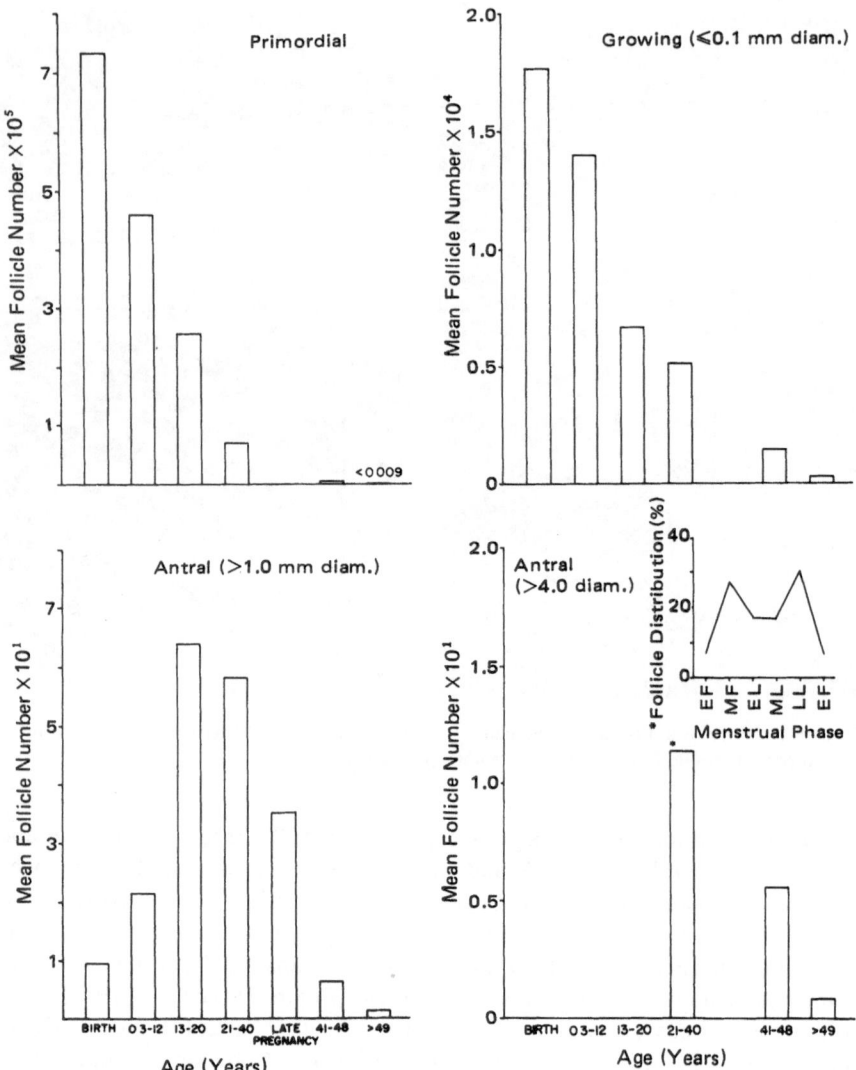

Figure 2. Number of follicles in women throughout the life cycle. From Nicosia (1983).

women takes place at the average age of 51.4 years (Hammond and Maxson, 1982).

Admittedly, the most characteristic feature of the postmenopausal ovary is the marked depletion of the ovarian reserve of follicles (Figure 2). However, a few primordial follicles can still be found well beyond the menopause (Costoff and Mahesh, 1975). Atretic follicles, cystic follicles, and luteinized follicular

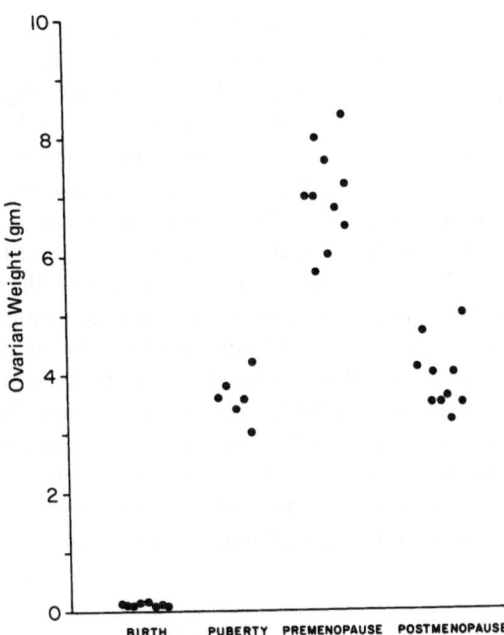

Figure 3. Ovarian weight at different phases of life. Data represent the wet weight of individual normal ovaries obtained at autopsy.

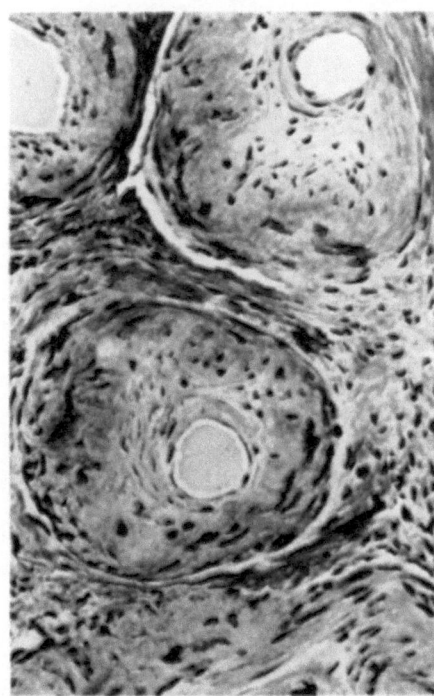

Figure 4. Hyalinized arteriolar sclerosis in postmenopausal ovary; ×220. (Reproduced at 80%).

cysts (Figure 5A, B) are fairly common perimenopausally but can also be found up to 10 years after the menopause (Bigelow, 1958; Boss *et al.*, 1965). In addition, approximately 23% of women older than 50 years may ovulate and produce a corpus luteum (Figure 6) (Novak, 1970). These corpora lutea may show various degrees of functional activity and together with atretic follicles are resorbed into the deep cortical and medullary regions of the ovary. Here, they may persist as fibrous corpora albicantia up to the eighth and ninth decades of life, with a well-preserved microvascular blood supply (Thung, 1961).

The prominence of the corticomedullary stroma is, perhaps, the second most distinguishing feature of the postmenopausal ovary. Diffuse sheets or nodular aggregates of stromal cells can be found among the blood vessels of the ovarian medulla (Figure 7A) and in the cortex. This last compartment may contain areas of hyalinized fibrous tissue or be totally replaced by stromal cells (Figure 7B). Usually, these cells have a fibroblastoid morphology (Figure 7C) (Reeves and Jacobs, 1980). However, they may often develop abundant intracellular lipids (Figure 7D) and cytochemical as well as ultrastructural features of steroidogenesis (Thung, 1961; Balboni, 1976; Fiemberg, 1981; Nicosia,

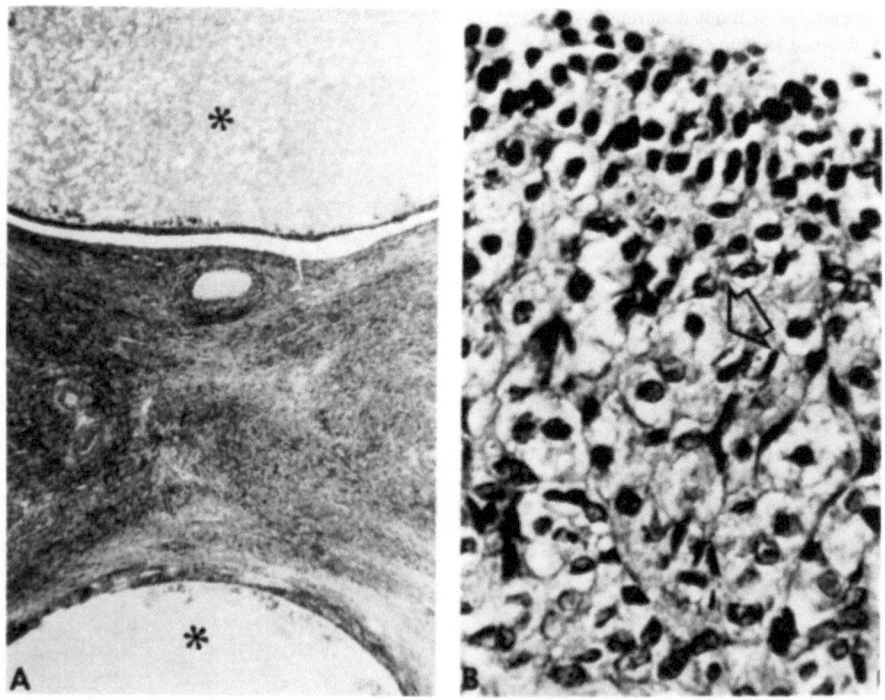

Figure 5. Ovary of a 50-year-old woman. (A) Cystic follicles (asterisks); ×110. (B) Wall of cystic follicle with luteinized theca cells (arrow); ×450. (Reproduced at 80%.)

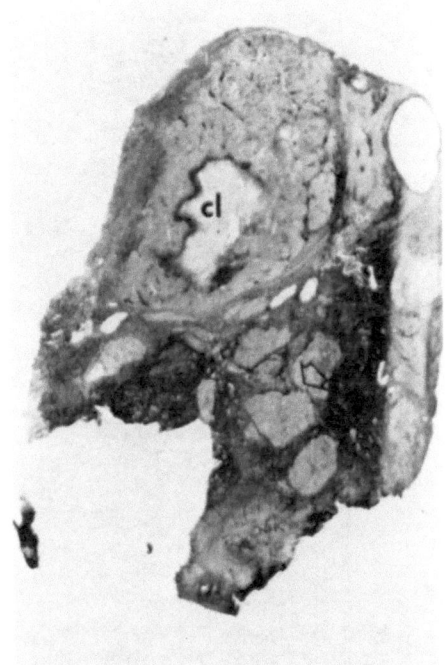

Figure 6. Recent corpus luteum (cl) in a postmenopausal woman. Hyalinized remnants of old corpora lutea are also present (arrow); ×3.5. (Reproduced at 80%.)

1983). This process is described as stromal luteinization. Other cells with steroidogenic properties may become prominent in the postmenopausal ovary. As indicated by their names, these cells are found alone or in endocrinelike islets near the hilus of the ovary and in intimate association with nonmyelinized nerve fibers (Figure 8A, B) (Laffargue *et al.*, 1978). It is unclear if hilus cells represent embryological remnants similar to the interstitial cells of the human testis or if they derive from ovarian stromal cells under the induction of sympathetic nerves (Laffargue *et al.*, 1978).

Other morphologic signs of activity may be present in the postmenopausal ovary. These include the formation of cortical granulomata and the proliferation of the surface epithelium and of the rete ovarii. Cortical granulomata represent small foci of epitheloid cells and lymphocytes (Hertig, 1944; Bigelow, 1958; Boss *et al.*, 1965). These structures are less frequently found in premenopausal ovaries (Herbold *et al.*, 1984) and their pathogenesis is not understood. They may originate from nodular aggregates of regressing hyperplastic stromal cells (Figure 9A, B). Atretic follicles can also be associated with macrophage like cells and lymphocytes (Figure 9C, D). It has also been suggested that cortical granulomata may be related to surface epithelium inclusion cysts (Boss *et al.*, 1965). These cysts derive from the invagination and entrapment of the surface

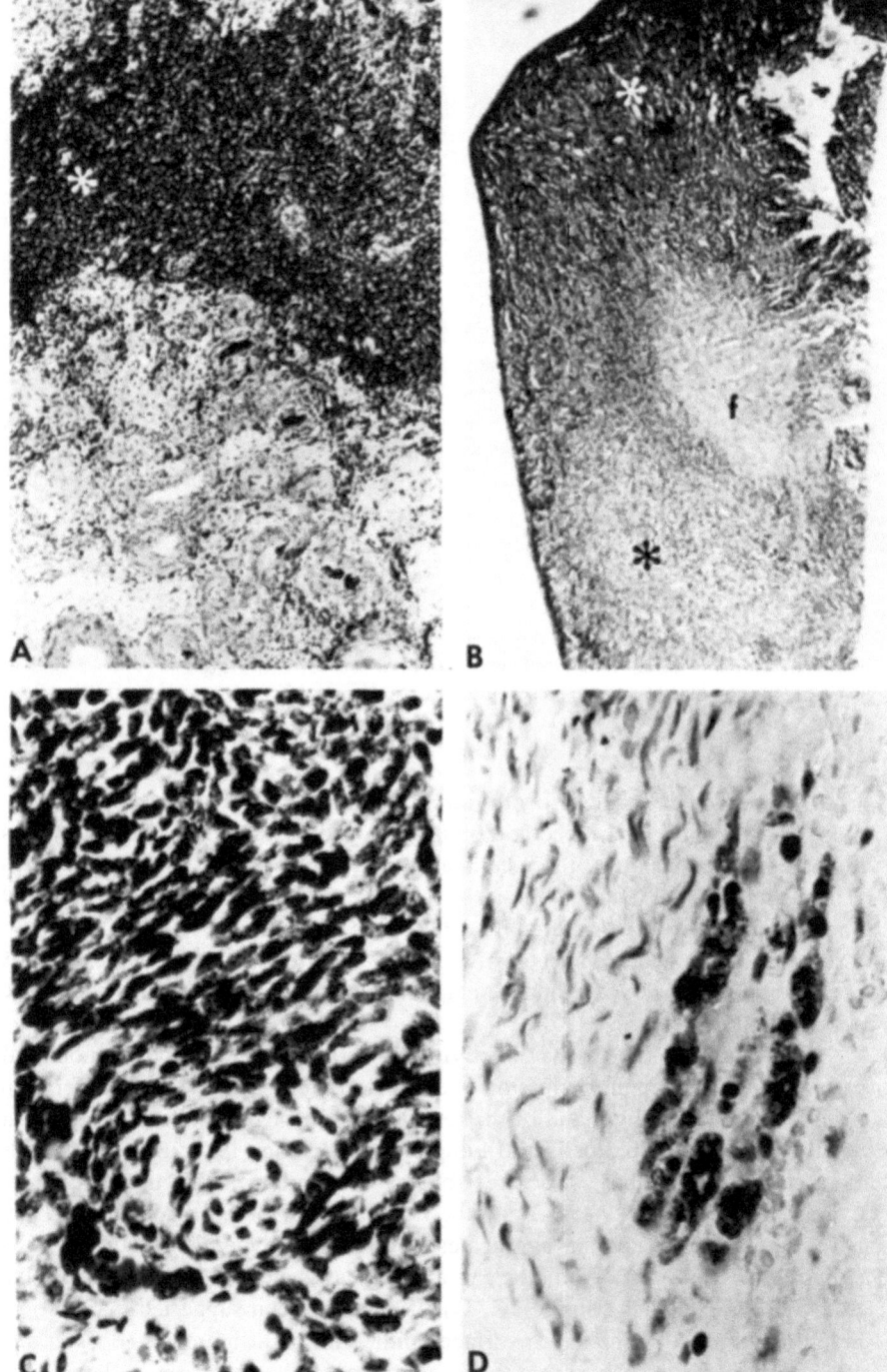

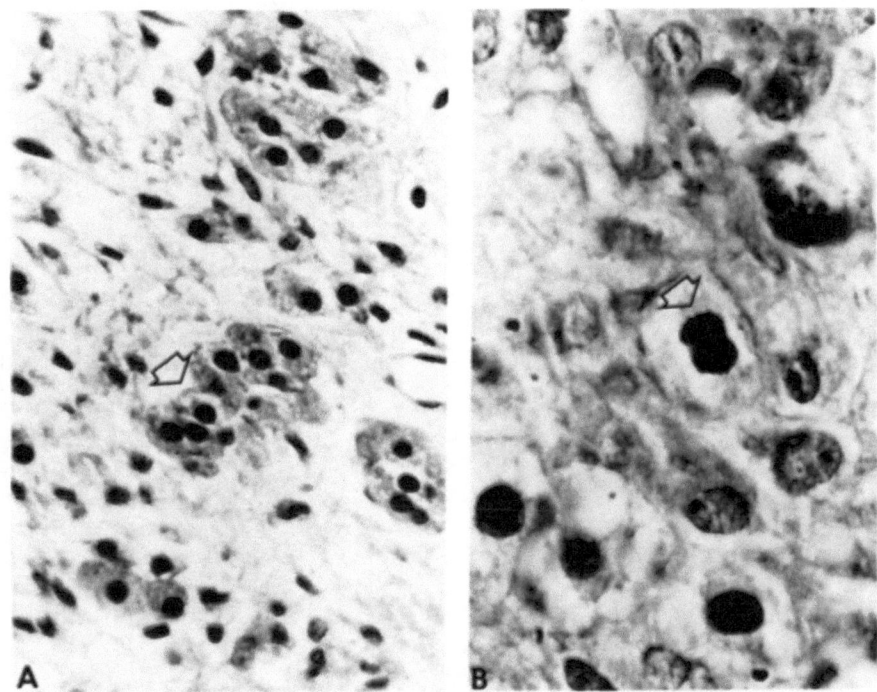

Figure 8. Ovary of a 62-year-old woman. (A) Scattered groups of hilus cells (arrow); ×450. (B) Mitosis (arrow) in an endocrine isletlike group of hilus cells; ×1100. [(A, B) Reproduced at 80%.]

epithelium into the ovarian cortex (Figure 10A). Although present throughout life, they are more frequently during the late reproductive years and after menopause (Thung, 1961; Blaustein, 1981; Bigelow, 1958; Boss *et al.*, 1965). Indirect evidence of higher activity in the postmenopausal surface epithelium is also provided by the occurrence of cell stratification (Figure 10B), papillary proliferations (Figure 10C), and cords (Figure 10D). Finally, careful evaluation of the postmenopausal ovary may frequently reveal hyperplastic agregates of ductlike tubules near the hilar region (Figure 11A). These tubules, better known as rete ovarii, are lined by a cuboidal to low columnar and occasionally ciliated epithelium similar to that of other paramesonephric derivatives (Figure 11B). The histogenesis of these structures is unclear although they can be frequently

---

Figure 7. Postmenopausal ovary. (A) Infiltrating sheet of stromal cells (asterisk) in ovarian medulla; ×110. (B) Diffuse cortical stroma (asterisks) and focal fibrosis (f); ×35. (C) Elongated stromal cells; ×450. (D) Luteinized stromal cells. Intracellular lipids are demonstrated as dark deposits after staining with oil red 0; ×450. [(A–D) Reproduced at 80%.]

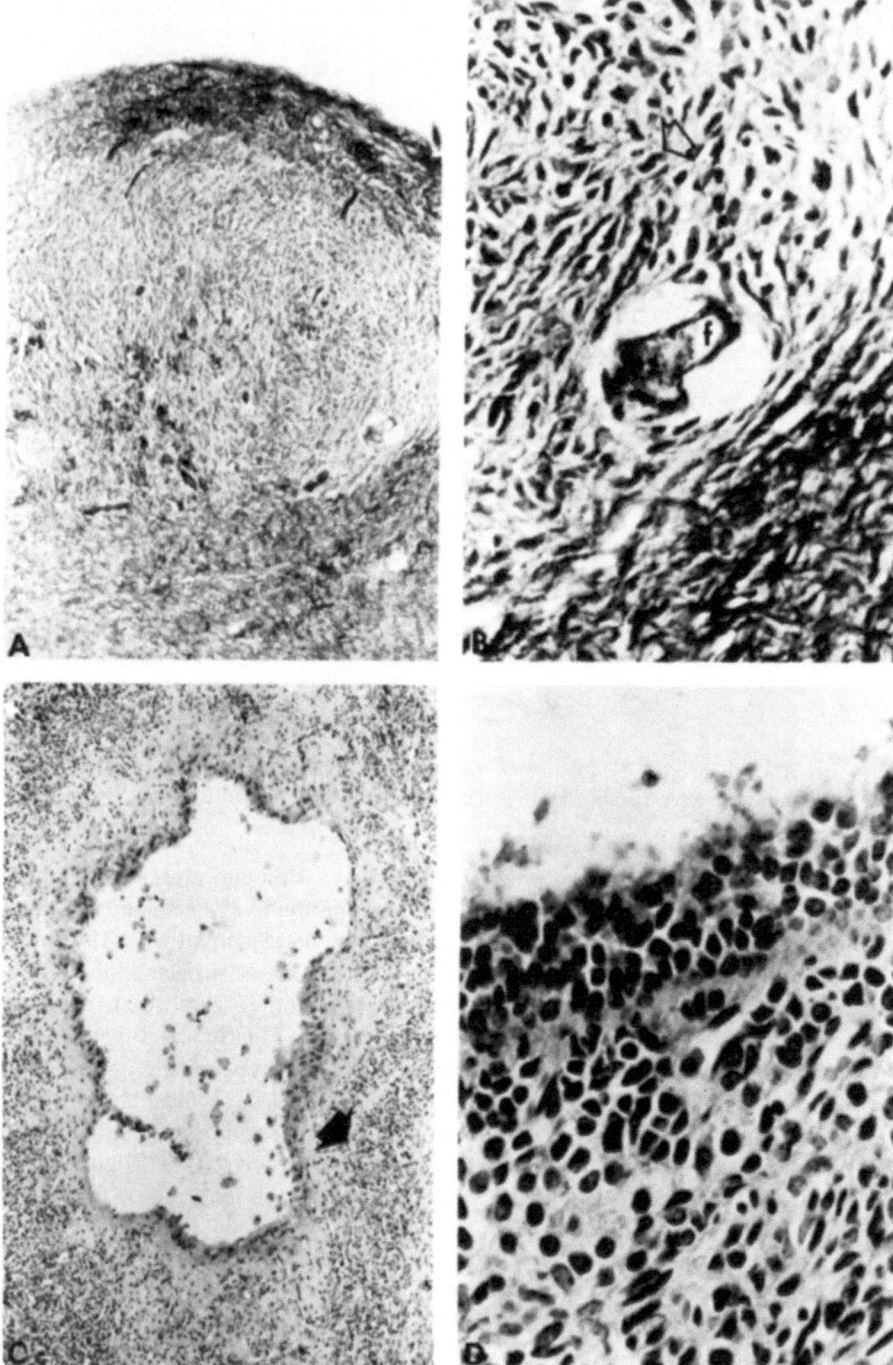

seen near to, or outgrowing from, the overlying surface epithelium (Thung, 1961).

Bigelow (1958) and Boss *et al.* (1965) have analyzed the incidence of the ovarian changes described above. These authors examined the ovaries of women between the age of 20 and over 76 years who had undergone hysterectomy for reproductive tract disorders (Bigelow's retrospective study) or had died of non-ovarian diseases (Boss' prospective autoptic study). Both studies showed that while stromal cell proliferation was common during the reproductive years, its highest incidence and prominence occurred after menopause prior to the eighth decade (Figure 12). Steroidogenic or "luteinized" stromal cells were present in approximately 35% of ovaries between the age of 56 and 70 years and their frequency remained high in more elderly women. Hilus cells were observed in 83% of all ovaries but were most prominent after menopause, particularly in women older than 70 years (Figure 12). An association was found between the degree of stromal proliferation and the incidence of morphological stromal luteinization and of prominent hilus cells (Figure 13). Stromal luteinization and proliferation were associated with clinical signs of androgen hypersecretion and their incidence appeared to be a function of parity, suggesting a pathogenetic role for gonadotropins. Finally, these semiquantitative studies indicated that follicular cysts, cystic follicles, and luteinized cysts were found more frequently 3 years after menopause (Figure 14). After menopause, there was also an increase in the incidence of cortical granulomata and surface epithelium cysts (Figure 14).

Structural-Functional Correlates of Ovarian Aging

The age-related increase in ovarian stroma may be due to the metaplastic transformation of theca cells from atretic follicles into stromal cells (Mossman and Duke, 1973; Hughesdon, 1978). It may also be attributed to the proliferation of stromal cells under the cyclic or, later in life, sustained influence of gonadotropins (Balboni, 1976; Nicosia, 1983). This morphogenetic view of ovarian aging is supported by the observation that the ovarian stroma is deficient in conditions associated with a numerical decrease or impaired development of follicles (Ross and Lipsett, 1978) and is augmented after physiologic or pathologic follicular atresia (Hughesdon, 1978; Reeves, 1980).

The functional consequence of a postmenopausally augmented stromal compartment has not been clearly assessed. Most likely, this compartment rep-

---

Figure 9. Postmenopausal ovaries. (A) Cortical granuloma; ×110. (B) Degenerating primordial follicle (f) and regressing stromal cells (arrow) at periphery of granuloma; ×450. (C) Atretic follicle surrounded by an inflammatory infiltrate (arrow); ×220. (D) This infiltrate is composed of round cells, mainly lymphocytes and macrophages; ×450. [(A–D) Reproduced at 80%.]

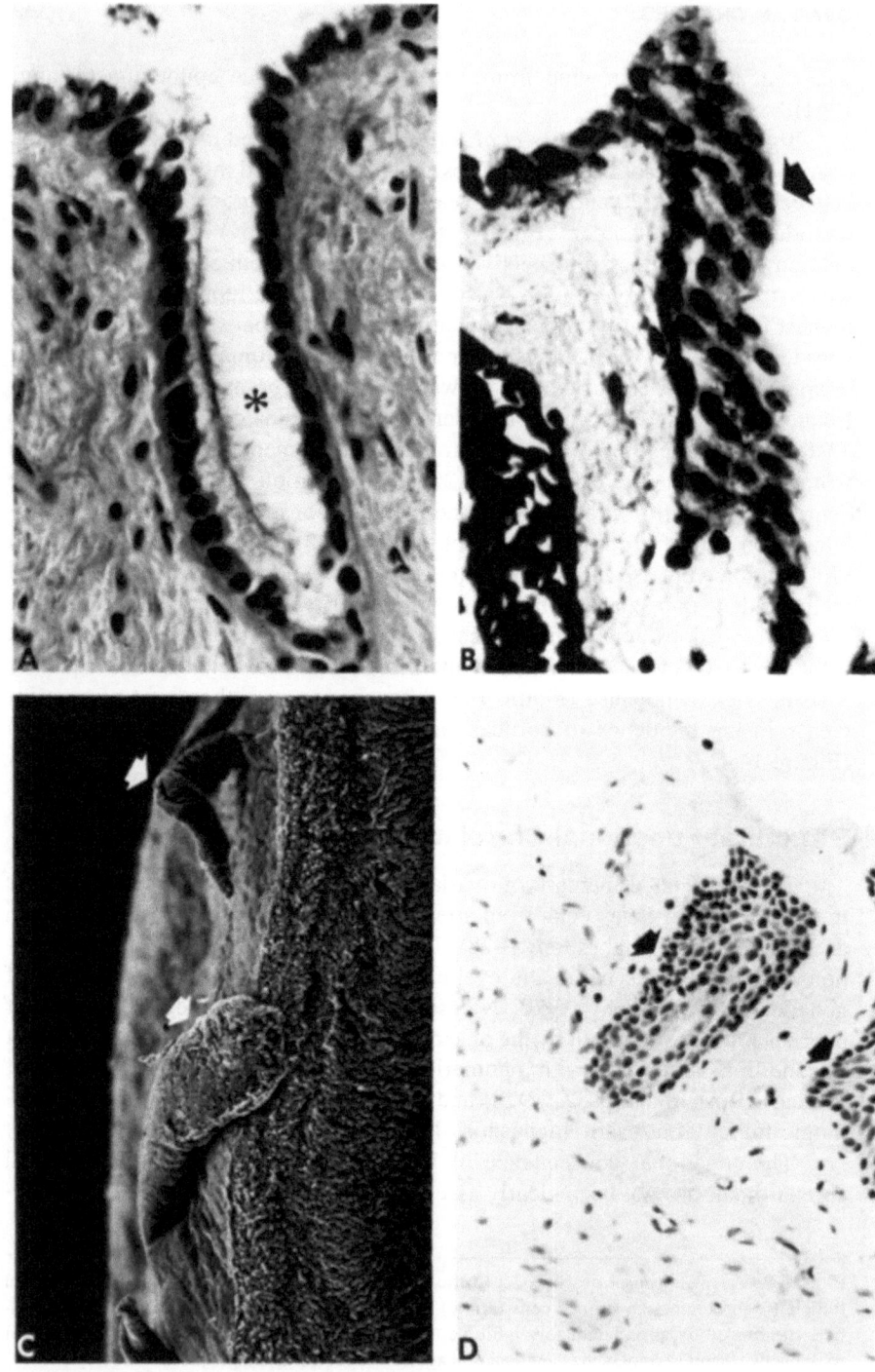

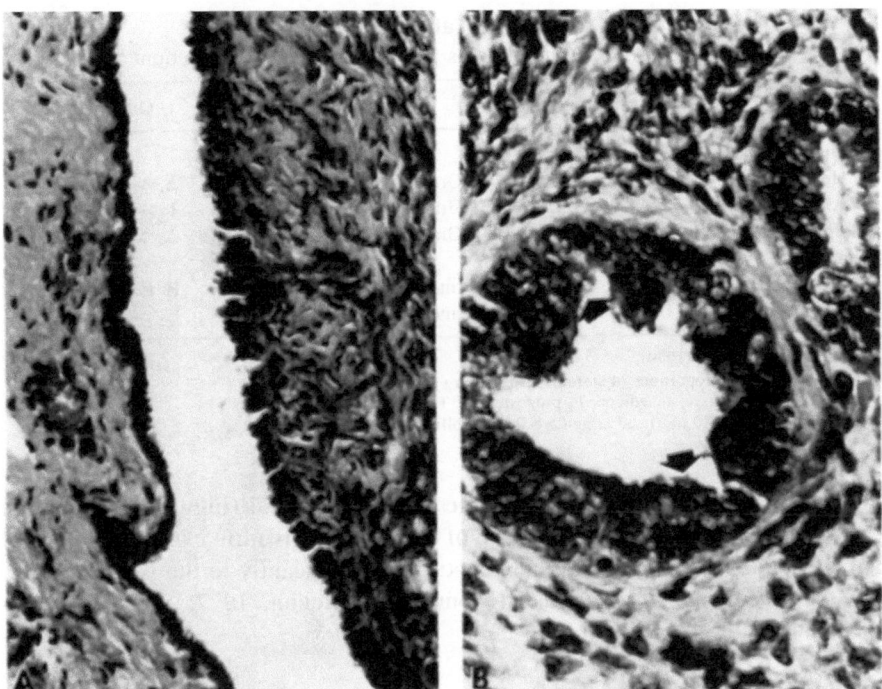

Figure 11. Rete ovarii in a postmenopausal ovary. (A) Ductlike tubule; ×220. (B) Lining epithelium with ciliated cells (arrows); ×450. [(A, B) Reproduced at 80%.]

resents a pool of potentially steroidogenic cells. In fact, it has been reported that up to 30% and 50%, respectively, of circulating androstenedione and testosterone may be produced by the postmenopausal ovary (Chang and Judd, 1981). Thus, the postmenopausal ovary may not be a totally exhausted endocrine gland but only an organ that has lost its rhythmic capacity to produce hormones and from a predominantly estrogen and progesterone producer has become a mostly androgen secretor. The structural basis of such a change has been recently reaffirmed by *in vitro* studies (Piana and Laffargue, 1975; McNatty *et al.*, 1979; Dennefors *et al.*, 1982). These studies have shown that most ovarian cells are capable of steroid secretion but that this capability is expressed in qualitative and quantitative different terms in various ovarian compartments (Table I). The ultimate consequence of multicellular ovarian steroido-

Figure 10. Ovarian surface epithelium in postmenopausal ovaries. (A) Invagination (asterisk); ×450. (B) Cell stratification (arrow); ×450. (C) Papillary processes (arrows); ×180. (D) Cortical cords (arrows); ×220. [(A–D) Reproduced at 80%.]

Table I

Major Ovarian Compartments and Putative Steroid Secretion[a,b]

Compartments	Major steroid[c]	Cell source	Other steroids[c]
Follicle[d]			
Healthy	E_2	Granulosa, theca	$\Delta_4 > P > E_1 \geqslant T$
Atretic	Δ_4	Granulosa, theca	$T > P \geqslant E_2 > E_1$
Corpus Luteum	P	Granulosa-lutein	$\Delta_4 > E_2 > E_1$
Stroma			
Premenopausal	Δ_4	Interstitial (stroma-thecal)	$P, E > T > E_1$
Postmenopausal	Δ_4	Interstitial (stroma-thecal)	$E_2 > T, P, E_1$

[a] From McNatty *et al.* (1979).
[b] Based on relative proportions of steroids secreted or synthesized *in vitro*.
[c] E_2, estradiol; Δ_4, androstenedione; P, progesterone; T, testosterone; E_1, estrone.
[d] Includes small (<8.0 mm) and large (>8.0 mm) follicles.

genesis is the creation of local (Table I) and peripheral (Figure 15) hormonal milieus characteristic to each phase of the reproductive life cycle. The adrenal glands and extraglandular tissues contribute significantly to the peripheral milieu in postmenopausal women (Monroe and Menon, 1977; Siiteri and McDonald, 1973).

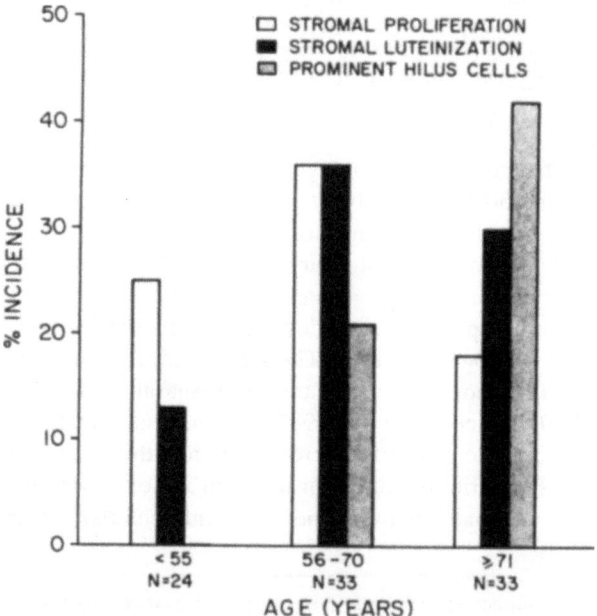

Figure 12. Age distribution of ovarian changes. Modified from Boss *et al.* (1965).

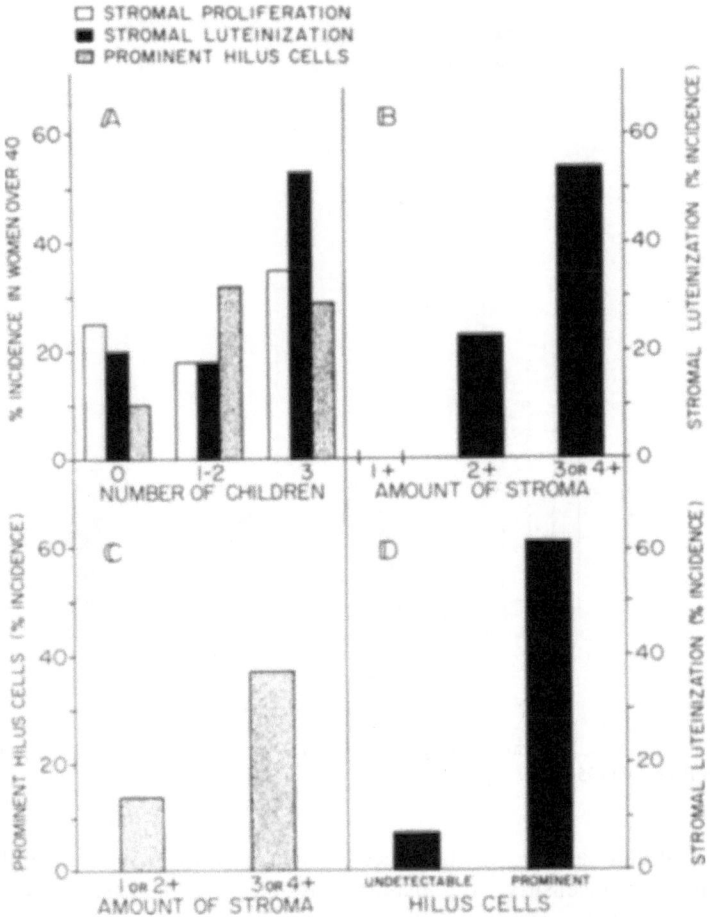

Figure 13. Correlation of ovarian changes with parity (A) and among themselves (B–D). Modified from Boss *et al.* (1965).

Conclusions and Suggestions for Future Studies

While quite different from its younger counterpart, the climacterial ovary is clearly not a quiescent organ. To support this assertion, a number of structural features have been reviewed in this chapter. These features are schematically outlined in Figure 16.

In spite of several decades of research in reproduction, our comprehension of the postmenopausal ovary and of the mechanisms leading to its formation is only partial. For instance, we need to understand better what regulates follicular atresia in the presence of adequate or elevated levels of gonadotropins in peri-

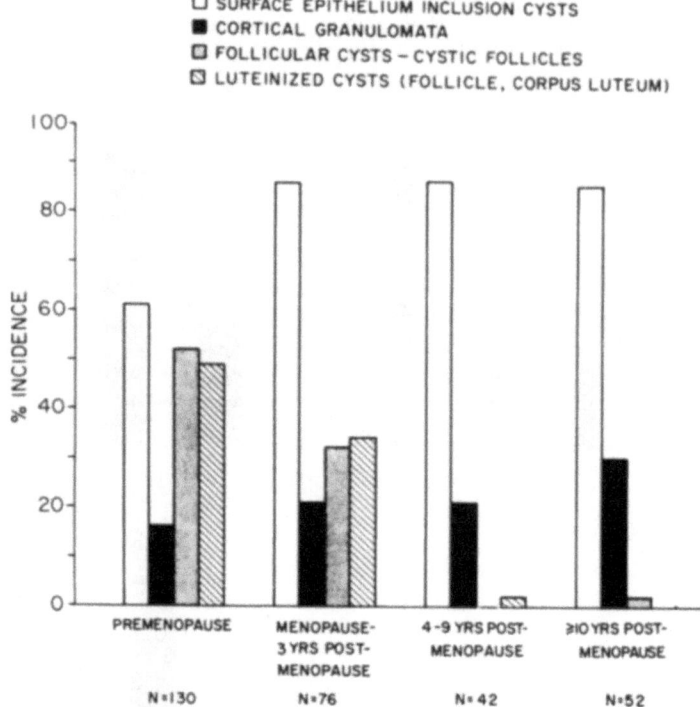

☐ SURFACE EPITHELIUM INCLUSION CYSTS
■ CORTICAL GRANULOMATA
☐ FOLLICULAR CYSTS – CYSTIC FOLLICLES
☐ LUTEINIZED CYSTS (FOLLICLE, CORPUS LUTEUM)

Figure 14. Distribution of ovarian changes before and at various intervals after menopause. Modified from Bigelow (1958).

menopausal and early postmenopausal women. Are the remaining follicles of the aging ovary insensitive to gonadotropins and is the increasingly androgenic intraovarian milieu a result or a cause of abortive follicular maturation? Is the proliferation and differentiation of stromal and hilus cells under neural or hormonal influences (Peluso *et al.*, 1976; Laffargue *et al.*, 1978)? Do cortical granulomata represent an immunological response to altered oocytes or stromal or surface epithelial cells (Russell *et al.*, 1982)? Finally, we need to know if surface epithelial and rete ovarii cells are sensitive to steroid and protein hormones. Many of these questions may not be easily pursued in humans for obvious reasons. Animal models may prove useful although a menopause similar to that occurring in women has not been described in mammals except, perhaps, in macaques (Graham *et al.*, 1979). In spite of some differences, the use of imbred mouse strains, such as the CBA and the $O_{20} \times DBA_f)_{F_1}$ mice, may also be appropriate (Loeb, 1948; Papadaki *et al.*, 1979).

Research on ovarian aging may transcend the goals of basic reproductive biology. In fact, one of the most frequent and lethal form of ovarian cancer

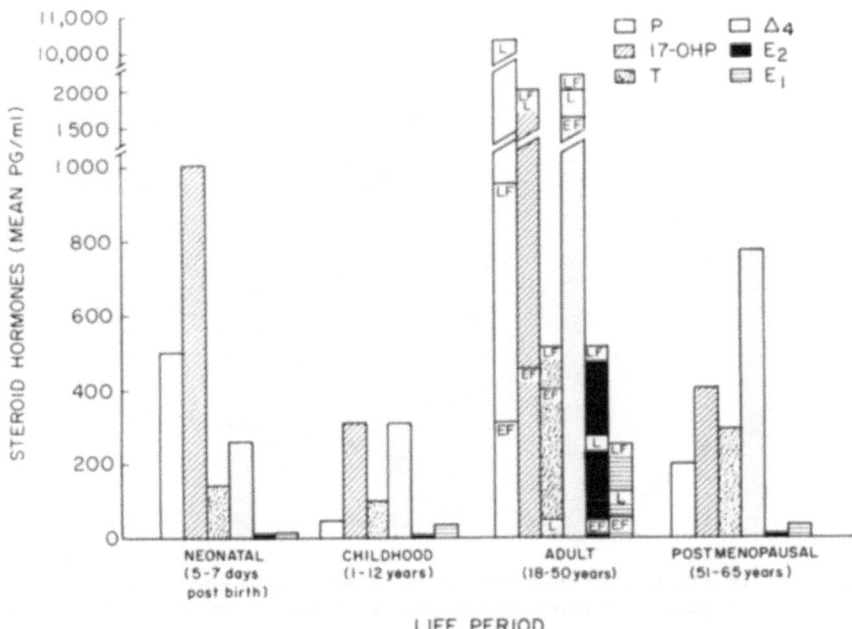

Figure 15. Peripheral steroid hormone levels throughout the life cycle. Data compiled from Bodlingmaier *et al.* (1973); Faiman *et al.* (1976); Forest (1979); Forest *et al.* (1976); Judd *et al.* (1974); McNatty *et al.* (1975); Sanyal *et al.* (1974); Vermeulen (1976).

arises in postmenopausal women from surface epithelial cells (Scully, 1977; Blaustein, 1981). The proliferative tendency of these cells after menopause has been described, and understanding its regulation may give insight into the genesis of ovarian surface cancer. Until a few years ago, the limited supply of surface epithelium present in normal mammalian ovaries has restricted the study of this tissue only to ultrastructural methods (Gondos, 1975; Anderson *et al.*, 1976; Motta *et al.*, 1980; Nicosia and Johnson, 1984). Recently, the development of isolation and culture procedures has allowed a more direct experimental manipulation of this most neglected ovarian tissue (Hamilton *et al.*, 1981; Adams and Auersperg, 1981; Nicosia *et al.*, 1984, 1985). These *in vitro* studies have provided direct evidence indicating that the growth of surface epithelial cells is stimulated by androgens (Hamilton *et al.*, 1980) and by protein hormones (Osterholzer *et al.*, 1985). Such data support morphologic and epidemiologic studies pointing to an association between surface epithelium proliferation, elevated gonadotropins, and prominence of a steroidogenically active ovarian stroma (Thung, 1961; Cramer and Welch, 1983).

It seems reasonable to predict that further investigation of surface epithelium changes as well as of other structural aspects of ovarian aging will con-

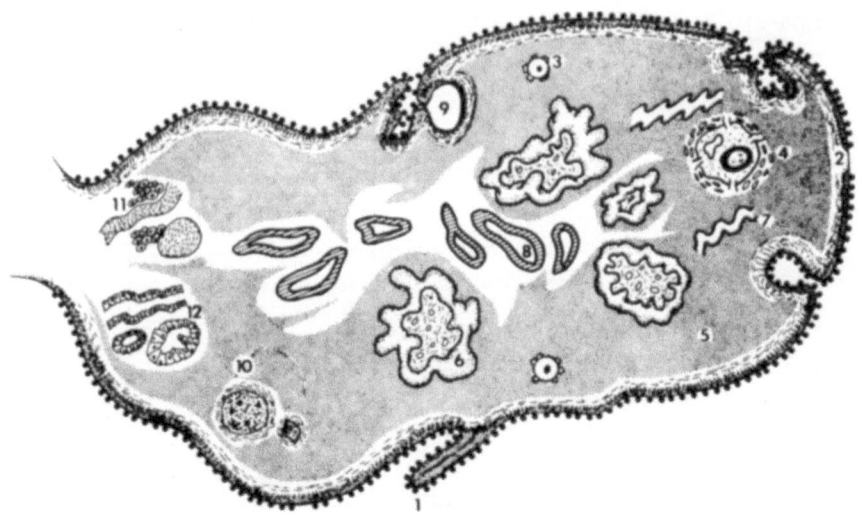

Figure 16. Schematic representation of the most salient changes in the postmenopausal ovary. 1, Papillary surface epithelium; 2, thickened tunica albuginea; 3, residual primordial follicle; 4, atretic follicle (recent); 5, diffuse stroma; 6, hyalinized remnant of corpus luteum; 7, atretic follicle (old); 8, sclerosed blood vessels, 9, surface epithelium cyst; 10, cortical granulomata; 11, hilus cells; 12, rete ovarii.

tribute not only to basic research but also to the quality of life of over half of the fastest growing segment of our population, the postmenopausal woman (Moment, 1978).

ACKNOWLEDGMENTS

The authors wish to thank Ms. Mary Jo Larson, Mr. Kenneth Ray, and Ms. Young Hyun for expert technical assistance in artistic and photographic work. The secretarial skills of Ms. Neisha Son is also gratefully acknowledged. This work was partly carried out using the facilities of a Tissue Core Laboratory sponsored by USPHS grant HD-06274.

References

Adams, A. T., and Auersperg, N., 1981, Transformation of cultured rat ovarian surface epithelial cells by Kirsten murine sarcoma virus, *Cancer Res.* **41**:2063–2072.

Anderson, E., Lee, G., Letourneau, R., Albertini, D. F., and Meller, S. M., 1976, Cytological observations of the ovarian epithelium in mammals during the reproductive cycle, *J. Morphol.* **150**:135–166.

Balboni, G. C., 1976, Histology of the ovary, in: *The Endocrine Function of the Human Ovary*, Volume 7 (V. H. T. James, M. Serio, and G. Giusti, eds.), Academic Press, New York, pp. 1–24.

Bigelow, B., 1958, Comparison of ovarian and endometrial morphology spanning the menopause, *Obstet. Gynecol.* **11**:487–513.

Blaustein, A., 1981, Surface (germinal) epithelium and related ovarian neoplasms, *Pathol. Annu.* **16**(1):247–294.

Bodlingmaier, J., Wagner-Barmack, M., Butenandt, O., and Knorr, D., 1973, Plasma estrogens in childhood and puberty under physiologic and pathologic conditions, *Pediatr. Res.* **7**:901–907.

Boss, J. H., Scully, R. E., Wegner, K. H., and Cohen, R. B., 1965, Structural variations in the adult ovary—Clinical significance, *Obstet. Gynecol.* **25**:747–763.

Chang, J. R., and Judd, H. L., 1981, The ovary after menopause, *Clin. Obstet. Gynecol.* **24**:181–191.

Costoff, A., and Mahesh, V. B., 1975, Primordial follicles with normal oocytes in the ovaries of postmenopausal women, *J. Am. Geriatr. Soc.* **23**:193–196.

Cramer, D. W., and Welch, W. R., 1983, Determinants of ovarian cancer risk. II. Interferences regarding pathogenesis. *JNCI* **7**:717–721.

Dennefors, B. L., Janson, P. O., Hamberger, L., and Knutsson, F., 1982, Hilus cells from human postmenopausal ovaries: Gonadotrophin sensitivity, steroid and cyclic AMP production, *Acta Obstet. Gynecol. Scand.* **61**:413–416.

Faiman, C., Winter, J. S. D., and Reyes, F. I., 1976, Patterns of gonadotropins and gonadal steroids throughout life, *Clinics Obstet. Gynec.* **3**:467–483.

Fiemberg, R., 1981, Thecosis, *Pathol. Annu.* **16**(2):239–271.

Forest, M. G., 1979, Function of the ovary in the neonate and infant, *Eur. J. Obstet. Gynecol. Reprod. Biol.* **9**:145–160.

Forest, M. G., dePeretti, E., and Bertrand, J., 1976, Hypothalamic-pituitary-gonadal relationships in man from birth to puberty, *Clin. Endocrinol.* **5**:555–569.

Gondos, B., 1975, Surface epithelium of the developing ovary, *Am. J. Pathol.* **81**:303–320.

Graham, C. E., Kling, R. O., and Steiner, R. A. 1979, Reproductive senescence in female non-human primates, in: *Aging in Nonhuman Primates* (D. M. Bowden, ed.), Van Nostrand Reinhold, New York, pp. 183–202.

Hamilton, T. C., Henderson, W. J., and Eaton, C., 1980, Isolation and growth of the rat ovarian germinal epithelium, in: *Tissue Culture in Medical Research* (R. J. Richards and K. T. Rajan, eds.), Pergamon, New York, pp. 237–244.

Hamilton, T. C., Davies, P., and Griffiths, K., 1981, Androgen and oestrogen binding in cytosols of human ovarian tumours, *J. Endocrinol.* **90**:421–431.

Hammond, C. B., and Maxson, W. S., 1982, Current status of estrogen therapy for the menopause, *Fertil. Steril.* **37**:5–25.

Herbold, D. R., Frable, W. J., and Kraus, F. T., 1984, Isolated noninfectious granulomas of the ovary, *Int. J. Gynecol. Pathol.* **2**:380–391.

Hertig, A. T., 1944, The aging ovary: A preliminary note, *J. Clin. Endocrinol.* **4**:581–582.

Hughesdon, P. E., 1978, Postnatal formation of ovarian stroma and its relation to ovarian pathology, *Int. J. Gynaecol. Obstet.* **16**:8–19.

Judd, H. L., Judd, G. E., Lucas, W. E., and Yen, S. S. C., 1974, Endocrine function of the postmenopausal ovary: Concentration of androgens and estrogens in ovarian and peripheral vein blood, *J. Clin. Endocrinol. Metab.* **39**:1020–1024.

Laffargue, P., Benkoel, L., Laffargue, F., Casamora, P., and Chamlian, A., 1978, Ultrastructural and enzyme histochemical study of ovarian hilar cells in women and their relationships with sympathetic nerves, *Hum. Pathol.* **9**:650–659.

Lang, W. R., and Aponte, G. E., 1967, Gross and microscopic anatomy of the aged female reproductive organs, *Clin. Obstet. Gynecol.* **10:**454–465.

Loeb, L., 1948, Aging processes in the ovaries of mice belonging to strains differing in the incidence of mammary carcinoma, *Arch. Path.* **46:**401–440.

McNatty, K. P., Hunter, W. M., McNeilly, A. S., and Sawers, R. S., 1975, Changes in the concentration of pituitary and steroid hormones in the follicular fluid of human graafian follicles throughout the menstrual cycle, *J. Endocrinol.* **64:**555–571.

McNatty, K. P., Makris, A., DeGrazia, C., Osathanondl, R., and Ryan, K. J., 1979, The production of progesterone, androgens, and estrogens by granulosa cells, thecal tissue, and stromal tissue from human ovaries in vitro, *J. Clin. Endocrinol. Metab.* **49:**687–699.

Moment, G. B., 1978, The Ponce de Leon trail today, in: *The Biology of Aging* (J. A. Behnke, C. E. Finch, and G. B. Moment, eds.), Plenum Press, New York, pp. 1–17.

Monroe, S. E., and Menon, K. M. J., 1977, Changes in reproductive hormone secretion during the climacteric and postmenopausal periods, *Clin. Obstet. Gynecol.* **20:**113–122.

Mossman, H. W., and Duke, K. L., 1973, *Comparative Morphology of the Mammalian Ovary*, University of Wisconsin Press, Madison.

Motta, P. M., VanBlerkom, J., and Makabe, S., 1980, Changes in the surface morphology of ovarian germinal epithelium during the reproductive cycle and in some pathological conditions. A correlative three-dimensional analysis by transmission, scanning and high voltage electron microscopy, *J. Submicrosc. Cytol.* **12:**407–425.

Nicosia, S. V., 1983, Morphological changes of the human ovary throughout life, in: *The Ovary* (G. B. Serra, ed.), Raven Press, New York, pp. 57–81.

Nicosia, S. V., and Johnson, J. M., 1984, Surface morphology of ovarian mesothelium (surface epithelium) and other pelvic and extrapelvic mesothelial sites in the rabbit, *Int. J. Gynecol. Pathol.* **3:**249–260.

Nicosia, S. V., Johnson, J. M., and Streibel, E. J., 1984, Isolation and ultrastructure of rabbit ovarian mesothelium (surface epithelium), *Int. J. Gynecol. Pathol.* **3:**348–360.

Nicosia, S. V., Johnson, J. M., and Streibel, E. J., 1985, Growth characteristics of rabbit ovarian mesothelial (surface epithelial) cells, *Int. J. Gynecol. Pathol.* **4:**58–74.

Novak, E. R., 1970, Ovulation after fifty, *Obstet. Gynecol.* **36:**903–910.

Osterholzer, H. O., Streibel, E. J., and Nicosia, S. V., 1985, Growth effects of protein hormones on cultured rabbit ovarian surface epithelial cells, *Biol. Reprod.* **33:** 247–258.

Papadaki, L., Beilby, J. O. W., Chowaniec, J., Coulson, W. F., Darby, A. J., Newman, J., O'Shea, A., and Wikes, J. R., 1979, Hormone replacement therapy in the menopause: A suitable animal model, *J. Endocrinol.* **83:**67–77.

Peluso, J. J., Steger, R. W., Jaszczak, S., and Hafez, E. S. E., 1976, Gonadotropin binding sites in human postmenopausal ovaries, *Fertil. Steril.* **27:**788–795.

Peters, H., Himelstein-Braw, R., and Faber, M., 1976, The normal development of the ovary in childhood, *Acta Endocrinol.* **82:**617–630.

Piana, L., and Laffargue, F., 1975, Les ovaries apres la menopause, *Inserm* **55:**195–204.

Reeves, G., 1980, Stromal and follicular compartments of the ovary. Their influence in the development and involution of the organ, in: *Endocrine Physiopathology of the Ovary* (R. I. Tozzini, G. Reeves, and R. L. Pineda, eds.), Elsevier/North Holland Biomedical Press, Amsterdam, pp. 137–150.

Reeves, G., and Jacobs, J., 1980, Ultrastructure of the stroma of the human ovary, in: *Endocrine Physiopathology of the Ovary* (R. I. Tozzini, G. Reeves, and R. L. Pineda, eds.), Elsevier/North Holland Biomedical Press, New York, pp. 121–136.

Ross, G. T., and Lipsett, M. B., 1978, Hormonal correlates of normal and abnormal follicle growth after puberty in humans and other primates, *Clin. Endocrinol. Metabol.* **7:**561–573.

Russel, P., Bannatyne, P., Shearman, R. P., Fraser, I. S., and Corbett, P., 1982, Premature hypergonadotropic ovarian failure: Clinicopathological study of 19 cases, *Int. J. Gynecol. Pathol.* **1**:185–201.

Sanyal, M. K., Berger, M. J., Thompson, I. E., Taymor, M. L., and Horne, H. W., 1974, Development of grafian follicles in adult human ovary. I. Correlation of estrogen and progesterone concentration in antral fluid with growth of follicles, *J. Clin. Endocrinol. Metab.* **38**:828–835.

Scully, R. E., 1977, Ovarian tumors, A. Review, *Am. J. Pathol.* **87**:686–720.

Siiteri, P. K., and MacDonald, P. C., 1973, Role of extraglandular estrogen in human endocrinology, in: *Handbook of Physiology: The Female Reproductive System*, Section 7, Volume 2, Pt. 1 (R. Greep, ed.), American Physiological Society, Bethesda, pp. 615–629.

Thung, P. J., 1961, Aging changes in the ovary, in: *Structural Aspects of Aging* (G. H. Bourne, ed.), Hafner, New York, pp. 109–142.

Timiras, P. S., and Meisami, E., 1972, Changes in gonadal function, in: *Developmental Physiology and Aging* (P. S. Timiras, ed.), MacMillan, New York, pp. 527–541.

Utian, W. H., 1976, The climacteric syndrome, in: *Consensus on Menopause Research* (P. A. van Keep, R. B. Greenblatt, and M. Albeaux-Fernet, eds.), MTP Press, Lancaster, p. 1.

Vermeulen, A., 1976, The hormonal activity of the postmenopausal ovary, *J. Clin. Endocrinol. Metab.* **42**:247–253.

12

NONGONADAL ESTROGEN AND ENDOCRINE FUNCTION DURING THE MENOPAUSE

PENTTI K. SIITERI

Introduction

Estrogens are an integral component of a complex array of hormonal signals elaborated by the brain, pituitary, and ovary that direct the reproductive process in all mammalian species. In addition to their essential actions on tissues of the female reproductive tract, however, estrogens have more subtle but important actions on many other organs and tissues such as liver, bone, fat, and the vascular system in both sexes. Together with the knowledge that many tissues other than the gonads and placenta actively synthesize estrogens it is now evident that the traditional views gained by study of reproductive biology must be broadened. Indeed, the great diversity of chemical structures having estrogenic activity in both the animal and plant kingdoms suggests that the estrogenic signal may have arisen very early in evolution. Considering the strong but yet ill-defined association of estrogens with neoplastic transformation of target cells, the redundancy of mechanisms for regulating or negating estrogen action (protein binding, progesterone, androgens, and glucocorticoids) may reflect adaptive responses required to cope with a necessary evil. In order to understand the consequences of menopause and aging on estrogen-responsive tissues it is necessary to consider (1) nonovarian estrogen production in relation to other hormones, particularly androgens, (2) estrogen transport, and (3) the cellular mechanism of estrogen action. Each of these subjects will be addressed in turn.

PENTTI K. SIITERI • Reproductive Endocrinology Center, Department of Obstetrics, Gynecology, and Reproductive Sciences, University of California, San Francisco, San Francisco, California 94143.

Physiological and psychological aspects of estrogen action, or lack thereof, are considered by colleagues elsewhere in this volume.

Estrogens from the Womb to Menopause

During pregnancy both the mother and developing fetus are exposed to much greater amounts of estrogens and progesterone than at any other period of life. The outpouring of estrogens and progesterone by the placenta in quantities that cannot be achieved pharmacologically is unique to the human female. Although the modern era of endocrinology began more than 50 years ago with the isolation and characterization of steroid metabolites from human pregnancy urine, we still do not understand the full biological significance of this remarkable phenomenon. The copious amount of enzymes required to convert the adrenal androgen dehydroepiandrosterone sulfate (DHEAS) accounts for the efficient synthesis of estradiol (E_2) and estriol by the placenta (Siiteri and MacDonald, 1966; Siiteri and Seron-Ferre, 1981). At the end of gestation total plasma estrogen levels are 100- to 1000-fold higher than those found in the cycling female or postmenopausal woman. Clearly, the molecular mechanism(s) of estrogen action on reproductive tract tissues during pregnancy must be modulated differently than during the postmenopausal period when they are exquisitely sensitive to very low levels of estradiol (see below). The balance between agonistic and antagonistic signals such as the well-known antiestrogenic properties of progesterone is clearly important in this regard. Estriol, which is always present in greater amounts than E_2 during human gestation, may also be an important antagonist to estrogen action, as proposed recently by Notides and his co-workers (Sasson and Notides, 1983a,b).

Following birth, estrogens are cleared from the infant's circulation over a period of 4–5 days, and levels remain low in adolescent girls since estrogen production by the ovaries is negligible. After adrenarche, estrone (E_1) and estradiol plasma levels slowly begin to rise with the onset of puberty (Lee et al., 1976; Apter, 1980). Although the ovaries are generally thought to be the source of this estrogen, peripheral conversion of adrenal androgens, particularly androstenedione to E_1 also may be important during this period. Since fat is a major site of this transformation (Schindler et al., 1972; Forney et al., 1981; MacDonald et al., 1978), more efficient conversion of androgens to estrogens in prepubertal girls whose body fat composition is relatively high likely explains the better correlation of body weight than chronological age with the timing of menarche (Frisch, 1984). Following maturation of the central pulse generator mechanisms, the enhanced output of peripheral estrogen in obese girls could increase sensitivity of the pituitary to hypothalamic gonadotropin-releasing hormone (GnRH). Consequently, the secretion of follicle-stimulating hormone (FSH) and luteinizing hormone (LH) in amounts sufficient to activate a full ovarian cycle of folicular development, ovulation, and corpus luteum formation can occur earlier in obese than in thin girls. According to this view, positive effects on the pituitary, rather than changes in negative feedback sensitivity to estrogens (gonadostat), may influence the timing of menarche. Furthermore, mechanisms by which this gradually maturing process may be disturbed are readily apparent. Excessive peripheral estrogen production resulting from marked obesity and/or elevated androstenedione production by the ovaries and/ or adrenals together with increased availability of E_2 to the pituitary owing to depressed levels of the transport protein, sex hormone-binding globulin (SHBG), in obese subjects (see below) could inhibit FSH secretion and lead to failure of ovulation. In extreme cases, hypersecretion of LH and excessive production of androgens by ovarian thecal cells would result in excessive atresia of ovarian follicles and the polycystic ovarian syndrome.

Normal maturation and function of the hypothalamic pituitary ovarian axis leads to cyclical ovarian secretion of E_2 and progesterone during the reproductive years. Peripheral production of estrogens under these conditions appears to play a minor role in stimulating most target tissues. However, during the early follicular phase of the menstrual cycle when ovarian estrogen secretion is very low, peripherally formed estrogen may be essential to maintain the ability of the pituitary gland to respond appropriately to GnRH. The reversible amenorrhea of women athletes who reduce their body fat content to extremely low levels during periods of high activity (Warren, 1980) could be explained by desensitization of the pituitary when peripheral estrogen production falls below a critical level.

Postmenopausal Estrogen Production

This topic has been reviewed extensively in recent years (Siiteri and MacDonald, 1973; Siiteri, 1982) and will be discussed only briefly here. Following a 1–2 year period of erratic ovulatory cycles, the ovaries become quiescent. Cyclical secretion of estradiol (E_2) and progesterone ceases and only small amounts of the androgens androstenedione and testosterone continue to be produced. In the absence of adequate estrogen production, ovarian testosterone may contribute to hirsutism and even virilization in postmenopausal women (Judd *et al.*, 1974). The rather abrupt drop in ovarian steroid output is accompanied by a slow decline in adrenal secretion of DHEAS. A major decline in plasma DHEAS level occurs in both men and women with advancing age (Abraham and Maroulis, 1975), but neither the mechanism(s) responsible for this change nor the biological consequences are clear at present. In contrast, adrenal secretion of androstenedione continues unabated and its conversion to E_1, in peripheral tissues accounts for most if not all of the estrogen that is produced in postmenopausal women (Grodin *et al.*, 1973). E_1 is a weak estrogen but it is converted to E_2 in many tissues. The amount of E_1 formed is variable, depending upon several factors, the most important of which is obesity. The correlation between degree of obesity and the efficiency of androstenedione to E_1 conversion *in vivo* was recognized in early isotopic studies (Grodin *et al.*, 1973; Siiteri and MacDonald, 1973; MacDonald *et al.*, 1978) and has been substantiated by many subsequent studies of plasma estrogen levels in postmenopausal women (Judd *et al.*, 1974a; Judd *et al.*, 1974b). It is now abundantly clear that except in rare situations such as liver disease or androgen-producing tumors (Siiteri and MacDonald, 1973), the estrogenic mileu in the postmenopausal female is largely dependent upon her body weight. The biological effects of estrogen or lack thereof on various target tissues are consistent with this mechanism. The long-recognized association of obesity and endometrial hyperplasia and carcinoma on the one hand and increased severity of osteoporosis in thin women on the other is explicable by the relationship between body weight and estrogen production (Laufer *et al.*, 1983). While many studies have demonstrated that adipose tissue and cells (Simpson *et al.*, 1981; Ackerman *et al.*, 1981) contain the aromatase enzyme, the true significance of this relationship is not clear. The simplest view is that estrogen production is proportional to the amount of adipose tissue, which is determined by complex dietary, metabolic, and genetic factors. Alternatively, a causal relationship may exist between estrogen synthesis in fat and the number of fat cells. Although the data in support of this notion is meager at present, several recent findings are in accord with this view. Estrogens have been reported to stimulate growth of preadipocytes in culture (Roncari and Van, 1978), and estrogen receptors have been found in peripheral fat deposits of the rat (Wade and Gray, 1978). Thus, the intriguing possibility exists that estrogen formed in stromal cells of adipose tissue stimulates adipocyte proliferation and obesity. Unpublished studies from this laboratory are in accord with this possibility.

During the breeding season both male and female squirrel monkeys (*Saimiri sciureus*) experience weight gain. In males body weight increases (fatting response) as much as 20–25% during this 3- to 4-month period. Endocrine studies have revealed that plasma levels of both androgens (androstenedione and testosterone) and estrogen (E_1 and E_2) increase dramatically during this period. The estrogens appear to be derived by peripheral mechanisms since the conversion of androstenedione to E_1 *in vivo* per unit weight is about 1000-fold greater than occurs in humans. Furthermore, genital skin fibroblasts from squirrel monkeys have much higher aromatase activity than found in rhesus monkey or human cells. In marked contrast, squirrel monkey cells lack the expected 5-α-reductase activity needed to form dihydrotestosterone (DHT). Thus, the intriguing possibility is raised that the fatting response is elicited by local estrogen production from circulating androgens in adipose tissue depots. Further studies are in progress now that we have the long-sought model for studying the physiologic function(s) of peripheral estrogen production.

Plasma Transport of Estradiol

Endocrinologists generally believe that only the free (protein unbound) fraction of steroid and thyroid hormones in plasma is able to enter target cells and exert biological effects. According to this view, the specific high-affinity binding proteins in blood, such as corticosteroid-binding globulin (CBG), SHBG, and thyroxine-binding globulin (TBG), function only to regulate the free concentrations of the hormones that they bind. We became interested in reevaluating this question when we found that the concentration of SHGB was severely depressed in obese postmenopausal women (Nisker *et al.*, 1980; Davidson *et al.*, 1981). The results suggested that the reduced binding capacity of SHBG would increase the percentage of free E_2 in plasma as had been noted for testosterone in hyperandrogenic states by many authors. Despite the elegant studies of Anderson (1974), however, little attention had been given to binding of E_2 in plasma. Clear evidence was presented for reciprocal variations in free E_2 and SHBG levels (Wu *et al.*, 1976), whereas others concluded that E_2 binding to SHBG was insignificant under physiologic conditions (Vigersky *et al.*, 1979). In view of these controversial results we developed a new approach to the measurement of free steroids in plasma that would yield physiologically relevant information. The procedure that evolved, isodialysis, combines the principles of dialysis, ultrafiltration, and double isotope methodology (Hammond *et al.*, 1980). This method permits the rapid determination of the percentage of free steroid in many small samples (0.2 ml) of whole plasma at 37°C. Using this method we obtained unequivocal evidence for binding of E_2 to SHBG and demonstrated that the percentage of free E_2 is inversely related to the plasma SHBG concentrations (Siiteri *et al.*, 1982). When postmenopausal women with and without endometrial cancer were studied, the expected increase in percentage of free E_2 in obese subjects was observed in both groups (Nisker *et al.*, 1980; Davidson *et al.*, 1981). Thus, the effects of increased estrogen production appears to be augmented in obese women by increased availability of plasma estrogens to target tissues. The failure to find differences between women with and without endometrial cancer was not surprising since estrogens are generally considered to be promoters of cancer rather than carcinogens themselves. Similar studies were carried out in women with breast cancer, and the influence of obesity was evident, particularly in postmenopausal women since elevated free E_2 levels were observed (Moore *et al.*, 1982). In addition, there appeared to be a small number of patients who had elevated free E_2 levels despite normal plasma SHBG (Siiteri *et al.*, 1981). These intriguing results suggested that other mechanisms such as competing ligands or an abnormal SHBG binding site might chronically elevate free estradiol levels in some women. Subsequent studies have failed to confirm these results (unpublished observations). It now appears that despite careful matching of samples the earlier findings were an artifact due to long-term storage of the serum samples, For reasons that are not clear, the percentage of free E_2 increases upon storage (3–8 years) without a decline in the SHBG binding capacity.

In further collaborative studies with Dr. Howard Judd, we compared androgen and estrogen levels in postmenopausal women with endometrial cancer and those with osteoporosis (Laufer *et al.*, 1983). No significant differences were found in androstenedione and testosterone levels, whereas the total E_1 and E_2 levels in the more obese (cancer) group were significantly higher than in the thin women with osteoporosis as expected. In addition, the percentage of free E_2 was higher in the plasma of the obese subjects as indicated earlier. As a consequence, the mean free E_2 level was 2.5- to 3-fold higher in obese than in thin postmenopausal women. Although at first glance these results were satisfying, they also raised some interesting questions. Since the steroid receptor concept (see below) suggests that the biological response to a hormone is proportional to the fractional degree of receptor occupancy, the less than threefold difference in E_2 concentrations between the extremes of estrogenicity in postmenopausal women was surprising. Second, the actual levels of unbound E_2 (ca. 10^{-12} M) were far below the concentrations required to saturate the estrogen receptor (ER) if the dissociation constant (K_D) is around 10^{-9} M, as estimated in many previous studies.

We have considered several alternatives that might explain these discrepancies. First, it has been suggested that albumin-bound steroids also may be available to target cells (Pardridge *et al.*, 1980). As demonstrated experimentally, this is possible if the transit time of blood through an organ is slow relative to the fast dissociation rate of E_2 or other steroids from albumin. Although the albumin-bound fraction, which varies from 20 to 90% depending upon the SHBG level (Siiteri *et al.*, 1982), is much larger than the free fraction, the concentration of available (free + albumin bound) E_2 (10^{-11}–10^{-10} M) would still be far less than that needed to saturate the estrogen receptor. We next considered the possibility that the affinity of E_2 for its receptor may have been underestimated in the past.

Receptor Binding Studies

The central dogma of steroid hormone action developed over the past 20 years holds that steroids diffuse into target cells and bind to cytoplasmic receptor proteins and that the resulting complex translocates to the nucleus where it interacts with specific gene regulatory sites on DNA to modulate messenger RNA synthesis. However, important issues such as the intracellular locus of the native receptors are still being debated. The binding of E_2 to its receptor is a more complex phenomenon than a simple biomolecular interaction of the steroid with a fixed binding site. The binding process appears to result in a change in size of the receptor as a result of dimerization of proteins having a molecular weight of about 65,000 daltons and sedimentation coefficient of 4 S to form a dimer having a molecular weight of about 120,000 and a sedimentation constant of 5 S. This transformation is associated with positive cooperativity of binding (Notides et al., 1981) and a change in properties (activation) that promotes binding of the steroid receptor complex to DNA. Other steroid receptors appear to undergo the activation but not the transformation phenomenon. Our early in vivo and in vitro studies demonstrated that transformation of the estrogen receptor occurs in the nuclear compartment of rat uterine cells and further that it has the characteristics of a cooperative process (Siiteri et al., 1973; Siiteri et al., 1974; Linkie and Siiteri, 1978). Recent reports have confirmed these findings (Kasid, et al., 1984) and also demonstrated that the native estrogen receptor is located in the cell nucleus (King and Greene, 1984; Welshons et al., 1984). It now appears that receptors found in cytoplasmic extracts are artifacts of tissue disruption. How the 5 S E_2 receptor acts to alter cell function still is not clear, although we previously speculated that alterations in chromatin structure may allow binding of RNA polymerases to specific gene regulatory sites that promote transcription of RNA (Linkie and Siiteri, 1978).

Recent observations in our laboratory suggest that this model is compatible with the physiologic picture in postmenopausal women discussed earlier. In order to obtain physiologically relevant estimates of the binding affinity of ER, we have adapted the isodialysis technique (Hammond et al., 1980) for studies of receptor binding. As recently pointed out (Siiteri, 1984), the use of an equilibrium assay in which true free ligand concentrations are actually measured rather than calculated avoids underestimates of binding affinity that arise from binding of ligand to nonreceptor components in impure preparations. As shown in Figure 1, the affinity of the rat uterine estrogen receptor for E_2 is about 100-fold higher than previously estimated. The value of K_D at 4°C is about 20 pM and does not change at elevated temperatures (25–37°C) providing that the receptor binding site is protected by the addition of reducing agents such as dithiothreitol (DTT) to the buffers. If this precaution is not taken, heating the receptor preparation leads to a slow increase in K_D to values about 30-fold higher (0.6 nM). This change could account for the apparent presence of low-affinity binding sites (Type II) observed by others (Markaverich and Clark, 1979) when the estrogen receptor is examined by exchange assays carried out at 25–30°C in the absence of thiol compounds. When E_2 binding in uteri from ovariectomized-adrenalectomized rats was examined, the results shown in Figure 2 were obtained. The domed appearance of the Scatchard plot is clear evidence for positive cooperativity of E_2 binding as observed with the bovine uterine estrogen receptor by Notides and his co-workers (Notides et al., 1981). We have ruled out several potential sources of error that can artifactually produce such curvilinear Scatchard plots. The high-affinity component estimated from the linear portion of the Scatchard plot has a K_D in the range of 0.01–0.02 nM,

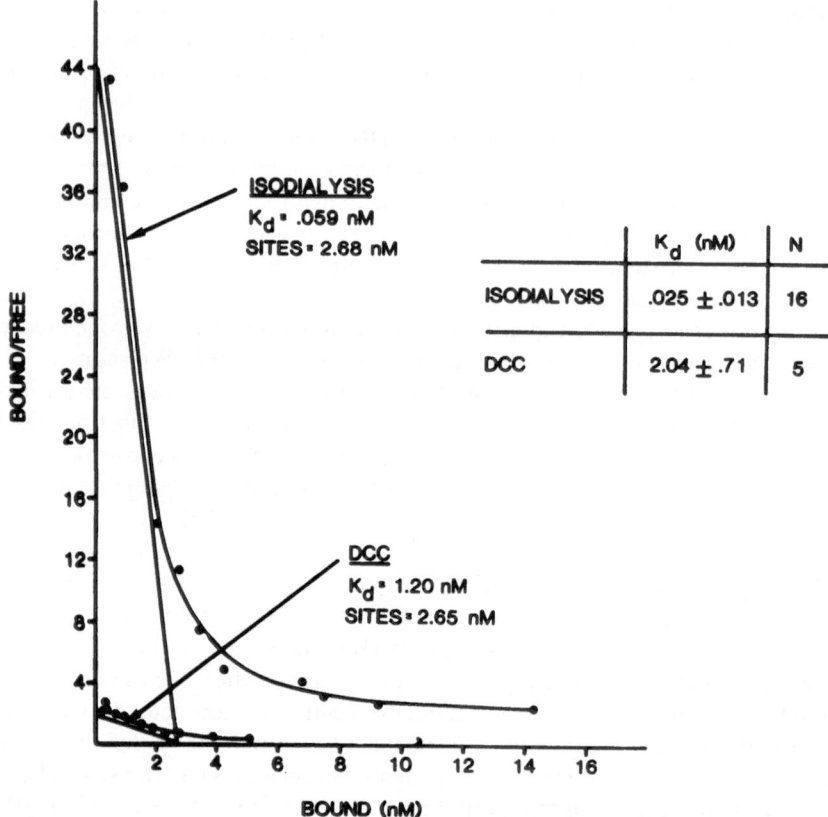

Figure 1. Comparison of the dissociation constants and binding capacities for the rat uterine E_2 receptor obtained by the DCC method and isodialysis. Adult castrated rats were treated for 2 days with E_2 (10 μg/day) and sacrificed 24 hr after the last injection. Cytosol was prepared as described in the text and E_2 receptors assayed by isodialysis. Specific E_2 binding components were determined using nonlinear, multiple regression computer analysis. Solid line, DCC method; broken line, isodialysis method. Inset shows a statistical analysis of the K_D values obtained by the two methods.

which approximates the concentration of available E_2 in plasma of postmenopausal women. The same phenomenon has been observed with estrogen receptor preparations from human breast tumors. As previously noted, positive cooperativity in receptor binding reduces the range of E_2 concentration needed to saturate the receptor (Notides *et al.*, 1981). This is illustrated in Figure 3. It is evident that a change of 3 orders of magnitude is required to achieve 90% saturation of the receptor if the binding process is noncooperative. However, using actual data obtained with rat uterine ER binding it can be seen that cooperativity reduces the range to about 10-fold. Since it is generally believed that

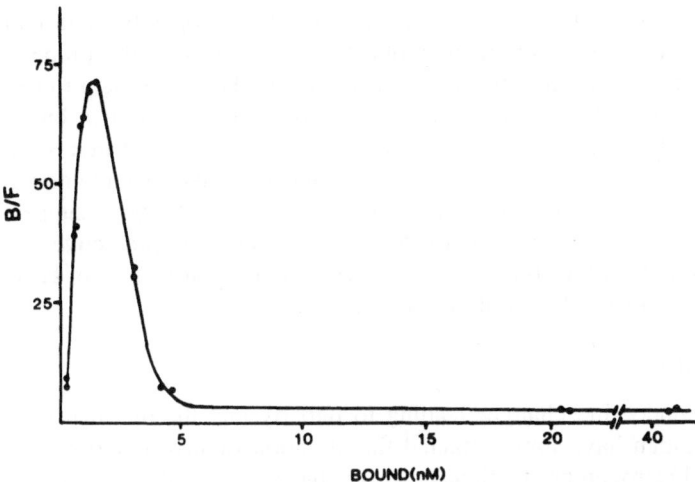

Figure 2. Domed Scatchard plot indicative of positive cooperativity in binding of [³H]-E₂ to the rat uterus estrogen receptor. Uterine cytosol was prepared from ovarectomized-adrenalectomized mature rats using Tris-EDTA buffer, pH 7.4. Analysis of binding was carried out using isodialysis at 4°C over 18 hr.

steroid effects are proportional to receptor occupancy, positive cooperativity may be operative in many systems since the fluctuations in plasma hormone levels rarely exceed a 10-fold range. Thus, in the absence of the antiestrogenic effects of progesterone, estrogens appear to act on reproductive tissues almost

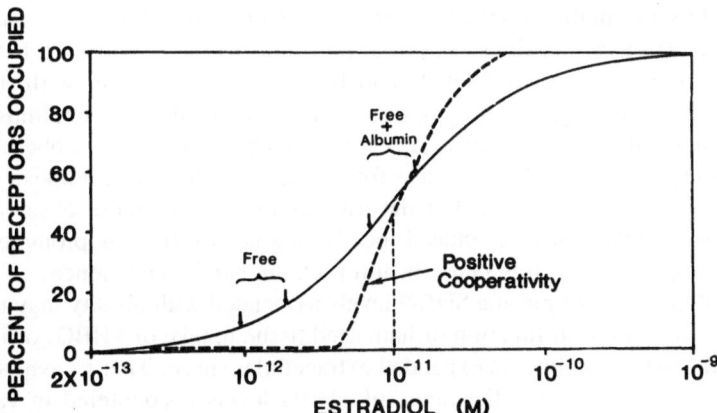

Figure 3. Theoretical and actual semilogarithmic saturation plots for E_2 binding to rat uterine receptor ($K_D = 2 \times 10^{-11}$ M). The curve labeled "positive cooperativity" was constructed from the data shown in Figure 2.

in an on-or-off mode as a consequence of positive cooperativity in E_2 binding. These findings have important implications for selecting the proper estrogen replacement therapy in postmenopausal women. They also help to explain the very low doses of various estrogen preparations needed to elicit biological effects in postmenopausal women. It is gratifying that current trends are toward lower doses of estrogen in light of the increased risk of endometrial cancer associated with estrogen doses used in the past. Recent reports suggesting that combined use of estrogens and progestins may even be protective against endometrial and perhaps breast cancer underscore the importance of basic studies on the mechanism of steroid hormone action.

Summary

The endocrine changes leading to puberty and the onset of ovulation in young women have long attracted the attention of investigators. The role of peripheral estrogen production, however, has received little consideration. The important influence of body fat composition rather than chronological age in the timing of puberty suggests that estrogen produced in fat may prime the pituitary to initiate discharge of gonadotropins in response to GnRH. Depressed levels of SHBG associated with obesity would act to increase exposure of the pituitary to low levels of E_2. Although speculative, this proposed relationship between body weight and estrogen activation of the reproductive process would ensure an optimum degree of female development in preparation for the metabolic demands of pregnancy. Following cessation of ovulatory function during the menopausal period of a woman's life, estrogen production falls to low but variable levels. The conversion of adrenal androstenedione to E_1 in fat and other peripheral tissues in the principal source of postmenopausal estrogens. The production of estrogen in adipose tissue is reflected by the relationship between plasma estrone and estradiol levels and body weight. Together with the increased availability of plasma E_2 in obese subjects this mechanism almost certainly explains the greatly increased risk for endometrial cancer in obese postmenopausal women. Direct evidence for estrogen formation in adipose tissue and cells have been obtained, but the true significance of these observations remains to be determined. If indeed locally produced estrogens promote obesity, new approaches to this major health problem can be envisioned.

The depression of plasma SHBG levels associated with obesity may reflect a disturbance in thyroid function or increased tissue uptake of SHBG, or it may simply reflect dilution into an expanded extracellular space. That E_2 availability can vary 4- to 5-fold over the range of SHBG levels encountered in various clinical conditions whereas total E_2 levels only vary 2- to 3-fold in postmenopausal women illustrates the unique properties of this system. Measurements of total plasma E_2 alone cannot accurately reflect its impact on target tissues. This is particularly important since relatively small increases in intracellular E_2 levels can activate the estrogen receptor. For example, a plasma level of 20 pg/ml E_2 in a postmenopausal woman weighing 300 lb may induce endometrial hyperplasia, whereas the same level in a 90-lb woman may be associated with atrophy of the genital tract and osteoporosis. This knowledge of the production, transport, and mechanism of estrogen action can guide the management of postmenopausal as well as young anovulatory women and, it is hoped, lead to better methods for controlling estrogen-dependent neoplasia.

References

Abraham, G. E., and Maroulis, G. B., 1975, Effect of exogenous estrogen on serum pregnenolone, cortisol, and androgens in postmenopausal women, *Obstet. Gynecol.* **45**:271–274.

Ackerman, G. E., Smith, M. E., Mendelson, C. R., MacDonald, P. C., and Simpson, E. R., 1981, Aromatization of androstenedione by human adipose tissue stromal cells in monolayer culture, *J. Clin. Endocrinol. Metab.* **53**:412–417.

Anderson, D. C., 1974, Sex-hormone-binding globulin, *Clin. Endocrinol.* **3**:69–96.

Apter, D., 1980, Serum steroids and pituitary hormones in female puberty: a partly longitudinal study, *Clin. Endocrinol.* **12**:107–120.

Davidson, B. J., Gambone, J. C., Lagasse, L. D., Castaldo, T. W., Hammond, G. L., Siiteri, P. K., and Judd, H. L., 1981, Free estradiol in postmenopausal women with and without endometrial cancer, *J. Clin. Endocrinol. Metab.* **52**:404–408.

Forney, J. P., Milewich, L., Chen, C. T., Garlock, J. L., Schwarz, B. E., Edman, C. D., and MacDonald, P. C., 1981, Aromatization of androstenedione to estrone by human adipose tissue in vitro. correlation with adipose tissue mass, age, and endometrial neoplasia, *J. Clin. Endocrinol. Metab.* **53**:192–199.

Frisch, R. E., 1984, Body fat, puberty and fertility, *Biol. Rev.* **59**:161–188.

Grodin, J. M., Siiteri, P. K., and MacDonald, P. C., 1973, Source of estrogen production in postmenopausal women, *J. Clin. Endocrinol. Metab.* **36**:207–214.

Hammond, G. L., Nisker, J. A., Jones, L. A., and Siiteri, P. K., 1980, Estimation of the percent free steroid in undiluted serum by centrifugal ultra filtration-dialysis, *J. Biol. Chem.* **255**:5023.

Judd, H. L., Judd, G. E., Lucas, W. E., and Yen, S. S. E., 1974a, Endocrine function of the postmenopausal ovary; concentration of androgens and estrogens in ovarian and peripheral vein blood, *J. Clin. Endocrinol. Metab.* **39**:1020–1024.

Judd, H. L., Lucas, W. E., and Yen, S. S. E., 1974b, Effect of oophorectomy on circulating testosterone and androstenedione levels in patients with endometrial cancer, *Am. J. Obstet. Gynecol.* **118**:793–798.

Kasid, A., Huff, K., Green, G. L., and Lippman, M. E., 1984, A novel nuclear form of estradiol receptor in MCF-7 human breast cancer cells, *Science* **225**:1162–1165.

King, W. J., and Greene, G. L., 1984, Monoclonal antibodies localize oestrogen receptor in the nuclei of target cells, *Nature* **307**:745–747.

Laufer, L. R., Davidson, B. J., Ross, R. D., Lagasse, L. D., Siiteri, P. K., and Judd, H. L., 1983, Physical Characteristics and sex hormone levels in patients with osteoporotic hip fractures or endometrial cancer, *Am. J. Obstet. Gynecol.* **145**:585–590.

Lee, P. A., Xenakis, T., Winer, J., and Matsenbaugh, S., 1976, Puberty in girls: Correlation of serum levels of gonadotropins, prolactin, androgens, estrogens, and progestins with physical changes, *J. Clin. Endocrinol.* **43**:775–784.

Linkie, D. M., and Siiteri, P. K., 1978, A re-examination of the interactions of estradiol with target cell receptors, *J. Steroid Biochem.* **9**:1071–1078.

MacDonald, P. C., Edman, C. D., Hemsell, D. L., Porter, J. C., and Siiteri, P. K., 1978, Effect of obesity on conversion of plasma androstenedione to estrone in postmenopausal women with and without endometrial cancer, *Am. J. Obstet. Gynecol.* **130:**448–455.

Markaverich, B. M., and Clark, J. H., 1979, Two binding sites for estradiol in rat uterine nuclei: relationship to uterotropic response, *Endocrinology* **105:**1458–1462.

Moore, J. W., Clark, G. M. G., Bulbrook, R. D., Hayward, R. D., Murai, J. T., Hammond, G. L., and Siiteri, P. K., 1982, Serum levels of free and total oestradiol in patients with breast cancer and in normal controls, *Int. J. Cancer* **29:**17–21.

Notides, A. C., Lerner, N., and Hamilton, D. C., 1981, Positive cooperativity of the estrogen receptor, *Proc. Nat. Acad. Sci. U.S.A.* **78:**4926–4930.

Nisker, J. A., Hammond, G. L., Davidson, B. J., Frumar, A. M., Takaki, N. K., Judd, H. L., and Siiteri, P. K., 1980, Serum sex hormone-binding globulin capacity and the percentage of free estradiol in postmenopausal women with and without endometrial carcinoma, *Am. J. Obstet. Gynecol.* **138:**637–642.

Pardridge, W. M., Mietus, L. J., Frumar, A. M., Davidson, B. J., and Judd, H. L., 1980, Effects of human serum on transport of testosterone and estradiol into rat brain, *Am. J. Physiol.* **239:**E103–E108.

Roncari, D. A. K., and Van, R. L. R., 1978, Promotion of human adipocyte precursor replication by 17 beta-estradiol in culture, *J. Clin. Invest.* **65:**503–508.

Sasson, S., and Notides, A. C., 1983, Estriol and estrone interaction with estrogen receptor. I. Temperature-induced modulation of the cooperative binding of [3H]estriol and [3H]estrone to the estrogen receptor, *J. Biol. Chem.* **258:**8113–8117.

Schindler, A. E., Ebert, A., and Frederick, E., 1972, Conversion of androstenedione to estrone by human fat tissue, *J. Clin. Endocrinol. Metab.* **35:**627–630.

Shlomo, S., and Notides, A. C., 1983, Estriol and estrone interaction with the estrogen receptor. II. Estriol and estrone induced inhibition of the cooperative binding of [3H]estradiol to the estrogen receptor, *J. Biol. Chem.* **258:**8118–8122.

Siiteri, P. K., 1982, Review of studies on estrogen biosynthesis in the human, *Cancer Res. (Supp. 1)* **42:**3269s–3273s.

Siiteri, P. K., 1984, Receptor binding studies, *Science* **223:**191–193.

Siiteri, P. K., and MacDonald, P. C., 1966, Placental estrogen biosynthesis during human pregnancy, *J. Clin. Endocrinol. Metab.* **26:**751.

Siiteri, P. K. and MacDonald, P. C., 1973, Role of estraglandular estrogen in human endocrinology, 1973, in: *Handbook of Physiology*, Sect. 7, Vol. 2, Part 1 (R. O. Greep and E. B. Astwood, eds.), American Physiological Society, Washington, D.C., pp. 615–629.

Siiteri, P. K., and Seron-Ferre, M., 1981, Some thoughts on the feto-placental unit and parturition in primates, in: *The Oregon Regional Primate Research Center Symposia on Primate Reproductive Biology*, Vol. 1 (M. J. Novy and J. A. Resko, eds.), Academic Press, New York, pp. 1–34.

Siiteri, P. K., Schwarz, B., Moriyama, I., Ashby, R., and Linkie, D., 1973, Estrogen binding in the rat and human, *Adv. Exp. Med. Biol.* **36:**97–112.

Siiteri, P. K., Schwarz, B. E., and MacDonald, P. C., 1974, Estrogen receptors and estrone hypothesis in relation to endometrial and breast carcinoma, *Gyn. Oncology* **2:**228–238.

Siiteri, P. K., Hammond, G. L., and Nisker, J. A., 1981, Increased availability of serum estrogens in breast cancer. A new hypothesis, in: *Banbury Report 8: Hormones and Breast Cancer* (M. C. Pike, P. K. Siiteri, and C. W. Welsch, eds.), Cold Spring Harbor, New York.

Siiteri, P. K., Murai, J. D., Hammond, G. L., and Nisker, J. A., Raymoure, W. J., and Kuhn, R. W., 1982, Serum transport of steroid hormones, *Recent Prog. Hormone Res.* **38:**457–510.

Simpson, E. R., Ackerman, G. E., Smith, M. E., and Mendelson, C. R., 1981, Estrogen formation in stromal cells of adipose tissue of women: induction by glucocorticosteroids, *Proc. Natl. Acad. Sci., U.S.A.* **78:**5690–5694.

Vermeulen, A., 1976, The hormonal activity of the postmenopausal ovary, *J. Clin. Endocrinol. Metab.* **42:**247–253.

Vigersky, R. A., Kong, S., Sauer, M., Lipsett, M. B., and Loriaux, D. L., 1979, Relative binding of testosterone and estradiol to testosterone-estradiol-binding globulin, *J. Clin. Endocrinol. Metab.* **48**:899–904.

Wade, G. N., and Gray, J. M., 1978, Cytoplasmic 17beta-[3H]estradiol binding in rat adipose tissues, *Endocrinology* **103**:1695–1701.

Warren, M. P., 1980, The effects of exercise on pubertal progression and reproductive function in girls, *J. Clin. Endocrinol. Metab.* **51**:1050–1057.

Welshons, W. U., Lieberman, M. E., and Gorski, J., 1984, Nuclear localization of unoccupied oestrogen receptors, *Nature* **307**:747–749.

Wu, C. H., Motonashi, T., Abel-Rahman, H. A., Flickinger, G. L., and Mikhail, G., 1976, Free and protein-bound plasma estradiol-17beta during the menstrual cycle, *J. Clin. Endocrinol. Metab.* **43**:436–445.

13

THE BASIS OF MENOPAUSAL VASOMOTOR SYMPTOMS

HOWARD L. JUDD

In women the most common and characteristic symptom of the climacteric is an episodic disturbance consisting of sudden flushing and perspiration, referred to as a "hot flush or flash." It has been observed in 65–76% of women who experience a natural or surgical menopause (Hannan, 1927; Neugarten and Kraines, 1965; Jaszmann *et al.*, 1969; Thompson *et al.*, 1973; McKinlay and Jeffreys, 1974). Of those having flushes, 82% will experience the disturbance for more than 1 year (Jaszmann *et al.*, 1969) and 25–85% will complain of the symptom for more than 5 years (Neugarten and Krains, 1969; Thompson *et al.*, 1973).

Profound changes occur in physiological function with these events, indicating that some major disturbance in basic function is responsible. At first, reports described a prodromal period between the onset of the subjective feeling and the first recordable change in physiological function (Figure 1). With development of better methods to assess changes of physiologic function, it has been found that some physiologic changes actually precede the onset of the subjective feeling. The first measurable sign of the attack is cutaneous vasodilation (Mashchak *et al.*, 1982). This has been quantitated using a digital plethysmograph, a skin thermosensor to record increases in skin temperature and thermography (Figure 2) (Maschak *et al.*, 1982; Molnar, 1975; Sturdee and Reece, 1979). The vasodilation, as measured by the plethysmograph, begins approximately 1 min before the onset of the subjective flush and continues for

HOWARD L. JUDD • Department of Obstetrics and Gynecology, School of Medicine, University of California, Los Angeles, Los Angeles, California 90024.

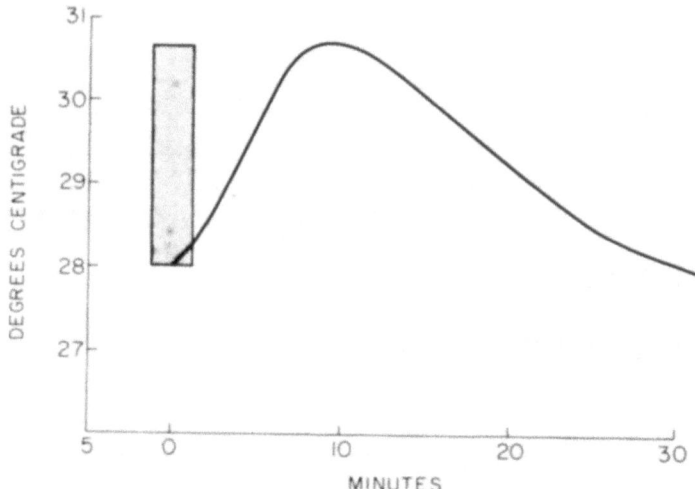

Figure 1. Mean characteristics and typical configuration of finger temperature fluctuations associated with hot flushes. The shaded area delineates the period from the mean beginning to completion of subjective flushing. From Meldrum *et al.* (1979), with permission.

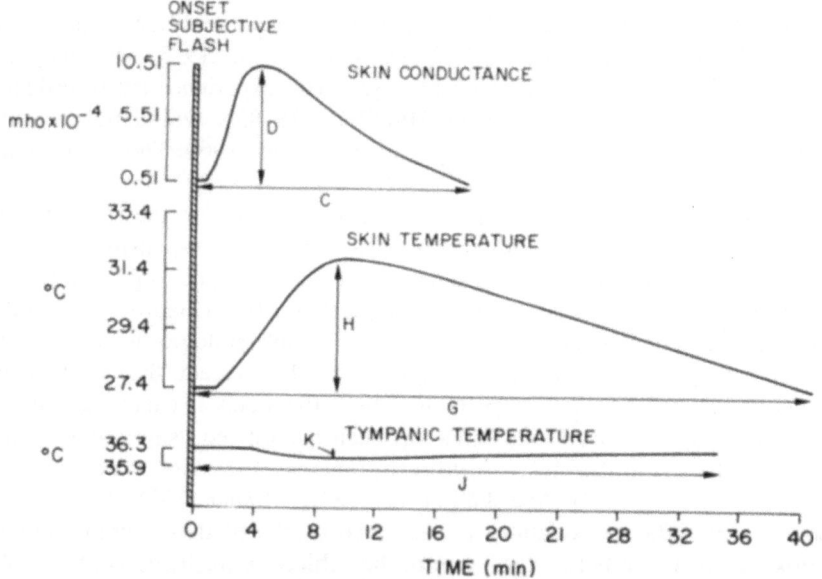

Figure 2. Characteristics of the changes in skin conductance, skin temperature, and tympanic membrane temperature based on observations of 25 hot flushes. All measurements are referenced to the signal by the patient at the onset of the subjective flush. From Tataryn *et al.* (1980), with permission.

about 8 min. The cutaneous vasodilation is generalized and not limited to the upper trunk and head. Measurable changes of skin temperature (a reflection of cutaneous vasodilation) have been observed on the fingers and toes (Figure 2) (Molnar, 1975; Tataryn et al., 1980).

The next measurable sign is a decrease in skin resistance, a measurement of perspiration (Figure 2). This begins on the average 45 sec after the onset of the subjective flush, reaches its maximum within 4 min and returns to baseline in 18 min.

As heat is lost from the body, by cutaneous vasodilation and perspiration, a decline in core temperature occurs (Figure 2). This commences about 4 min after the onset of the subjective symptoms, and returns to baseline in approximately 30 min (Meldrum et al., 1979; Tataryn et al., 1980). The average decrease in core temperature, as measured at the tympanic membrane, is 0.2°C. Alterations of pulse rate also occur during flushing, with increases of 13–20% being reported (Sturdee et al., 1978). Fluctuations of the baseline recording of the electrocardiogram are seen, probably reflecting changes in skin resistance (Molnar, 1975; Sturdee et al., 1978). Alterations of either heart rhythm or blood pressure have not been observed with flushes (Molnar, 1975; Sturdee et al., 1978).

These changes in physiological function do not correspond identically to the subjective symptoms. As mentioned previously, the subjective symptoms begin approximately 1 min after the onset of the first recordable change in cutaneous vasodilation and only last for an average of 4 min (Meldrum et al., 1979; Mashchak et al., 1982) (Figure 1). Thus, the physiological signs continue many minutes after the subjective symptoms have ceased.

The exact mechanism responsible for hot flushes is not known. Early investigators believed that an imbalance of the autonomic nervous system was somehow responsible, and this could occur because of central or peripheral instability of this system. Based on the physiological data presented above, it would appear that the menopausal hot flush is the result of a defect in central thermoregulatory function. There are three indications to support this conclusion.

First, the two major physiological changes associated with hot flushes are the result of different peripheral sympathetic functions. Excitation of sweat glands is by sympathetic cholinergic fibers (Venables, 1967), while cutaneous vasoconstriction is under the exclusive control and tonic α-adrenergic action on cutaneous vessels. It is difficult to envision a peripheral mechanism that would result in sympathetic cholinergic action on sweat glands and blockade of α-adrenergic tone on cutaneous vessels. However, these are the two basic mechanisms triggered by central thermoregulatory centers to lower core temperature. Thus, a hot flush appears to be a normal thermoregulatory event, occurring at an inappropriate time.

Second, during a hot flush, central temperature decreases following cutaneous vasodilation and perspiration. If hot flushes were the result of some peripheral mechanism, then one would expect the body's regulatory mechanisms to prevent a decrease in core temperature.

The third indication is the change in behavior associated with these symptoms. Women have a conscious desire to try and cool their bodies. They will remove clothing, throw off bedcovers, and stand by open windows and doors. All these feelings occur in the face of a normal central temperature. An analogous dissociation between perception and central temperature is found at the onset of a fever, when the individual feels cold or a "chill" prior to any change of central temperature. Because of this "chill," subjects will modify behavior to conserve heat. This assists in elevating the central temperature, thus the fever.

Most investigators working in the field of temperature regulation consider a fever to be the result of an elevation of the set point of central thermoregulatory centers, particularly those in the rostral hypothalamus (Snell and Atkins, 1968; Bligh, 1973; Kluger, 1978). Pyrogen elevates the central set point and the febrile organism, whether it be shellfish or human, actively raises the central body temperature, using both physiological (cutaneous vasoconstriction and shivering) and behavioral (curling in a ball, drinking hot liquids, etc.) mechanisms (Cooper et al., 1964; Reynolds et al., 1974). This continues until the core temperature reaches the new set point.

Employing all these observations it is suggested that the menopausal hot flush is triggered by a sudden downward setting of the central, hypothalamic thermostats (Figure 3). Subsequently, heat loss mechanisms are activated to bring the core temperature in line with the new set point resulting in the fall of core temperature.

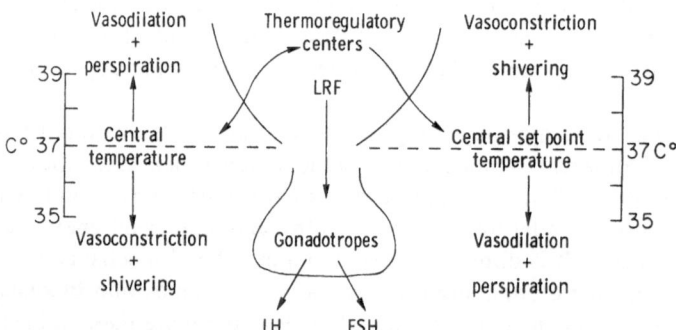

Figure 3. Proposed mechanism of hot flushes is a sudden downward setting of central set point temperature in hypothalamic thermoregulatory centers. Since central temperature would be higher, this would trigger vasodilation and perspiration (hot flush) to dissipate heat. From Judd (1983), with permission.

This raises the question, what then is the trigger for initiating these thermoregulatory episodes at inappropriate times? Since hot flushes occur after the spontaneous cessation of ovarian function or following ovariectomy, it has been presumed that the underlying mechanism responsible is endocrinological and has something to do with enhancement of pituitary gonadotropin secretion or reduction of ovarian estrogen secretion.

Studies have now indicated that both are involved. Of particular interest are the reports correlating the occurrence of hot flushes with pulsatile luteinizing hormone (LH) release. Tataryn et al. (1979) measured hot flushes in six women using continuous finger temperature recordings (Figure 4). Blood samples were drawn at 15-min intervals before and at 5-min intervals following the commencement of each flush. During 48 hr of study 34 flushes were recorded. Fluctuations of LH levels were also observed. Pulses of LH were defined as increases of hormone concentration of at least 20% over a nadir. Based on this definition 31 pulses of LH secretion were seen and 26 (84%) had a close temporal relationship to the occurrence of hot flushes. There was a strong correlation between LH levels and finger skin temperature in five of the six subjects. Independently, Casper et al. (1979) made similar observations. Since these reports several groups of investigators have confirmed this association (Mashchak et al., 1982). These results strongly suggest there is a close temporal relationship between the pulsatile release of LH and the occurrence of hot flushes. Tataryn et al. (1979) and Casper and co-workers (1979) also accessed the relationship of pulsatile follicle-stimulating hormone (FSH) release with hot flushes with one finding and the other not observing an association.

The close temporal relationship between pulsatile LH release and hot flushes suggests LH or other factors that initiate pulsatile LH release are in-

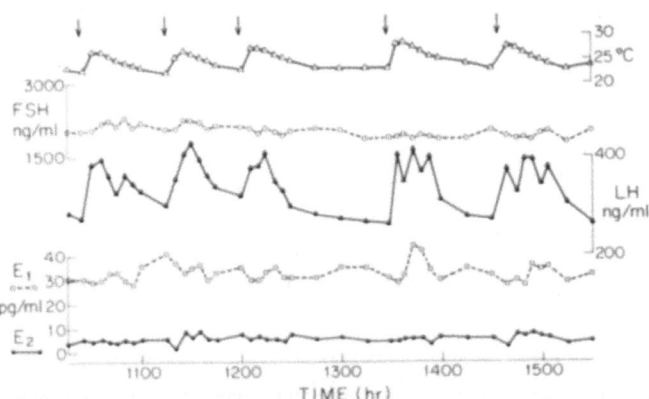

Figure 4. Serial measurements of finger temperature and serum FSH, LH, estrone, and estradiol in an individual subject. Arrows mark the onset of the temperature rises. From Tataryn et al. (1979), with permission.

volved with triggering these thermoregulatory events. It is doubtful that LH, itself, or increased pituitary activity is responsible, since subjectively measured hot flushes have been described in patients following a surgical hypophysectomy (Mulley *et al.*, 1977; Larsen, 1977). The observations of Meldrum *et al.* (1981) also support this conclusion. These investigators measured hot flushes objectively in two patients with hypoestrogenism secondary to surgically induced pituitary insufficiency (Figure 5). In each patient, gonadotropins were low with no pulsatile release. However, hot flushes were observed in both women. The episodes were similar to those experienced by normal postmenopausal subjects in regard to mean interval between flushes, magnitudes of the skin temperature and resistance changes, and relationships of these changes to the onset of subjective flushes.

Further support that the relationship between these two events is not related to pulsatile LH release, itself, is the finding of Casper and Yen (1982) that administration of a potent agonist of gonadotropin hormone-releasing hormone (GnRH-a) to postmenopausal women obliterates the pulsatile release of LH, but not the occurrence of hot flushes. A similar observation also has been made in premenopausal subjects given the same analog (De Fazio *et al.*, 1983). In subjects with normal menstrual cycles, the daily administration of agonist for one month blocked the pulsatile release of LH from the pituitary, presumably through pituitary desensitization (Figure 6). Ovarian function was also obliterated with estrone and estradiol levels falling to concentrations seen in postmenopausal women. Within 3 weeks of initiation of agonist administration hot flushes

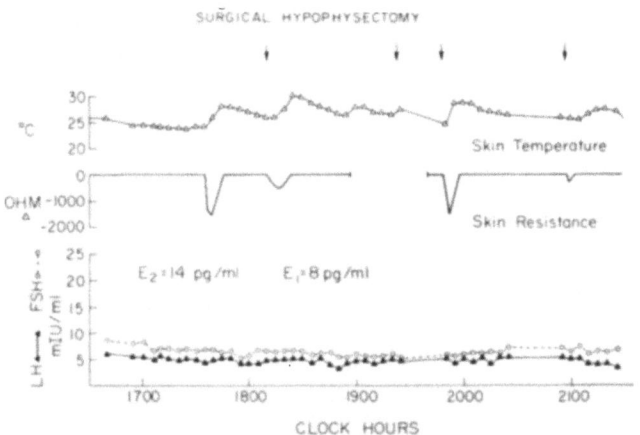

Figure 5. Serial measurements of the skin temperature, skin resistance, and serum LH and FSH levels in a woman following partial resection, cryotherapy, and irradiation of a chromophobe adenoma of the pituitary. Arrows mark the onsets of subjective flushes. From Meldrum *et al.* (1981), with permission.

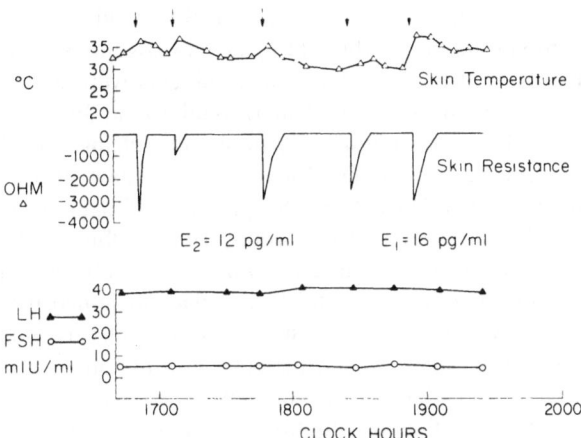

Figure 6. Induction of flush episodes following pituitary gonadotropin desensitization by treatment with GnRH agonist in a premenopausal woman. From DeFazio *et al.* (1983), with permission.

were induced in three of five subjects (60%). This incidence of occurrence was similar to the 76% incidence observed in 25 premenopausal women who underwent surgical castration. The characteristics of the flushes induced with GnRH-a were also similar to the events observed in the women following surgical oophorectomies.

Based on these findings, it seems likely that a suprapituitary mechanism must initiate hot flushes and is somehow influenced by the hypothalamic factors responsible for pulsatile LH release. In rats and monkeys the hypothalamic hormone, gonadotropin-releasing hormone (GnRH), fluctuates in the hypophysial portal vein blood, and these fluctuations are responsible for pulsatile release of LH from the pituitary (Carmel *et al.*, 1976; Eskay *et al.*, 1977).

In monkeys the site governing the pulsatile release of GnRH is within the arcuate nucleus of the hypothalamus (Krey *et al.*, 1975). The secretion of GnRH is modulated by several factors present in the hypothalamus that influence GnRH-secreting neurons, including norepinephrine, dopamine, endogenous opioids, and prostaglandins. These have all been shown to influence gonadotropin release from the pituitary, presumably through effects on GnRH release from the hypothalamus (LeBlanc *et al.*, 1976; Pang *et al.*, 1977; Linton *et al.*, 1979). Thus, GnRH or the factors that influence its release may somehow alter the set point of the thermoregulatory centers to trigger hot flushes. Since an animal model of hot flushes has not been identified, classical lesioning experiments of hypothalamic function cannot be performed to define which of the above factors trigger flushes. Thus, experiments must be confined to the study of women with spontaneously occurring lesions in hypothalamic function.

To address this issue, Gambone *et al.* (1982a) studied subjects with isolated gonadotropin deficiency, who had received exogenous estrogens for hormone replacement for at least 1 year. These subjects were used as a model of a spontaneously occurring defect of GnRH synthesis and/or release. This assumption was based on the findings of pathological changes in the hypothalamus in patients with this defect (DeMorsier and Gauthier, 1963) and the stimulation of gonadotropin release from the pituitary of patients with repetitive injections of GnRH (Crowley and McArthur, 1980). Objectively recorded hot flushes were observed in these subjects. Since these subjects with defects in GnRH synthesis and/or release had flushes, it was presumed that GnRH itself is probably not involved in triggering these thermoregulatory episodes.

These investigators also studied patients with hypothalamic amenorrhea. These women were considered to be models of a defect in neurotransmitter input to GnRH neurons. This assumption was based on the observations that patients with this syndrome have limited or no pulsatile release of gonadotropins (Yen *et al.*, 1973) and have evidence of increased dopaminergic and opioid activity in the hypothalamus (Quigley *et al.*, 1980). This abnormal neurotransmitter activity is thought to contribute to the reduced release of GnRH from the hypothalamus resulting in amenorrhea and hypogonadism.

The studies of the patients with hypothalamic amenorrhea were conducted within 1 year of disease onset and were limited to patients with hypoestrogenism equivalent to that seen following surgical castration. In these subjects no objectively measured hot flushes were recorded. This finding in these hypoestrogenic women suggests that the mechanism responsible for hypothalamic amenorrhea inhibits the occurrence of hot flushes. Since abnormal neurotransmitter input to GnRH neurons has been reported in this disease, this finding suggests that the hypothalamic factors responsible for GnRH release, not GnRH itself, may somehow alter the set point of thermoregulatory centers of the hypothalamus resulting in the initiation of hot flushes. The close proximity of some of the GnRH neurons with the thermoregulatory centers in the preoptic anterior hypothalamus is consistent with this concept (Lomax and Knox, 1973; Krey *et al.*, 1975; Kobayashi *et al.*, 1978; Reaves and Hayward, 1979). However, experimental evidence has shown that thermoregulatory responses to neurotransmitters are not precisely localized to the anterior hypothalamus. Destruction of the preoptic anterior hypothalamic nuclei in rats enhances the hypothermic response to norepinephrine injected into the third ventricle, indicating that the site of action of this catecholamine in regard to thermoregulation may be in other areas of the hypothalamus (Satinoff and Cantor, 1975).

The observations that catecholamines play roles in both central thermoregulatory function and GnRH release is also consistent with the hypothesis that neurotransmitter metabolism in the hypothalamus is responsible for the simultaneous occurrence of pulsatile LH release and hot flushes (Cox and Lomax,

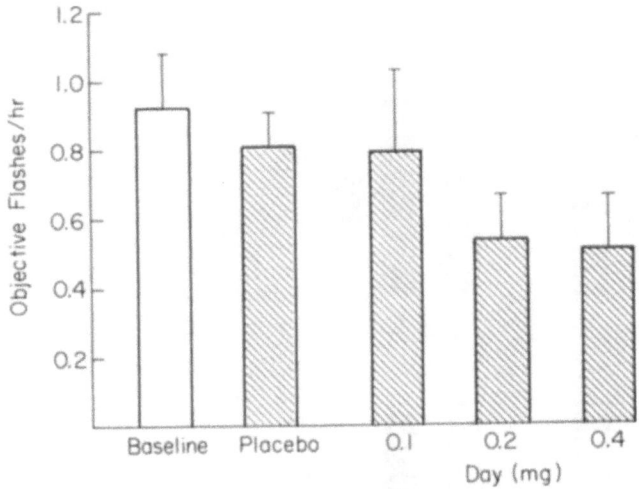

Figure 7. Effects of clonidine and placebo given to ten postmenopausal women with severe hot flushes. Note significant reductions with 0.2 and 0.4 mg doses of clonidine. From Laufer *et al.* (1982), with permission.

1977; Crowley *et al.*, 1978; Simpkins and Kalra, 1979). The finding that cloni- dine partially blocks the occurrence of objectively measured flushes also sup- ports this possibility (Laufer *et al.*, 1982) (Figure 7). The effect of clonidine on flushes could be exerted through peripheral or central mechanisms. Cloni- dine is an α-adrenergic receptor agonist that stimulates postsynaptic α-adrener- gic receptors in the depression site of the vasomotor center of the medulla ob- longata (Houston, 1981). In addition, it may also influence suprabulbar structures such as α-adrenergic receptors in the hypothalamus.

Opiatelike peptides have been observed to induce facial flushing when ad- ministered to normal human subjects. The effects of opiates on thermoregula- tion are similar to those physiologic changes occurring with hot flushes (Stubbs *et al.*, 1978). Infusion of naloxone, an opiate receptor agonist, has been re- ported to reduce subjective flushing and the frequency of LH pulses (Lightman *et al.*, 1981). To test this hypothesis further, DeFazio *et al.* studied 16 women who were experiencing frequent severe hot flushes following natural or surgical menopause. The subjects had not received hormone therapy for at least 1 month before study. Following 8-hr baseline recordings, subjects were randomized into two equal groups and the recordings were repeated on a subsequent day during infusion of either saline or naloxone, 22.2 μg/min. Blood samples were drawn through an intravenous catheter at 15 and 0 min prior to the infusion, and every 15 min during the last 4 hr of study.

Figure 8 shows the mean rates of objective and subjective hot flashes re-

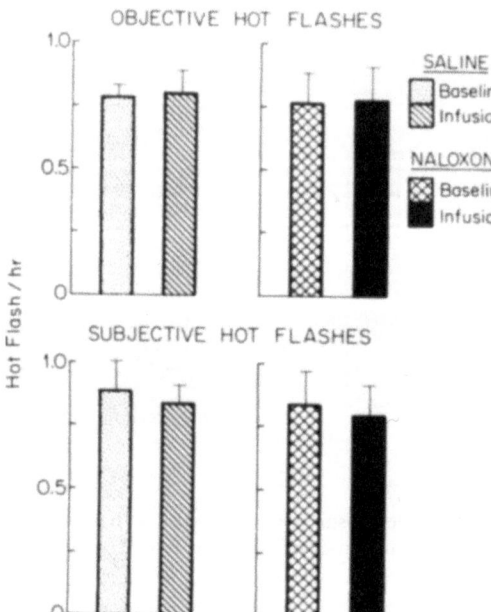

Figure 8. Effects of naloxone and saline infusions on subjectively and objectively measured hot flushes. From DeFazio *et al.* (1984), with permission.

corded in the two groups before and following infusions of naloxone or saline. The rates of objective and subjective episodes following saline infusion were similar to the respective values obtained during baseline recordings. Similarly, the rates of objectively and subjectively recorded hot flushes during naloxone infusion were unchanged from those recorded prior to treatment.

Figure 9 shows the onsets of subjective symptoms (arrows), and continuous recordings of finger temperature and skin resistance in a subject before and during naloxone infusion. No appreciable changes were observed in the occurrence of subjectively or objectively recorded hot flushes with naloxone administration.

This study failed to confirm an effect of naloxone infusion on the rate of either subjective or objectively measured hot flushes. Naloxone infusion in normal women during the late follicular phase has been shown to cause a gradual progressive increase in basal LH levels and LH pulse amplitude, whereas infusion during the luteal phase resulted in prompt and marked episodic LH variation (Quigley and Yen, 1980). No effect was noted during the early follicular phase. These data suggest that endogenous opiates inhibit gonadotropin secretion and exert this effect mainly or solely under the influence of relatively high levels of circulating estradiol and progesterone. The putative involvement of endogenous opiates in the initiation of the hot flush would appear to be para-

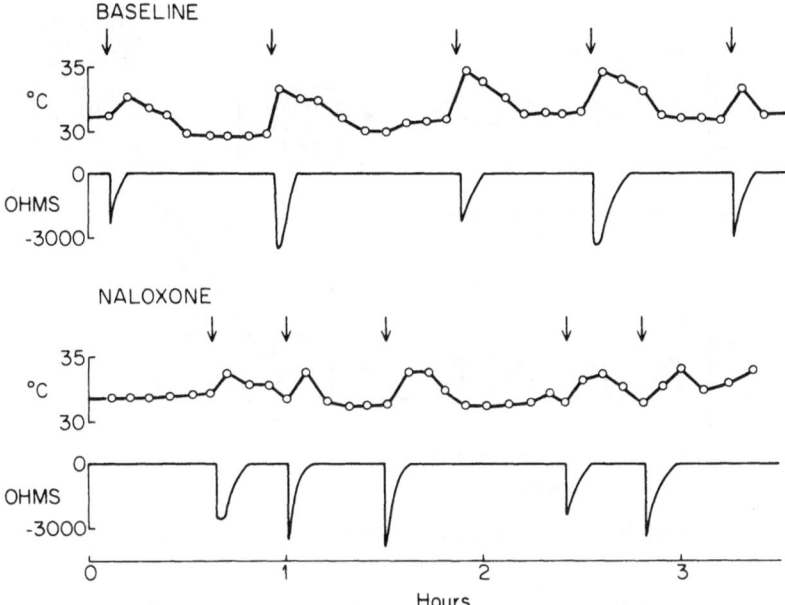

Figure 9. Effects of naloxone infusion on objectively measured hot flushes and LH and FSH levels in a symptomatic postmenopausal woman. Arrows mark occurrence of subjective hot flush. From DeFazio *et al.* (1984), with permission.

doxical, since release of an inhibitor of gonadotropin secretion would be unlikely to be temporally associated with the initiation of LH pulsatile increase.

This lack of effect of opiate receptor blockade suggests a very low or absent input of hypothalamic opioid peptides on gonadotropin secretion in women lacking ovarian function and is consistent with the undetectable levels of β-endorphin measured in hypophyseal portal blood in ovariectomized monkeys (Wehrenberg *et al.*, 1982). These data are all consistent with the conclusions that hypothalamic β-endorphin secretion is dependent upon stimulation by normal levels of ovarian sex steroids and that estrogen and progesterone-deficient postmenopausal women have decreased or absent input of endogenous opiates on gonadotropin secretion.

In summary, menopausal hot flushes apparently represent normal thermoregulatory events triggered centrally by altered hypothalamic activity (Figure 10). A likely candidate is norepinephrine. Following loss of ovarian function, hypothalamic norepinephrine activity is enhanced with the removal of ovarian steroids that exert negative feedback on GnRH and gonadotropin release. With loss of negative feedback pulsatile GnRH and gonadotropin release is enhanced. This presumably is associated with increased norepinephrine input to GnRH

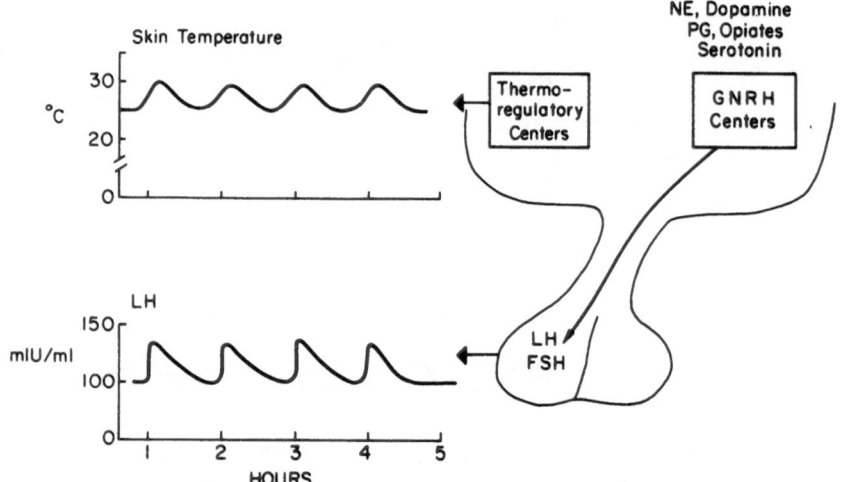

Figure 10. Proposed model of mechanism of menopausal hot flush.

neurons. Somehow this enhanced norepinephrine activity triggers normal thermoregulatory episodes leading to the symptom complex of hot flushes. This working hypothesis provides new concepts to test and possibly insights into new treatment methods.

References

Bligh, J., 1973, *Temperature Regulation in Mammals and Other Vertebrates*, North-Holland Publishing, Amsterdam.

Carmel, P. W., Araki, S. and Ferin, M., 1976, Pituitary stalk portal blood collection in rhesus monkeys: Evidence for pulsatile release of gonadotropin-releasing hormone (GnRH), *Endocrinology* **99**:243–248.

Casper, R. F., and Yen, S. S. C., 1982, Menopausal flushes: Effect of pituitary gonadotropin desensitization by a potent luteinizing hormone-releasing factor agonist, *J. Clin. Endocrinol. Metab.* **53**:1056–1058.

Casper, R. F., Yen, S. S. C. and Wilkes, M. M., 1979, Menopausal flushes: A neuroendocrine link with pulsatile luteinizing hormone secretion, *Science* **205**:823–825.

Cooper, K. E., Cranston, W. I., and Snell, E. S., 1964, Temperature regulation during fever in man, *Clin. Sci.* **27**:345–356.

Cox, B., and Lomax, P., 1977, Pharmacologic control of temperature regulation, *Annu. Rev. Pharmacol. Toxicol.* **17**:341–353.

Crowley, W. F., Jr., and McArthur, J. W., 1980, Simulation of the normal menstrual cycle in Kallman's Syndrome by pulsatile administration of luteinizing hormone-releasing hormone (LHRH), *J. Clin. Endocrinol. Metab.* **51**:173–175.

Crowley, W. R., O'Donohue, T. L., Wachslicht, H., and Jacobowitz, D. M., 1978, Effects of estrogen and progesterone on plasma gonadotropins and on catecholamine levels and turnover in discrete brain regions of ovariectomized rats, *Brain Res.* **154**:345–357.

DeFazio, J., Meldrum, D., Laufer, L., Vale, W., Rivier, J., Lu, J., and Judd, H., 1983, Induction of hot flashes in premenopausal women treated with a long acting GnRH agonist, *J. Clin. Endocrinol. Metab.* **56:**445–448.

DeFazio, J., Verheugen, C., Chetkowski, R., Nass, T., Judd, H. L., and Meldrum, D. R., 1984, The effects of naloxone on hot flashes and gonadotropin secretion in postmenopausal women, *J. Clin. Endocrinol. Metab.* **58:**578–581.

DeMorsier, G., and Gauthier, G., 1963, La dysplasie olfacto-genitale, *Pathol. Biol. (Paris)* **11:**1267–1272.

Eskay, R. L., Mical, R. S., and Porter, J. C., 1977, Relationship between luteinizing hormone releasing hormone concentration in hypophysial blood and luteinizing hormone release in intact, castrated and electrochemically-stimulated rats, *Endocrinology* **100:**263–270.

Gambone, J., Meldrum, D., Laufer, L., Chang, J., Lu, J., and Judd, H., 1982, Further delineation of hypothalamic dysfunction responsible for menopausal hot flashes. Proc. 29th Annual Meeting Soc. Gynecol. Invest., Dallas, Texas. March 24–27, 1982, Ab. #60.

Hannan, J. H., 1927, *The Flushings of the Menopause,* Bailliere, Tindall and Cox, London, pp. 1–22.

Houston, M., 1981, Clonidine hydrochloride: Review of pharmacologic and clinical aspects, *Prog. Cardiovasc. Dis.* **23(5):**337–350.

Jaszmann, L., Van Lith, N. D., and Zaat, J. C. A., 1969, The perimenopausal symptoms, *Med. Gynecol. Sociol.* **4:**268–276.

Judd, H. L., 1983, Pathophysiology of menopausal hot flushes, in: *Neuroendocrinology of Aging* (J. Meites, ed.), Plenum Press, New York, pp. 173–202.

Kluger, M. J., 1978, The evolution and adaptive value of fever, *Ann. Sci.* **66:**38–43.

Kobayashi, R. M., Lu, K. H., Moore, R. Y., and Yen, S. S. C., 1978, Regional distribution of hypothalamic luteinizing hormone-releasing hormone in proestrous rats: effects of ovariectomy and estrogen replacement, *Endocrinology* **102:**98–105.

Krey, L. C., Butler, W. R., and Knobil E., 1975, Surgical disconnection of the medial basal hypothalamus and pituitary function in the rhesus monkey. I. Gonadotropin secretion, *Endocrinology* **96:**1073–1087.

Larsen, I. F., 1977, Hot flashes after hypophysectomy, *Br. Med. J.* **2:**1356.

Laufer, L. R., Erlik, Y., Meldrum, D. R., and Judd, H. L., 1982, Effect of clonidine on hot flashes in postmenopausal women, *Obstet. Gynecol.* **60:**583–586.

Leblanc, H., Lachelin, G. C. L., Abu-Fadil, S., and Yen, S. S. C., 1976, Effects of dopamine agonists on LH release in women, *J. Clin. Endocrinol. Metab.* **44:**728–732.

Lightman, S. L., Jacobs, H. S., Maguire, A. K., McGarrick, G., and Jeffcoate, S. L., 1981, Climacteric flushing: clinical and endocrine responses to infusion of naloxone, *Br. J. Obstet. Gynecol.* **88:**919–924.

Linton, E. A., Bennet, G. W., and Whitehead, S. A., 1979, Prostaglandins and the release of LHRH from hypothalamic synaptosomes, *Neuroendocrinology* **28:**394–401.

Lomax, P., and Knox, G. V., 1973, The sites and mechanisms of action of drugs affecting thermoregulation, *The Pharmacology of Thermoregulation.* San Francisco 1972, Karger, Basel, pp. 146–154.

Mashchak, C. A., Kletzky, O. A., Artel, R., and Mishell, D. R. Jr., 1982, Postmenopausal vasomotor perfusion changes investigated by digital plethysmography and plasma catecholamine levels. Proc. 29th Annual Meeting Soc. Gynecol. Invest., Dallas, Texas, March 24–27, 1982. Ab #277.

McKinlay, S., and Jefferys, M., 1974, The menopausal syndrome, *Br. J. Prev. Soc. Med.* **28:**108–115.

Meldrum, D. R., Erlik, Y., Lu, J. H. K., and Judd, H. L., 1981, Objectively recorded hot flashes in patients with pituitary insufficiency, *J. Clin. Endocrinol. Metab.* **52:**684–687.

Meldrum, D. R., Shamonki, I. M., Frumar, A. M., Tataryn, I. V., Chang, R. J., and Judd, H.

L., 1979, Elevations in skin temperature of the finger as an objective index of postmenopausal hot flashes: standardization of the techniques, *Am. J. Obstet. Gynecol.* **135**:713–717.

Molnar, G. W., 1975, Body temperatures during menopausal hot flashes, *J. Appl. Physiol.* **38**:499–503.

Mulley, G., Mitchell, J. R. A., and Tattersall, R. B., 1977, Hot flushes after hypophysectomy, *Br. Med. J.* **2**:1062.

Neugarten, B. L., and Kraines, R. J., 1965, Menopausal symptoms in women of various ages, *Psychosom. Med.* **27**:266–273.

Pang, C. N., Zimmermann, E., and Sawyer, C. H., 1977, Morphine inhibition of the preovulatory surges of plasma luteinizing hormone and follicle stimulating hormone in the rat, *Endocrinology* **101**:1726–1732.

Quigley, M. E., Sheehan, K. L., Casper, R. F., and Yen, S. S. C., 1980, Evidence for increased dopaminergic and opioid activity in patients with hypothalamic hypogonadotropic amenorrhea, *J. Clin. Endocrinol. Metab.* **50**:949–954.

Quigley, M. D., and Yen, S. S. C., 1980, The role of endogenous opiates on LH secretion during the menstrual cycle, *J. Clin. Endocrinol. Metab.* **51**:179–181.

Reaves, T. A., and Hayward, J. M., 1979, Hypothalamic and extrahypothalamic thermoregulatory centers, in: *Body Temperature: Regulation, Drug Effects and Therapeutic Implications* (P. Lomax and E. Schonbaum, eds.), Dekker, New York, pp. 39–70.

Reynolds, W. W., Casterlin, M. E. and Covert, J. B., 1974, Behavioral fever in teleost fishes, *Nature* **259**:41–42.

Satinoff, E., and Cantor, A., 1975, Intraventricular norepinephrine and thermoregulation in rats, in: *Temperature Regulation and Drug Action* (P. Lomax, E. Schonbaum, and J. Jacob eds.), Karger, Basel, pp. 103–110.

Simpkins, J. W., and Kalra, S. P., 1979, Central site(s) of norepinephrine and LHRH interaction, *Fed. Proc.* **38**:1107.

Snell, E. S., and Atkins, E., 1968, *The Mechanisms of Fever, the Biological Basis of Medicine* (E. E. Bittar and N. Bittar, eds.), Academic Press, New York, pp. 397–419.

Stubbs, W. A., Jones, A., Edward, C. R. W., Delitala, G., Jeffcoate, W. J., Ratter, S. J., Besser, G. M., Bloom, S. R., and Alberti, K. G. M. M., 1978, Hormonal and metabolic responses to an enkephalin analogue in normal man, *Lancet* **2**:1225–1227.

Sturdee, D. W., Wilson, K. A., Pipili, E., and Crocker, A. D., 1978, Physiological aspects of menopausal hot flush, *Br. Med. J.* **2**:79–80.

Sturdee, D. W., and Reece, B. L., 1979, Thermography of menopausal hot flushes, *Maturitas* **1**:201–205.

Tataryn, I. V., Meldrum, D. R., Lu, K. H., Frumar, A. M., and Judd, H. L., 1979, LH, FSH, and skin temperature during the menopausal hot flash, *J. Clin. Endocrinol. Metab.* **49**:152–154.

Tataryn, I. V., Lomax, P., Bajorek, J. G., Chesarek, W., Meldrum, D. R., and Judd, H. L., 1980, Postmenopausal hot flushes: a disorder of thermoregulation, *Maturitas* **2**:101–107.

Thompson, B., Hart, S. A., and Durno, D., 1973, Menopausal age and symptomatology in general practice, *J. Biol. Sci.* **5**:71–82.

Venables, R., 1967, *Methods in Psychophysiology* (C. C. Brown, ed.), Williams and Wilkins, Baltimore, pp. 1–26.

Wehrenberg, W. B., Wardlaw, S. L., Frantz, A. G., and Ferin, M., 1982, β-endorphin in hypophyseal portal blood: variations throughout the menstrual cycle, *Endocrinology* **111**:879–881.

Yen, S. S. C., Rebar, R., Vandenberg, G., and Judd, H., 1973, Hypothalamic amenorrhea and hypogonadotropinism: Responses to synthetic LRF, *J. Clin. Endocrinol. Metab.* **36**:811–816.

14

FEMALE SEXUALITY DURING AND FOLLOWING MENOPAUSE

DIANE S. FORDNEY

Introduction

The most important finding about sexual expression in aging women is the appalling lack of study and information about it. For the 30 million people over the age of 60, approximately 85% of whom are women who will live to an average age of 70 years, such omission is another index of neglect (Strean, 1983). Wharton (1981) published an annotated bibliography on sex and aging covering slightly over 1100 references, which include public addresses, meeting presentations, book chapters, books, and foreign as well as English language articles. It is fair to estimate that only 25% of those references are to research studies—of all levels of quality—in sexuality. The majority of articles are on men only rather than women or paired male and female populations. My own literature search discovered less than 100 articles studying the sexuality of aging women, and many of those were simple restatements of other articles rather than original research studies. Nonetheless, mythologized statements about an older woman's sexuality recur over and over. The best intentioned of these are statements that a woman has no obligatory loss in function with aging (Brown *et al.*, 1978; Butler and Lewis, 1978; Diamond and Karlen, 1981; Mancini, 1983; Scheingold and Wagner, 1974; Simpson and McKinney, 1970). However, a deestrogenized woman will have definite anatomic and physiological changes that adversely affect her functioning sexually, and there are postmenopausal changes that appear to be uncorrected by estrogen replacement (Dia-

DIANE S. FORDNEY • Health Sciences Center, College of Medicine, University of Arizona, Tucson, Arizona 85724.

mond and Karlen, 1981; Marten, 1981; Masters and Johnson, 1966; Scheingold and Wagner, 1974). The worst myths are that when she develops dysfunction with increasing age the dysfunctions are almost always psychogenic and that a majority of older women have very little or no sexual interest. As to the former, there is absolutely no information about subtle decrements in a woman's vascular, neurologic, metabolic, and endocrine function accompanying aging and disease processes in aging, and a global attribution to psychological ill health is both unfair and unwarranted. Certainly, for aging males (40 and over) the literature of the last 5 years would seem to attribute almost all erectile failure, the most significant male dysfunction, to an organic state. The reality probably rests somewhere between for both sexes. Statements concerning little or no desire in aging women blithely disregard prominent factors about women partnered with inactive men, the high disproportion in numbers of women to available men by age 55, and the social factors of poverty and deprivation these "excess" women endure. As in other areas of sexual interest, arousal, behavior, and function, it would seem that loss of sexual desire is more the result of other problems than a *de facto* cause.

Behavior

Until we garner more objective information, professionals are reduced to being well-intentioned "witch" doctors. The dearth of information about women's sexuality and aging has multiple causes applicable to both the topic of sexuality and that of aging. There was an embarassing lack of interest in the problems of aging in our society until the last 15 years. There was an almost total lack of sex research on women of any age until the middle 1960s. Both have been taboo, with strong social sanctions against investigation. Embarassment by the investigator is a part of that process and more importantly has resulted in almost total failure of interdisciplinary collaboration among sexologists, gerontologists, and those studying related issues. Moreover, there is a continuing low priority in developing or supporting research in the areas of sex and of aging. The complexities of social factors, loss issues, physiology, disease, and iatrogenic effects of treatment of disease have presumably also contributed to avoidance of appropriate studies (Flint, 1981; Persky, *et al.*, 1976; Scheingold and Wagner, 1974). Finally, aging women as well are reluctant to discuss their sexual needs and problems openly, another cultural liability. If they encounter a perceived life threat, they often avoid sexual expression entirely under the assumption that the behavior will increase the risk of harm, pain, or death (Thurer and Thurer, 1982). These characteristics are powerful and prevalent in our aging population and perpetuate silence, fear, and deprivation.

Numerous authors have enumerated the constant attrition of gratification and losses occurring to our "young-old," 55–65 years of age, and our "elderly," 65 and over (Mancini, 1983; Newman and Nichols, 1960; Rubin, 1976; Simpson and McKinney, 1970). Both these populations share certain mythologies that any sexual activity is less credible, less normal, and less important for the aging. It is even less acceptable for women than for men (LaTorre and Kear, 1977; White, 1982). Surprisingly, the youngest groups queried (19–35 years) find sexual behavior of the elderly more acceptable than do either the middle-aged (35–50 years) or older groups (50–60 years) (LaTorre and Kear, 1977). More importantly, irrespective of age, education, or status, there is widespread ignorance about sexual behavior in the elderly, and a general expectation that it will cease somewhere in the sixth decade. Such an opinion has the power of a self-fulfilling prophecy and a true taboo, creating a fear of ostracism in those who would violate the taboo (Flint, 1981; Gaitz, 1974; Pattison, 1983).

The human being is sexually unique by virtue of behavior that is primarily pleasure oriented and not procreative during all life stages and without respect to fertility. Nonprimate mammalian species usually have estrous cycles in which proceptivity, attractiveness of females to males, and receptivity, acceptance of males by females, occur only in close approximation to ovulation (Young, 1961). Nonhuman primates demonstrate true menstrual or "hidden" ovulation cycles as well as sexual activity not dictated by fertility proximity (Nadler, 1977). However, they also show definite increases in sexual behavior allied to hormonal and ovulatory conditions (Nadler, 1977; Young, 1961). These patterns are affected by social learning and dominance factors as well as natural or laboratory stress factors, but the underlying procreative base is clear. Human beings, however, have never been clearly established to have such a fertility-based drive because of dominating affective and social patterns governing human sexual behavior (Udrey and Morris, 1968). For aging there are virtually no animal models, as only "pets," usually, survive into old age. Laboratory studies of postmenopausal rhesus monkeys reveal both decreased proceptive and receptive behavior unless the female animal receives estrogen replacement (Young, 1961). While this may have some limited applicability to women, animal models for sexual activity with aging have been of limited usefulness.

A final difficulty in the study of sexuality in aging women is that their patterns of interest and behavior are still seriously overridden by men's interest and behavior, at least for heterosexual women, in all age groups. Throughout all life stages men consistently express higher rates of interest, initiation, and activity than do women, who tend to develop active sexual interest and behavior later and then generally follow their partners into primarily coital patterns (Butler and Lewis, 1978; Gebhardt, 1972; Kinsey et al., 1948; Kinsey et al., 1953). Table I is a composite of data from several sociologic surveys and demonstrates sex differences in solitary and partnered activity. When Christensen and Johnson (1973) studied women aged 50–69, whose patterns were never male dominated, they discovered that 70% were in some way sexually active, 79% practicing masturbation, 62% coitus, and 18% homosexual activity. These rates are much higher than shown in studies of ever-married (divorced and widowed) women of the same age groups (Christiansen and Gagnon, 1965; Gebhardt, 1972). Newman and Nichols (1974) reported on 250 women aged 60–93 years, and 45% of the partnered and only 7% of the unpartnered were coitally active. If only those women over 75 were reported, the partnered women's rate fell to 20%. Christensen and Gagnon (1965) displayed a decline in coital activity from 80% at age 50 to 25% by age 70, using coitus at least once a week as criteria for coitally active in partnered women, and from 37% at age 50 to zero by age 65 in unpartnered women. During a longitudinal study over 6 years of 500 married couples in two age groups, one under 56 and another over 65 years, men's intercourse activity remained stable for 62% of the younger and 42% of the older groups (George and Weiler, 1981). Women's, however, remained stable for only 42% of the younger and 33% of the older. Virtually no one showed an increase in activity over the 6 years, and while men who remained active, but at a decreased rate, comprise less than 10% in both age groups, women's decreases equaled 10% in the younger and 30% in the older groups. For those, who declined from active intercourse to not active, the rate was not significant for younger people of either sex, but was 22% for older males and 33% for older females. Only older males showed an increase in coital activity (for 10%), and while the groups of no intercourse activity at the onset or end of the 6 years was very small for younger men and women, it was 18% for older males and 33% for older women (George and Weiler, 1981). An appropriate interpretation would be that while 20% of men became coitally inactive after age 65 and 18% were inactive by the time they reached 65, a total of 38%, over 66% of women are coitally inactive by or after age 65. Thus, while half the female sample could be coitally inactive because their partners were, the other half of women who became inactive do so because they have no coital outlet. Since masturbation rates are very low for women in the senior age groups, 30% at highest, and homosexuality is rarely practiced, the majority of elderly women have no sexual outlet of any kind. A sobering reality is that differential death rates account for much of the sexual deprivation inactivity in older women. At ages 40 to 45 the ratio of all women to all men in that age group is 233/100 or 2.3 to 1. By age 65, it is 6 to 1, and by age 70, 8 to 1 (Blumstein and Schwartz, 1983). Part of this discrepancy is due to the fact that women in their twenties marry men an average of 4 years older, and by age 65, 50% are widows. They also live 7.7 years longer than their husbands (Butler and Lewis, 1983). For men who are either divorced or widowed, more than 90% will remarry. Widows are less likely to remarry than divorced women of comparable age and by ages 40 and 50 only 32% and 12%, respectively, of divorcees will

Table I

Solitary and Partnered Sexual Activity by Age[a,b]

	Age	Gender	Nocturnal orgasm	Masturbation	Petting	Coitus	Homosexual
Activity prior to onset of intercourse	15–19	Men	99+%	99+%	99+%	—	20%
		Women	<15%	30%	50%	—	<10%
Activity after onset of intercourse	19–45	Men	N.D.[c]	80%	80%	80%	5–15%
		Women	N.D.	50%	100%	80%	2–5%
	50+	Partnered men	N.D.	50%	N.D.	<50%	5–15%
		Partnered women	N.D.	20%	N.D.	<30%	2%
		Single women	N.D.	30%	N.D.	<10%	N.D.

[a] Composite data from Christensen and Gagnon (1965), Gebhardt (1972), and Ginsey et al. (1948, 1953).
[b] Intended to display patterns of activity only, as sampling techniques and sizes varied among the represented studies
[c] Not determined.

remarry (Blumstein and Schwartz, 1983). Since 54% of men over age 45 will remarry women 8 or more years younger, they are in reality unavailable to the same-aged female population (Bernardo *et al.*, 1983).

Postmarital intercourse rates reflect a similar problem. Half as many widows as divorcees have postmarital intercourse, from a high for divorcees aged 41–45 years of 68.8% to a low for the same group of 2% over age 65. The frequency of coital activity declined from just under once per week to once per month prior to becoming coitally inactive (Gebhardt, 1972; Newman and Nichols, 1974). In the one year study describing the relationships of sexual activity and sexual interest, Charatan (1982) reported that sexual interest declined about 5 years after cessation of activity in women and 5–10 years in men.

It is well documented that the most sexually active and functional men at younger ages remain the most sexually functional and active men at advanced age if healthy (Flint, 1981; Kinsey *et al.*, 1948; Marten, 1981; Masters and Johnson, 1966; Simpson and McKinney, 1970). The correlation holds for their female partners as well (Flint, 1981; Kinsey *et al.*, 1953; Marten, 1981). In addition, a woman who maintains regular coital contact maintains the highest level of interest as she ages, irrespective of partners (Christensen and Gagnon, 1965; Gebhardt, 1972; Kinsey *et al.*, 1953). In at least one study, the only significant prediction of her sexual interest and activity was the presence of an interested partner (Charatan, 1982).

Physiology

There *are* reported postmenopausal changes in sexual anatomy and physiology. The mean age of menopause in this country is now 51 years. At this occurrence, ovulation and progesterone production cease. Estrogen production by the ovary declines gradually over a 2- to 5-year period with natural menopause and precipitously if castration causes artificial menopause. Sixty percent of women have no overt estrogen-deficiency sexual symptoms until 5 years after natural menopause, while 40% have mild to moderate symptoms and 10% have severe symptoms of vaginal atrophy and resultant vaginitides, atrophic urethritis and cystitis, arteriolar vascular instability, serious sleep disorders, and cardiovascular disease with the limitations on sexual activity these conditions cause do not appear till 10 years after menopause with no estrogen replacement (Butler and Lewis, 1983; Simpson and McKinney, 1970).

The sexual effects seen in the deestrogenized women and the aged woman are numerous. Anatomically, these are decreased vaginal mucosal thickness (from 25 to 2 cells), decreased vaginal acidity, decreased elasticity with collagen loss, and decreased non-sexually stimulated vaginal secretion production as well as decreased sexually stimulated vaginal secretion production and a longer period of time required to produce it. This anatomic cluster of changes predispose to pain, irritation, and injury with coitus and vaginal infections (Butler and Lewis, 1978; Diamon and Karlen, 1981; Masters and Johnson, 1966; Scheingold and Wagner, 1974; Simpson and McKinney, 1970). With the exception of collagen loss, replacements of estrogen can reverse these changes. There are other deficiencies noted in aging woman that are only partially or possibly not affected by estrogen levels. These include a decrease in the vulvar sex skin color change with sexual arousal, decreased clitoral engorgement, decreased labia minora engorgement, decreased voluntary and involuntary prestimulated and stimulated pubococcygeal muscle tone, decreased size and volume of the orgasmic platform, decreased intensity and number of pubococcygeal contractions with orgasm, decreased uterine bladder and rectal contractions with arousal and orgasm, decreased breast sexual enlargment, decreased peripheral skin sex flush, and increased amounts of stimulation required to achieve maximal response (Butler and Lewis, 1983; Diamon and Karlen, 1981; Masters and Johnson, 1966; Simpson and McKinney, 1970). While there is no reported change in the ability of an older woman to achieve orgasm, even multiple orgasm, or a measurable increase in the refractory period between orgasms, the overall sexual response is less intense, less strong, and requires more stimulation to achieve (Masters and Johnson, 1966). Data are preliminary and no large-scale studies or surveys have been done. The exact role of estrogen presence or absence in production or alleviation of observed changes is not understood. Recent information on estrogen mediation of peripheral sensory and motor nerves suggests that both the decreased sensation and occasional hypersensitivity reported by postmenopausal women may be induced by estrogen deficiency. Additionally, there is evidence of decreased autonomic control of parasympathetic nerve fibers over the normal sexual vasocongestive response (Sorrell and Sorrell, 1984). There are reports that the above decremental changes are all less likely to occur in women continuing frequent coital activity. Perhaps the best of these reports was a controlled study by Bachman and Leiblum (1983) in which sexual activity, estrogen levels, testosterone, and gonadotropins were correlated to vaginal and labial indices, and only frequency of coitus was seen to correlate with "younger" anatomic and physiologic parameters. There is a problem, however, with both a higher mean weight within the sexually active groups and no correction for peripherally bound estrogen. One must exercise caution in stating observed sexual changes are so minor as to be insignificant or, conversely, are highly significant. In the male, erectile failure, which is a failure of the earliest phase of sexual arousal, almost always predicates no coitus and no orgasmic response because of cessation of continued penile stimulation. No study has observed whether orgasm could occur without erection in the male although this is a physiologic possibility. In the female, decreased early sexual arousal, which results in decreased vaginal lubrication, does not necessarily eliminate either coitus or orgasm. Similarly, while aging men show an increased incidence of retarded ejaculation, also known as ejaculatory incompetence, equivalent studies on possibly retarded orgasm have not been done with post-menopausal women.

The endocrine state of the female has been briefly discussed as it pertains to estrogen's anatomic effects. Although the literature is contradictory on this point, most experts agree that estrogen has very little if any effect on sexual desire, function, or response beyond its direct effect on the elimination of pain with vaginal atrophy (Abplanalp *et al.*, 1979; Bachman and Leiblum, 1981; Benedek and Rubinstein, 1939; Persky, 1983; Verwoelt, 1969). Klaiber (1979) administered estrogen to severely depressed menopausal women who had been unresponsive to other therapies and reported significant improvements in mental health and sexual desire. Improved mental health and sexual desire were not independent variables and it is somewhat more likely than not that return to more normal emotional states allowed the resurgance of sexual desire. The effect of progesterone on sexual desire and function has not been seriously investigated at any age (Hamberg, 1966; Persky, 1983; Persky *et al.*, 1976; Udrey and Morris, 1968). One could theorize that degradation to androgenic compounds might enhance desire and hence result in decreased desire in menopausal women. One study on menstruating women demonstrated a midluteal increase in sexual initiation and activity with correspondingly increased progesterone levels, but progesterone effect has not been reported in menopausal women (Persky *et al.*, 1976). Another study also documented a small increase in initiation and coital behavior for women given progesterone during and before ovulation as a dependent variable tied to lessened anxiety and depression (Hamburg, 1966). Androgens are known to be potent stimulators of desire in both men and women although they have no direct effect on function per se (Bancroft, 1980; Bancroft and Wu, 1983; Benkert *et al.*, 1979; Brown *et al.*, 1978; Money and Erhardt, 1972; Persky, 1983). Testosterone, androstenedione, dihydroepiandrosterone, and dihydrotestosterone all increase desire, initiatory behavior, and responsivity as well as increased coital frequency and sexual satisfaction in women (Bancroft, 1980; Money and Erhardt, 1972; Persky, 1983). None of the androgens improve vaginal lubrication, vasocongestive response, or orgasmic capacity (Bancroft, 1980). Since androgens are primarily produced by the adrenal, one would expect little change with menopause until extreme old age—unless progesterone production contributes significantly to either the androgen pool or to the biologic effect.

Data presented on possible physiologic effects in aging have been limited to otherwise healthy women. When one looks at the effects of illness on sexual function, a common condition involving an increasing percentage of people with advancing age, the sparse existing data is, once more, primarily on men. Older women who also have diabetes show markedly decreased sexual desire and secondary anorgasmia. Only Jensen (1981) correlated anorgasmia with the presence of peripheral neuropathy. Studies of survivors of gynecologic and breast cancer showed marked diminution in general sexual satisfaction (85% to 48%) from before diagnosis to the onset of symptoms and diagnosis, decreased frequency of coitus to cessation in 50%, and decreased frequency of regular orgasm (from 58% to 15%). On a 2-year follow-up, counseled subjects did significantly better by these measures, achieving essentially predisease levels, than did those who received only an initial interview. A comparable group who had neither an interview nor counseling continued to show deterioration, with 25% who reported general sexual satisfaction, 68% who reported cessation of all sexual activity, and 12% with regular orgasm (Harris et al., 1982; Waxenberg et al., 1960). This study convincingly demonstrates the profound psychosocial effects of serious disease when physical type, medical treatment, and permanent physiological damage are controlled. While there are no comparable data on women, 60% of males undergoing coronary diagnostic or corrective surgical procedure do not resume coitus because of fear. Their partners are adversely affected as well (Scheingold and Wagner, 1974; Thurer and Thurer, 1982). Many of the drugs used to alleviate cardiovascular disease severely impair the autonomic nervous system and cause sexual dysfunction in men, although no parallel studies on women exist (Buffman et al., 1981). The precise effects on sexual activity of chronic disabilities such as low back pain and degenerative and rheumatoid arthritis are not known, but have potential for the diminution of sexual behavior directly and by treatment effects. Similarly, some psychoactive drugs used to treat depression and anxiety, as well as these aging-prevalent emotional disorders themselves, can produce sexual dysfunction and resultant disinterest (Buffman et al., 1981; Klein, 1965).

Discussion

We have both the behavioral and technological methodology to obtain valid objective information on sexual responsivity, behavior, capacity, and desire in older women. Static and longitudinal well-designed survey instruments and psychometric testing can be done on samples with high validity. Social, aging, and experimental variables can be assessed independently in adequate samples. Access to random populations of aging women is more readily available than to either younger adults or minor children (Butler and Lewis, 1978).

Vaginal photoplethysmography, labial thermistors, and muscle transducers have all been used to measure physiological response in younger women (Chambless et al., 1982; Graber, 1982; Henson et al., 1979). Noninvasive sonographic venous and arterial blood flow assessment, measurements of nerve potentials, and sophisticated hormone evaluations are used to study medical conditions and sexual functioning in men but not, as yet, in women (Nadler, 1977; Brown et al., 1978). Again, longitudinal, disease state, and pre- and posttreatment effects can be evaluated effectively, but have not been for either female sexual or female aging phenomena.

The gathering of sexual behavior data in conjunction with other physiologic or medical researches could greatly enhance our knowledge and is essential. Obviously, this requires education of the researchers and demystification of the subject.

Marten (1981) delineated prerequisites for continued heterosexual functioning with advancing age. These include the integrity of anatomical and physiological processes involved in sexual learning and response, continued exposure to significant erotic stimuli, an erotic reaction to such stimuli, the cooperation of an interested partner, and a desire for sexual participation and sexual acts free of emotional disturbances such that the activity is promptly repeated. Three major processes contribute to sexual apathy. A lifelong low level of activity can, with a small decrease in motivation, lead to cessation. Social isolation dampens motivation. The failure to respond to previously effective stimuli secondary to an attendant physiologic state is common. Unfortunately, a majority of our women will face the absence of three or more of the prerequisites required for continued sexual functioning and the intrusion of at least one of the conditions leading to apathy.

Touch deprivation is one of the most grievous losses of the aged, and sexual inactivity is an important part of that (Butler and Lewis, 1978; Mancini, 1983). Current conventional social and medical systems do little to alleviate that loss and in some instances advance it. Examination of the capacities and alternatives of older women is overdue.

References

Abplanalp, J., Rose, R., and Donnelly, A., 1979, Psychoendocrinology of the menstrual cycle: II. The relationship between enjoyment of activities, moods, and reproductive hormones, *Psychosom. Med.* **41**:605.

Bachman, G., and Leiblum, S., 1981, Sexual expression in menopausal women, *Med. Aspects Hum. Sexuality* **15**:96b–96h.

Bancroft, J., 1980, Endocrinology of sexual function, *Clin. Obstet. Gynecol.* **7**:253–281.

Bancroft, J., and Wu, F., 1983, Changes in erectile responsiveness during androgen replacement therapy, *Arch. Sexual Behav.* **12**:59–66.

Benedek, T., and Rubinstein, B., 1939, The correlations between ovarian activity and psychodynamic process II: The menstrual phase, *Psychosom. Med.* **1**:461–466.

Benkert, O., Witt, W., Adam, W., and Leitz, A., 1979, Effectiveness of testosterone on sexual potency and the hypopituitary–gonadal axis of impotent men, *Arch. Sexual Behav.* **8**:471–480.

Bernardo, F., Vera, H., and Bernardo, D., 1983, Age-discrepant marriages, *Med. Aspects Human Sexuality* **17**:57–76.

Blumstein, P., and Schwartz, P., 1983, *American Couples: Money, Work, Sex,* Vol. 1, William Morrow and Co, New York.

Brown, W., Monti, P., and Corriveau, D., 1978, Serum testosterone and sexual activity and interest in men, *Arch. Sexual Behav.* 7:97–103.

Buffman, J., Smith, D., Moser, C., Apter, M., Buxton, M., and Davison, J., 1981, Drugs and sexual function, *Sexual Problems in Medical Practice* (H. Leif, ed.), American Medical Association, Monroe, Wisconsin, pp. 211–242.

Butler, R., and Lewis, M., 1978, The second language of sex, *Sexuality and Aging*, Volume 1 (R. Skolnick, ed.), University of Southern California Press, Los Angeles.

Butler, R., and Lewis, M., 1983, Sexual frustrations of older women, *Med. Aspects Hum. Sexuality* 17:65–78.

Chambless, D., Stern, T., Sultan, F., Williams, A., Goldstein, A., Lineberger, M., Lifshitz, J., and Kelly, L., 1982, The pubococcygeus and female orgasm: A correlational study with normal subjects, *Arch. Sexual Behav.* 11:479–490.

Charatan, F., 1982, Geriatric sexuality, *Med. Aspects of Hum. Sexuality* 16:68v–68dd.

Christensen, C., and Gagnon, J., 1965, Sexual behavior in a group of older women, *J. Geriatr. Psychol.* 20:351–356.

Christensen, C., and Johnson, W., 1973, Sexual patterns in a group of older never-married women, *J. Geriatr. Psychol.* 6:80–98.

Diamond, M., and Karlen, A., 1981, The sexual response cycle, *Sexual Problems in Medical Practice*, Volume 1 (H. Leif, ed.), American Medical Association, Monroe, Wisconsin, pp. 44–50.

Flint, M., 1981, Cross-cultural factors that affect age of menopause, in: *Consensus on Menopause Research, Proceedings of the First International Congress on Menopause*, University Park Press, Baltimore, pp. 73–85.

Gaitz, C., 1974, Sexual activity during menopause, *Med. Aspects Hum. Sexuality* 8:67–69.

Gebhardt, P., 1972, Postmarital coitus among divorces and widows, *Sex and Society* (R. Hodge, ed.), Markham Publishing Co., Chicago, pp. 142–154.

George, L., and Weiler, S., 1981, Sexuality in middle and later life, *Arch. Gen. Psychol.* 38:919–923.

Graber, B., 1982, The circumvaginal musculature, a literature review, *Circumvaginal Musculature and Sexual Function* (B. Graber, ed.), S. Karger, Basel, pp. 1–7.

Hamberg, D., 1966, Effects of progesterone on behavior, *Res. Publ. Assoc. Res. Nerv. Ment. Dis.* 43:251–253.

Harris, R., Good, R., and Pollack, L., 1982, Sexual behavior of gynecologic cancer patients, *Arch. Sexual Behav.* 11:503–510.

Henson, B., Rubin, H., and Henson, D., 1979, Women's sexual arousal assessed by three genital measures, *Arch. Sexual Behav.* 8:459–469.

Jensen, S., 1981, Diabetic sexual dysfunction: A comparative study of 160 insulin-treated diabetic men and women and an age-matched control group, *Arch. Sexual Behav.* 10:493–504.

Kinsey, A., Pomeroy, W., and Clyde, M., 1948, *Sexual Behavior in the Human Male*, Volume 1, W. B. Saunders, Philadelphia, pp. 226–235.

Kinsey, A., Pomeroy, W., and Clyde, M., 1953, *Sexual Behavior in the Human Female*, Volume 1, W. B. Saunders, Philadelphia, pp. 353, 714F, 716F.

Klaiber, E., Broverman, D., Vogel, W., and Kobajashi, Y., 1979, Estrogen therapy for severe persistent depressions in women, *Arch. Gen. Psychol.* 36:550–554.

Klein, D., 1966, Delineation of two drug-responsive anxiety syndromes, *Psychopharmocology* 5:397–408.

LaTorre, R., and Kear, K., 1977, Attitudes towards sex in the aged, *Arch. Sexual Behav.* 6:203–214.

Mancini, J., 1983, Strengthening marital relationships of older adults, *Med. Aspects Hum. Sexuality* 17:77–96.

DIANE S. FORDNEY

Marten, C., 1981, Factors affecting sexual functioning in 60–79 year old married males, *Arch. Sexual Behav.* **10**:399–420.

Masters, W., and Johnson, V., 1966, *Human Sexual Response*, Vol. 1, Little, Brown, and Co., pp. 223–276.

Money, J., and Erhardt, A., 1972, *Man and Woman: Boy and Girl, Differentiation and Dimorphism of Gender Identity from Conception to Maturity*, Volume 1, Johns Hopkins University Press, Baltimore.

Nadler, R., 1977, Sexual behavior of captive orangutans, *Arch. Sexual Behav.* **6**:457–475.

Newman, G., and Nichols, C., 1960, Sexual activities and attitudes in older persons, *JAMA* **173**:33–35.

Newman, G., and Nichols, C., 1974, Sexual activity in older persons, *Perspectives on Human Sexuality—Psychological, Social and Cultural Research Findings* (N. Wagner, ed.), Behavioral Publications, New York, pp. 501–504.

Pattison, E., 1983, When an adult's parents remarry, *Med. Aspects Human Sexuality* **17**:60b–60u.

Persky, H., 1983, Psychosexual effects of hormones, *Med. Aspects Hum. Sexuality* **17**:74–97.

Persky, H., Khan, M., and O'Brien, C., 1976, Reproductive hormone levels, sexual activity and moods during the menstrual cycle, *Psychosom. Med.* **38**:62–63.

Rubin, I., 1976, The sexless older years: A socially harmful stereotype, *Growing Old in America* Volume 1 (B. Hess, ed.), Transaction Books, New Brunswick, New Jersey.

Scheingold, L., and Wagner, N., 1974, *Sound Sex and the Aging Heart*, Volume 1, Behavioral Publications, New York, pp. 33–56.

Simpson, I., and McKinney, E., 1970, *Normal Aging*, Volume 1, Duke University Press, Durham, North Carolina.

Sorrell, P., and Sorrell, L., 1984, Male sexual dysfunction linked to wife's menopause, *Sexuality Today* **7(33)**:1–3.

Strean, H., 1983, *The Sexual Dimension*, Volume 1, Free Press, New York, pp. 176–196.

Thurer, R., and Thurer, S., 1982, Sex after coronary bypass surgery, *Med. Aspects Hum. Sexuality* **7**:68f–68r.

Udrey, J., and Morris, N., Distribution of coitus in the menstrual cycle, *Nature* **220**:593–596.

Verwoelt, A., Pfeiffer, E., and Wang, H., 1969, Sexual behavior, *Senescence Geriatrics* **24**:137–144.

Waxenberg, S., Finkbeinter, J., Drellich, M., and Sutherland, A., 1960, Changes in sexual behavior in relation to vaginal smears of breast cancer patients after oopehorectomy and adrenalectomy, *Psychosom. Med.* **22**:435–442.

Wharton, G. F., III, 1981, *Sexuality and Aging: An Annotated Bibliography*, Volume 1, Metuchen Press, Metuchen, New Jersey.

White, C., 1982, A scale for the assessment of attitudes and knowledge regarding sex in the aged, *Arch. Sexual Behav.* **11**:491–502.

Young, W., 1961, The hormones and mating behavior, *Sex and Internal Secretions*, Volume II (W. Young, ed.), Williams and Wilkins, Baltimore, pp. 1173–1239.

V

MANAGEMENT OF THE MENOPAUSE

15

HEALTH STATUS AND HEALTH CARE UTILIZATION BY MENOPAUSAL WOMEN

SONJA M. McKINLAY and JOHN B. McKINLAY

Introduction

According to the popular view and many experts, the menopause, or so-called "change of life," is thought to represent a major cultural, psychological, and physiological milestone for women during the middle years. It signifies the end of reproduction in societies such as the United States, where sexuality and reproduction are considered evidence of personal success and fulfillment. It is a prominent biological marker for an aging process in cultures that extol youthfulness. The menopause is even perceived as a major negative life event of the same magnitude as, for example, job loss or loss of a spouse. Consequently, it is thought to be accompanied by increases in morbidity and health care utilization, commensurate with the increases observed for such other negative life events.

Associated with these perceptions surrounding the menopause, this normal physiological event has been viewed as a "syndrome" (McKinlay and Jefferys, 1974; Townsend and Carbone, 1980; Greenblatt, 1974), and, more recently, as a newly discovered "deficiency disease" (McCrea, 1983). Women experiencing this normal end of reproduction are thought to experience regretfulness, to evidence signs of clinical depression (involutional melancholia), to present with a broad range of accompanying symptoms, to be high utilizers of physicians' services, and generally to consume a disproportionate share of medical

SONJA M. McKINLAY and JOHN B. McKINLAY • Cambridge Research Center, American Institutes for Research, Cambridge, Massachusetts 02138 and Boston University, Boston, Massachusetts 02215.

resources. The "typical" menopausal woman, then, is perceived as growing old, beset with psychosocial complaints, experiencing major physical changes, losing cultural significance, evidencing regretfulness and depression, and generally being a burden on medical care resources (Wilson, 1966; Reuben, 1969). Such a characterization would appear to receive reinforcement both from pharmaceutical advertisements in professional medical journals (Chapman, 1979; Prather and Fidell, 1975) and from patient images used and developed during the course of medical education (Scully and Bart, 1973; Howell, 1974).

While remaining the predominant view, it has evolved from biased clinical encounters and is actually without epidemiological foundation. Studies of health status and health service utilization in adult populations show no evidence of increased morbidity or health service use in this mid-aged population beyond expected increases with age (Anderson and Anderson, 1979; Maurana *et al.*, 1981). Most of the work to date that focuses on the menopause tends to be based on hospital/clinic populations or general practice settings that, by definition, represent patients presenting complaints (Shepherd *et al.*, 1966; Skegg *et al.*, 1977). This view is equally attributable to the practice of reporting only positive findings or association between the menopause and, for example, CHD, osteoporosis, or depression (Kuller *et al.*, 1984).

As a move towards correcting the present imbalance in scientific knowledge concerning the menopause, this paper focuses on three interrelated themes, which may be summarized as follows:

- A vast majority of mid-aged women do *not* express regret at reaching menopause, do *not* report more symptoms or poorer health status, and do *not* evidence increased use of medical services. The "typical" menopausal woman is neither sick nor a high user of medical care.
- Women who experience an artificial menopause (cessation of menses through surgery) constitute an atypical subgroup, which differs from the vast majority of women in important ways unrelated to the menopause. These are the women who have been extensively studied and discussed in the professional and popular literature.
- It is this atypical subgroup that physicians tend to see (as patients) and from whom they understandably derive their statistically biased image of the "typical" menopausal woman.

Methods

The data reported in this study were obtained through mailed questionnaires and telephone interviews completed by 8050 women aged 45–55 years (as of December 31, 1981) and residing in Massachusetts at the time of the survey (September 1981 to May 1982). Simple random samples of eligible women were selected in each of 38 towns and cities in a two-stage cluster sampling design, with a response rate of 77%.

Self-report data were obtained on a range of sociodemographic, health, and employment variables using, in all cases, the most reliable methodologies available. The *sociodemographic* variables considered include age, education (in three categories), number of regular contacts per week with friends or relatives not living with the respondents, and household composition. Occupation and per capita income were considered as socioeconomic indicators but regarded as inappropriate for this analysis. Occupation was relatively invariant in this population of mid-aged women, even though 67% were currently working for pay (54% of these were in clerical or service jobs). Per capita income could not be assessed for a sizeable proportion (14%) of women due to missing information on total household income. Many women simply do not know what their husbands earn. Educational status, which is strongly associated with per capita income in this sample, was therefore chosen as the most reliable socioeconomic indicator for this population. The number of contacts with relatives or friends was used as a crude measure of social network size, which has been shown to correlate strongly with health behavior in other populations (McKinlay, 1981; Broadhead and Kaplan, 1983). Household composition rather than marital status was thought to provide more sensitive information on the respondent's nurturing role (including children as well as other, usually elderly, adult relatives) as a potential source of stress for mid-aged women. At the same time, this variable includes most information on marital status, which has been well-documented as related to health and health behavior (Verbrugge, 1983; Haynes and Feinleib, 1980; Haynes *et al.*, 1984; Berkman and Syme, 1979). Elsewhere it has been reported that employment status is unrelated to health and health behaviors in this same population (Jennings *et al.*, 1984), and it was, therefore, not included here. A small subgroup of 721 women, not in the labor force primarily for health reasons, was identified in the earlier analysis as contributing all of the apparent difference in health status between employed and unemployed women. This subgroup, however, although more likely to have an early surgical menopause, was too small to affect results by menopausal status, in the sample.

Five measures of *health status* were included in this investigation. Self-assessed health in relation to one's peers is the primary measure, which is highly correlated with all others and has been shown to be a reliable indicator of actual health status (Maddox and Douglass, 1973). Women were also asked if they currently had, or were receiving treatment for, any of the following common chronic conditions, which usually require clinical diagnosis: diabetes (4%), high blood pressure (19%), asthma (4%), allergies of eczema (14%), heart disease (4%), ulcer (4%), arthritis or rheumatism (23%), and cancer (1%). The third health status variable is the number of restricted activity days in the prior 2 weeks due to some aspect of the respondent's health. This measure is comparable to that used in the National Center for Health Statistics' ongoing Health Interview Survey. Finally, respondents were asked to indicate whether or not each on a list of common symptoms had been experienced in the past 2 weeks. Eliminating those listed that related to either menstruation or the menopause (hot flushes/flashes, cold sweats, menstrual problems, fluid retention), the remaining symptoms were divided into two groups. Common physical symptoms included diarrhea and/or constipation, persistent cough, upset stomach, backaches, headaches, sore throat, and aches/stiffness in the joints. A group of symptoms of a less obviously physical origin, especially in the absence of other physical signs or diagnosed disease, included dizzy spells, lack of energy, irritability, feeling blue or depressed, shortness of breath, trouble sleeping, and loss of appetite. This latter group was labeled as primarily psychological in origin, given the low prevalence of heart disease and cancer in this sample.

Five measures of *utilization* of formal and informal health care resources were included. Prescribed and over-the-counter (OTC) medications reportedly taken in the prior 2 weeks were classified separately. Respondents were also asked if they had consulted either a health professional (nurse, nurse practitioner, physician assistant, physician, pharmacist, etc.) and/or relative or friend concerning a problem with their health in the prior 2 weeks. Consultation with one or more health professionals indicated formal health service use, while consultation with a friend or relative indicated use of informal or lay helping resources. Finally, reported breast surgery for cysts or benign tumors was used as an additional indicator of formal health service use. The use of a 2-week time frame for the first four of these variables has been shown by NCHS to ensure reliable recall.

Menopausal status was reliably determined from a combination of questions on current menstrual status and changes in the last 12 months. *Natural menopause* was considered to have occurred if no menses were reported for 12 months, in the absence of surgery that would terminate menstruation. Twelve months of amenorrhea is the widely accepted definition used in European studies since the 1950s (see, for example, Magursky *et al.*, 1975) and is recommended by Treloar (1974) on the basis of his prospective study of normal menstrual patterns. Women are classified as *perimenopausal* if menses have been reported in the last 12 months, but with periods of amenorrhea and/or changes in regularity or flow. Again, this is consistent with prior research (Jaszmann *et al.*, 1969; Magursky *et al.*, 1975; Treloar, 1974). Women are *still menstruating* if they report regular menses in the last 3 months. *Artificial menopause* is considered to have occurred if menses were surgically stopped by *either* a hysterectomy (with or without removal of the ovaries) (26%) *or* a bilateral oopherectomy (2%).

A final variable included in this analysis was the response to an *attitudinal question* on how respondents felt about the time when menstrual periods stopped altogether. Four precoded options were provided: "relief," "regret," "mixed feelings," and "no particular feelings at all."

The analytical approach followed a logical sequence of steps involving primarily contingency table methods and step-wise discriminant modeling. Initial cross-tabulations of menopausal status by each of the health status and health behavior variables indicated that women with an artificial menopause reported health status and health-related behavior that were markedly different from those reported by the other three menopausal groups. Perimenopausal women showed selected differences compared to women still menstruating or who had experienced a natural menopause. Given the very different age distributions in each menopausal group (see Figure 1), Fisherian discriminant functions (Mosteller and Tukey, 1977) were estimated, using a step-wise procedure (SAS Institute, 1982), between: (a) artificially and naturally menopausal women; and (b) perimenopausal and still menstruating women. These estimated functions indicated which sociodemographic variables required control in final comparisons and which health status and health behavior variables contributed significant differences in the two comparisons. Differences in combined proportions, adjusted for sociodemographic factors, were tested for significance in the final comparisons (Cochran, 1954).

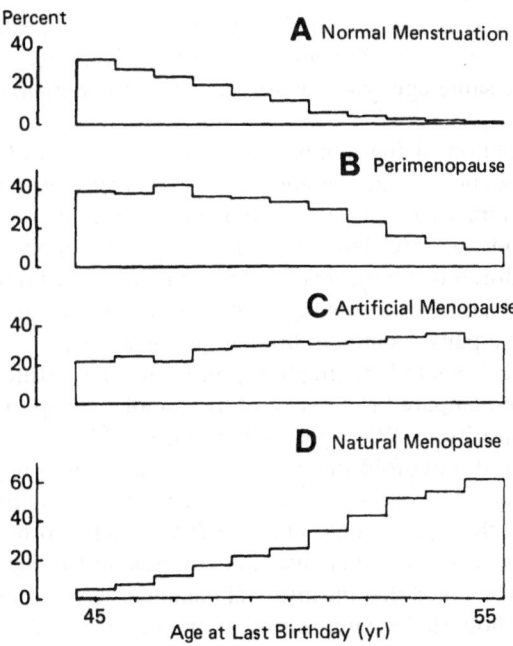

Figure 1. The distribution of menopause status by age.

Results

Figure 1 presents the distribution of the four menopausal status groups by age for the sample cases with complete data. As expected, the proportions of still menstruating and perimenopausal women decrease with age, the latter group more slowly. The proportion reporting a natural menopause increases rapidly in this cohort, while the proportion reporting a surgical menopause increases somewhat before age 50, then remains constant at about 30%, indicating that the majority of women undergo this surgery before the age of 50 and, therefore, before the occurrence of a natural menopause (median age 51.4 ± 0.19 years in this sample).

A summary of the distributions of selected sociodemographic characteristics is presented in Table I. As anticipated from Figure 1, the mean ages in each of the four menopausal status groups are significantly different with increments of 1.5 years, going from still menstruating to perimenopausal to artificially and finally to naturally menopausal women, in that order.

The proportion reporting more than 12 years of education, however, does not decrease with age as expected from survey and census data in adult populations (U.S. Department of Commerce, 1980). The decline in education with increasing age is evident when the artificial menopause group is excluded. This latter group however, is not only *younger* than naturally menopausal women by 1.5 years, but is less well-educated than the otherwise oldest, least educated group of menopausal women. Such a reversal of the expected trend indicates that women with an artificial menopause have received markedly less education than women the same age who are approaching or experiencing a natural menopause.

The distribution of the number of weekly contacts with friends and relatives is independent of both age and education, with nonsignificant chi-square values. When compared to all other women in the sample, however, those with an artificial menopause are significantly more likely to report ten or more weekly contacts. This difference indicates that frequent contact with relatives or friends is, in some way, associated directly with a surgical menopause but is otherwise unrelated to menopausal status in this sample of women.

The atypical sociodemographic profile in the artificially menopausal subgroup, when compared to the rest of the sample, is repeated with respect to household composition. When this subgroup is excluded, the expected trend with age is toward a twofold increase in the proportion living alone (reflecting an increase in widowed, divorced or separated women), an increase in the proportion living with a spouse only (11% to 20%), and a corresponding decrease in the proportion with both a spouse and children still living at home (59% to 45%). It should be noted that the artificial menopause subgroup reports the same distribution for household composition as the older, more educated, naturally menopausal women. This equivalence is partly due to the higher rate of divorce

Table I

The Distribution of Selected Sociodemographic Characteristics by Menopausal Status[a]

Characteristic	Menopausal Status			
	Still menstruating	Perimenopausal	Artifically menopausal	Naturally menopausal
(1) Mean age (S.E.)	48.0 (0.07)	49.5 (0.06)	51.1 (0.06)	52.6 (0.05)
(2) Percent >12 years education	44.3	39.6	29.6	31.7
(3) Percent 10+ weekly contacts	57.8	57.8	63.4	58.5
(4) Household composition				
Living w/spouse and children	59.3	55.3	45.3	44.7
Living w/spouse only	11.2	14.7	21.2	20.5
Living alone	4.1	6.4	8.6	9.0
Total (100%)	994	2113	2223	2421

[a] All differences are significant at $\alpha = 0.001$. All tests included adjustments for age or age and education differences as appropriate.

in artifically menopausal women (12% compared to 9% in the rest of the sample), offsetting a lower proportion of never-married women in this group (4% compared to 7% in naturally menopausal women). The comparably low proportion of artificially menopausal women with both children and a spouse at home appears to contradict the higher parity in this group (92% with at least one child compared to 88% among naturally menopausal women). A possible explanation (not verifiable in this data set) is that the artificially menopausal women began childbearing considerably earlier than did women with a natural menopause.

When these sociodemographic characteristics (age, education, number of contacts with relatives/friends, and household composition) were entered stepwise into the two discriminant functions described above, age and number of contacts with friends and relatives were the two significant discriminators between artificially and naturally menopausal women (Table II). The variables significantly discriminating between the two younger groups were, as expected, age and educational status, although almost all the R^2 value (0.69) was contributed by age alone, with education adding little information to the discriminant. The lack of comparability in these two discriminants underscores the atypical sociodemographic profile of the artificial menopause group.

These sociodemographic variables were then held constant in the discriminants while all five health status measures were added, step-wise. From Table II, it is evident that health status variables add little to the discrimination of the two pairs of menopausal groups. Moreover, two of the three discriminatory variables added are different in the two comparisons. While both comparisons include physical symptoms as the primary discriminator, the younger premenopausal groups differ primarily on measures of acute health status, while the older groups differ on more chronic health measures.

The differences between all four menopausal groups are summarized in Table III, which shows that, irrespective of the sociodemographic differences already discussed, the artificial menopause subgroup reports consistently worse health on all five measures. The differences are particularly marked when this subgroup is compared to the rest of the sample with respect to measures of long-term health status (perceived general health and reported chronic conditions). The reporting of recent symptoms and restricted activity days also shows increases in this atypical subgroup, but they are not as marked and, as expected, are accompanied by increased reporting among the perimenopausal women. The increases in the latter group are associated with increases in menstrual flow problems and the hot flashes and sweats that are typical of this transitional phase preceding cessation of menses (see Table IV). The measure showing the least differences among the four groups is the number of psychological symptoms, although all three measures based on reporting in the prior 2 weeks show elevation in the perimenopausal and menopausal women compared to those still

Table II

Summary of Stepwise Fisherian Discriminant Analyses

	Comparison			
	Perimenopausal vs. still menstruating		Artifically vs. naturally menopausal	
Variable group	Variables entered (in order)	R^2	Variables entered (in order)	R^2
A. Sociodemographic	1. Age 2. Educational status	0.69	1. Age 2. Number of relatives/friends	0.47
B. Health status	3. Number of physical symptoms 4. General health status 5. Number of psychological symptoms	0.70	3. Number of physical symptoms 4. Number of chronic conditions 5. Number of restricted activity days	0.48
C. Health care	No variables Entered (none met minimum criterion)		6. Number of prescribed medications 7. Breast surgery 8. Consulting with relative or friend	0.49

Table III

The Distribution of Selected Measures of Health Status, by Menopausal Status[a]

Health status measure	Menopausal status			
	Still menstruating	Perimenopausal	Artifically menopausal	Naturally menopausal
Percent health worse than peers	3.8	6.3	10.6	7.9
Percent reporting two or more chronic conditions	9.7	14.9	21.3	16.6
Percent reporting one or more restricted activity days	11.7	15.8	18.2	12.6
Percent reporting two or more physical symptoms	34.8	46.7	52.8	41.8
Percent reporting two or more psychological symptoms	35.9	46.9	48.9	43.5
Total (100%)	989	2113	2118	2430

[a] All differences are significant at $\alpha = 0.01$. All comparisons are adjusted for age and education differences

Table IV

The Percent Reporting Flushes/Sweats and Menstrual Problems/Fluid Retention, by the Number of Other Physical Symptoms and Menopausal Status

	Still menstruating		Perimenopausal		Artificially menopausal		Naturally menopausal	
	<2	≥2	<2	≥2	<2	≥2	<2	≥2
Percent reporting flushes/sweats	7.2	26.2	30.4	45.2	33.5	54.1	33.4	55.1
Percent reporting menstrual problems/fluid retention	17.5	43.0	28.4	50.8	—	—	—	—
Total (100%)	645	343	1121	981	1037	1157	1404	1001

Number of other physical symptoms in women

menstruating naturally. This pattern indicates that the experience of menopause has some immediate but short-term effect on reported health but is certainly not of the magnitude reported by the artificial menopause subgroup and is primarily associated with expected increases in menstrual flow problems, hot flashes, and the accompaning sweats. These findings provide some cross-cultural corroboration for the studies by Flint (1975) (of 483 Indian women of the Rajput caste) and Maoz and his colleagues (1977) (of five ethnic groups of women in Israel). Although women from different cultures may exhibit differences in attitudes towards the menopause, they are similar in their reporting of symptomatology, especially hot flashes/sweats and menstrual cycle changes.

The transient nature of the increases in symptomatology among perimenopausal women is reflected in the lack of a corresponding increase in the use of formal health care by this group, as summarized in Tables II and V. In marked contrast, the relatively poorer health of the surgical menopause subgroup is accompanied by a twofold increase in prescribed medication (primarily estrogen replacement therapy), a 75% higher rate of reported breast surgery, a 30–50% increase in consultations with a health professional, and a 10–40% increase in lay consultation.

Both the perimenopausal women and those who are artificially menopausal show slight but insignificant increases in the use of OTC medications, as well as correspondingly higher rates of informal consultation with relatives or friends concerning a health problem.

The health behavior variables that significantly added to the discriminant (artificial versus natural menopause) were, in order of entry, the number of prescribed medications, breast surgery, and consultation with a relative or friend (Table II). Almost all the increase (2%) in R^2 (0.49) was contributed by the first variable, as anticipated. In comparing perimenopausal and menstruating women, none of the health behavior variables considered added significantly to the discriminant, confirming the lack of impact of the menopause on health behavior, evident in Table V. Only the use of OTC medications even approached significance in the discriminant.

Given the twofold higher rate of prescribed medication use by women with an artificial menopause, a more detailed analysis of this relationship was undertaken and the results are summarized in Figure 2. The two primary types of prescribed medication were estrogens (for hormone replacement therapy) and tranquilizers such as valium and librium. Because the primary motivation for estrogen therapy is castration (bilateral oopherectomy), the artificial menopause group was further classified according to whether or not both ovaries had been removed.

As expected, those most likely to be currently using estrogens (alone or combined with a tranquilizer) were castrated women (one third of this group). However, even among hysterectomized women with at least one ovary intact,

Table V

Distribution of Selected Measures of Health Care Utilization Behavior, by Menopausal Status[a]

Health care utilization measure	Menopausal status			
	Still menstruating	Perimenopausal	Artificially menopausal	Naturally menopausal
Percent reporting one or more prescribed medications	11.6	15.5	34.8	16.1
Percent reporting one or more OTC medications	77.0	83.0	84.7	78.7
Percent reporting breast surgery	9.2	11.6	17.8	11.0
Percent consulting one or more health professionals	19.1	22.3	28.8	21.0
Percent consulting a relative/friend	20.4	26.6	29.0	22.2
Total (100%)	981	2098	2197	2400

[a] All differences are significant at $\alpha = 0.001$. All comparisons are adjusted for age and education or age and number of relatives/friends.

nearly 16% were also using estrogens. These rates are in marked contrast to the remaining groups in which the rate of estrogen use varies from just over 6% (natural menopause group) to less than 2% (among menstruating women).

It is notable that there is no elevation of tranquilizer use in either surgical menopause group if it is not combined with estrogen prescription, the rates remaining remarkably constant at about 10% across all menopausal groups. However, when combined with estrogen prescription, tranquilizers are more likely to be used by women with an artificial menopause. This pattern could indicate that physicians who are likely to prescribe estrogen are more likely to prescribe other drugs. Alternatively, it could reflect demands from a subset of women who are high drug users.

The remaining differences of interest are the rates of formal and informal consultation of relatives and friends concerning a health problem. From Table V it appears that the perimenopausal women show some increase in informal consultation only, while the artificial menopause group shows higher rates of both formal and informal consultation. The relationship of these two outcomes

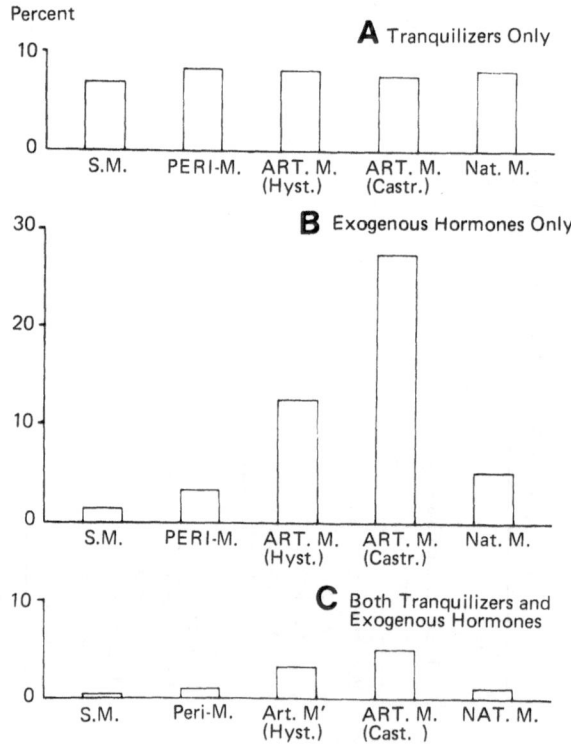

Figure 2. Prescribed exogenous hormone and tranquilizer use, by menopausal status. [ART. M. (Hyst.)] Those with hysterectomy and at least one ovary intact; [ART. M. (Castr.)] those with a bilateral oopherectomy (castration).

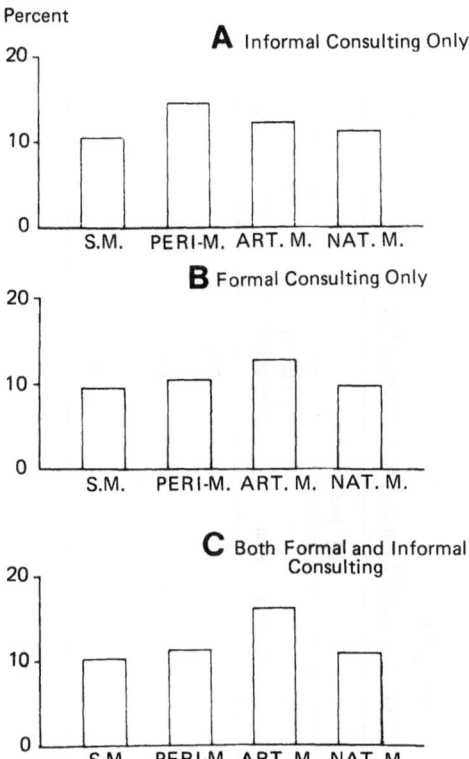

Figure 3. Consultation concerning health, with professionals (formal) and/or with relatives/friends (informal), by menopausal status.

was further explored by considering their joint distribution by menopausal status (Figure 3). From this figure it is evident that the perimenopausal and artificially menopausal women exhibit quite different consulting behavior patterns, as suggested in Table V. The younger, perimenopausal group shows a modest increase in lay consultation only with no increase in either the use of health professionals only, or of both sources. In contrast, the artificially menopausal women show a clear increase in the use of both formal and informal consultation, rather than independent increases in the use of each source separately.

Finally, Table VI summarizes women's feelings about cessation of menses, by menopausal status. Two findings should be highlighted in this table. *First*, only a very small proportion express regret (less than 3% overall), with only a slight increase among artificially menopausal women. *Second*, a clear majority (over 75%) express either relief or no particular feelings at all. The higher rates of mixed feelings reported by the two younger, premenopausal groups appear to change to more positive or neutral feelings in the older, men-

Table VI

Percentage Distribution of Attitude to Cessation of Menses, by Menopausal Status[a]

Attitude	Menopausal status			
	Still menstruating	Perimenopausal	Artificially menopausal	Naturally menopausal
Relief or no feelings at all (combined)	73.5	74.0	79.5	80.8
Mixed feelings	23.5	23.6	17.0	17.0
Regret	3.0	2.4	3.4	2.2
Total (100%)	935	2037	2130	2409

[a] $\chi^2(9) = 137.2$, $p < 0.001$ (including all four attitude categories).

opausal groups. This difference probably reflects an understandable reduction in anxiety as this event is experienced.

Discussion

For the vast majority of women, the so-called "change of life" actually produces very little change at all. Except for the experience of some temporarily bothersome symptoms (flashes, sweats, and menstrual irregularity), perimenopausal women (those currently experiencing the menopause) report no major increase in illness conditions and evidence no increase in health service utilization. Such findings are consistent with earlier studies reporting no increases in symptoms associated with the menopause (Neugarten and Kraines, 1965; McKinlay and Jeffreys, 1974; Greene, 1976; Hallstrom, 1973; Winokur, 1973). Our data indicate that women experiencing the menopause do increase their frequency of contact with relatives and friends (lay referral system) to some extent, which is understandable since most people discuss everyday events in their lives with their significant others. As expected, the frequency of contact with relatives and friends after a natural menopause returns to its lower premenopausal level. This use of friends and relatives (informal lay health system) by women experiencing a natural(nonsurgical) menopause is consistent with our findings that this is not a major health event in their everyday lives. It can be understood and coped with through lay resources, rather than health care services.

Just over a quarter of women in Massachusetts experience an artificial menopause (i.e., surgical cessation of menses). These are a distinctly atypical subgroup who differ from the majority of women sociodemographically, and perceive their health as worse than their peers, report a higher incidence of chronic conditions, experience more restricted activity days, and report more physical symptoms. They tend to be younger, of lower socioeconomic status (as measured by educational level), and more likely to be divorced than are their naturally menopausal counterparts. Understandably, they are significantly more likely to consult health professionals for these problems, undergo breast surgery, consume prescription drugs, and consult relatives and friends concerning their health problems.

For these atypical, artificially menopausal women, reported morbidity appears to be quite independent of menopause itself. That is, they report higher rates of diagnosed chronic conditions (e.g., diabetes, asthma, allergies, and cancer) that are independent of the menopause. For whatever reason (whether a direct result of their poorer health status or whether due to their consequent increased exposure to medical care), the artificially menopausal subgroup appears to be at increased risk to medical intervention. This is strongly suggested by their higher rate of breast surgery and increased use of estrogen replacement therapy and tranquilizers, in the absence of bilateral oopherectomy. It should be noted that the artificially menopausal women are higher users of *both* the formal health care system and their informal lay system of friends and relatives, perhaps signifying the more serious morbid episodes they encounter. Some studies of utilization behavior suggest that the use of one system tends to substitute for the other. For example, high use of relatives and friends is associated with low use of formal health services (McKinlay, 1973; McKinlay, 1981; Tennstedt, 1984). For the artificially menopausal women, however, the lay resources of friends and relatives are used in addition to, rather than instead of, formal medical care.

Health care workers are often criticized for their patient sterotypes, for example, the image of the "typical" menopausal woman (Ruzek, 1978). It is possible that this clearly erroneous image derives from frequent contact with the selected subgroup of atypical women whose major differences from the majority of woman experiencing the menopause have been described in this chapter. This subgroup clearly forms the core of the cinical caseload and remains the population most often studied and referenced (McKinlay and McKinlay, 1973).

The substantive findings presented here highlight two fairly obvious, but important, methodological requirements for further work in this area. *First,* epidemiological studies of representative samples of apparently healthy women experiencing normal aging are required to offset and complement the statistically biased portrait that understandably derives from highly selective clinical encounters with predominantly sick individuals (in, for example, general practice or clinic settings). It is axiomatic in epidemiological research that more can be learned about the sick by studying the well than by studying the sick themselves.

Secondly, only static associations can be presented here (an unavoidable limitation of cross-sectional designs), which raise rather than resolve questions concerning causality that can only be definitively addressed by longitudinal or prospective data. For example, the analysis presented here associates a surgical menopause with poorer health and higher health care utilization. However, the most likely causal sequence of these three factors remains to be determined. The results of this study indicate that poorer health *may* be independent of (and, therefore possibly, precede) a surgical menopause. However, it is not clear whether or not increased exposure to the health care system resulting from poorer health intervenes to increase the risk of surgical menopause.

Which of these possible causal explanations is the most likely can only be decided by appropriate analyses of longitudinal data. Such data are being obtained from the Massachusetts Women's Health Study, which is following a large sample of 2600 women every 9 months. These subsequent data will permit definitive answers to the derivative questions raised by the cross-sectional results presented in this study.

ACKNOWLEDGMENTS

The research reported herein is supported by Grant No. AG-03111 from the National Institute on Aging, National Institutes of Health. The authors would like to acknowledge the contribution of Cristine A. Crandall to the preparation of this chapter.

References

Anderson, R., and Anderson, O. W., 1979, Trends in the use of health services, in: *Handbook of Medical Sociology* (H. E. Freeman, S. Levine, and L. G. Reeder, eds.), Prentice Hall, Englewood Cliffs, pp. 371–391.

Berkman, L. F., and Syme, S. L., 1979, Social networks, host resistance and mortality: A nineyear follow-up study of Alameda County residents, *Am. J. Epidemiol.* **109**:186–204.

Broadhead, W. E., and Kaplan, B. H, 1983, The epidemiologic evidence for a relationship between social support and health, *Am. J. Epidemiol.* **117**:521–537.

Chapman, S., 1979, Advertising and psychotropic drugs: The place of myth in ideological production, *Soc. Sci. Med.* **13A**:751–764.

Cochran, W. G., 1954, Some methods for strengthening the common χ^2 tests, *Biometrics* **10**:417–451.

Flint, M. P., 1975, The menopause: Reward or punishment, *Psychosomatics* **16**:161–163.

Greene, J. G., 1976, A factor analytic study of climacteric symptoms, *J. Psychosom. Res.* **20**:425–430.

Greenblatt, R., 1974, *The Menopausal Syndrome*, Medcom Press, New York.

Hallstrom, J., 1973, *Mental Disorders and Sexuality in the Climacteric*, Orstadius Boktrgstreu A. B., Goteberg, Sweden.

Haynes, S. G., and Feinleib, M., 1980, Women, work and coronary heart disease: Prospective findings from the Framingham Heart Study, *Am. J. Public Health* **70**:133–141.

Haynes, S. G., Eaker, E. D., and Feinleib, M., 1984, The effect of employment, family and job stress on coronary heart disease in women, in: *The Changing Risk of Disease in Women: An Epidemiological Approach* (E. B. Gold, ed.), Collamore Press, Lexington, pp. 37–48.

Howell, M. C., 1974, What medical schools teach about women, *New Engl. J. Med.* **291**:304–307.

Jaszmann, L., VanLith, N. D., and Zaat, J. C. A., 1969, The age at menopause in the Netherlands, *Int. J. Fertil.* **14**:106–117.

Jennings, S., Mazaik, C., and McKinlay, S., 1984, Women and work: an investigation of the association between health and employment status in middle-aged women, *Soc. Sci. Med.* **19**:423–431.

Kuller, L. H., Meilahn, E. N., and Costello, E. J., 1984, Relationship of menopause to cardiovascular disease, *Behav. Med. Update* **5**:35–47.

Maddox, G. L., and Douglass, E., 1973, Self-assessment of health: A longitudinal study of elderly subjects, *J. Health Soc. Behav.* **14**:87–93.

Magursky, V., Mesko, M., and Sokolik, L., 1975, Age at menopause and onset of the climacteric in women of Martin District, *Int. J. Fertil.* **20**:17–23.

Maoz, B., Antonovsky, A., Apter, A., Wijsenbeek, H., and Datan, N., 1977, The perception of menopause in five ethnic groups in Israel, *Acta Obstet. Gynecol. Scand. Suppl.* **65**:69–76.

Maurana, C. A., Eichhorn, R. L., and Lonnquist, L. F., 1981, *The use of Health Services: Indices and Correlates*, Government Printing Office, Washington, D.C.

McCrea, F., 1983, The politics of menopause: The discovery of a deficiency disease, *Soc. Problems* **31**:111–123.

McKinlay, J. B., 1973, Social networks, lay consultation and help-seeking behavior, *Soc. Forces* **51**:(3):275.

McKinlay, J. B., 1981, Social network influence on morbid episodes and help-seeking behavior, in: *The Relevance of Social Science for Medicine* (L. Eisenberg and A. Kleinman, eds.), D. Reidel Publishing, The Hague, Netherlands, pp. 77-107.

McKinlay, S. M, and Jefferys, M., 1974, The menopausal syndrome, *Br. J. of Prev. Soc. Med.* **28**:108-115.

McKinlay, S. M., and McKinlay, J. B., 1973, Selected studies of the menopause: A methodological critique, *J. Biosoc. Sci.* **5**:533-555.

Mosteller, F., and Tukey, J. W., 1977, *Data Analysis and Regression*, Addison-Wesley, Reading, Massachusetts.

Neugarten, B. L., and Kraines, R. J., 1965, Menopausal symptoms in women of various ages, *Psychosom. Med.* **27**:266-273.

Prather, J., and Fidell, L., 1975, Sex differences in the content and style of medical advertisements, *Soc. Sci. Med.* **9**:23-26.

Reuben, D., 1969, *Everything You Always Wanted to Know About Sex, But Were Afraid to Ask*, David McKay, New York.

Ruzek, S. B., 1978, *The Woman's Health Movement*, Praeger Publishers, New York.

SAS Institute, 1982, *The User's Guide*, 1982 Edition, SAS Institute, Raleigh, North Carolina.

Scully, D., and Bart, P., 1973, A funny thing happened on the way to the orifice: Women in gynecology textbooks, *Am. J. Sociol.* **78**:1045-1051.

Shephard, M., 1966, *Psychiatric Illness in General Practice*, Oxford University Press, London.

Skegg, D. C., *et al.*, 1977, Use of medicine in general practice, *Br. Med.J.* **1**:1561-1563.

Tennstedt, S., 1984, *Informal Care of Frail Elders in the Community*, Unpublished doctoral dissertation, Boston University, Boston.

Townsend, J. M., and Carbone, C. L., 1980, Menopausal Syndrome: Illness or social role—a transcultural analysis, *Culture Med. Psychiatry* **4**:229—248.

Treloar, A. E., 1974, Menarche, menopause and intervening fecundabiity, *Human Biol.* **46**:89-107.

Verbrugge, L. M., 1983, Multiple roles and physical health of women and men, *J. Health Soc. Behav.* **24**:16-30.

Wilson, R., 1966, *Feminine Forever*, M. Evans, New York.

Winokur, G., 1973, Depression in the menopause, *Am. J. Psychiatry* **130**:92-93.

U.S. Department of Commerce, Bureau of the Census, 1980, *Statistical Abstract of the U.S.: 1980*, 101st Edition, Government Printing Office, Washington, D.C.

16

ESTROGEN DEPRIVATION
The Physiology, Pathophysiology, and Informed Management of Female Menopause or Castration

NATHAN KASE

Introduction

As a woman progresses through the fifth decade of life, the timing and function of the menstrual cycle change. The residual few thousand follicle units remaining in the ovaries at this age apparently are less sensitive to gonadotropin stimulation. This notion is supported by the progressive increase in follicle-stimulating hormone (FSH) and the reduced estrogen concentrations characteristic of the premenopause (Sherman *et al.*, 1976). Limited follicle maturation leads either to decreased cycle interval (shortened follicle maturation phase at first, followed by an inadequate luteal phase) or to variable lapses of cyclicity (oligoanovulation) with oligomenorrhea.

With the passage of time, as the number of follicles decreases and their resistance increases, less estrogen is produced by the ovaries. Eventually, despite higher levels of FSH, but not of luteinizing hormone (LH), the reduced level of estrogen secretion is no longer sufficient to proliferate a quantity of endometrium capable of producing visible menstruation. At this point, generally 50–52 years of age, the woman has reached menopause. Thereafter, the FSH level rises 10-fold to 20-fold above the usual cycle levels, reaching a maximum 1 to 3 years after menopause and persisting through the ninth decade (Vermeulen, 1976). At this stage, LH concentrations that had remained within

NATHAN KASE • Department of Obstetrics, Gynecology, and Reproductive Sciences, Mount Sinai Medical Center, New York, New York 10029.

normal cycle limits premenopausally now increase significantly; FSH and LH elevations are reliable chemical manifestations of the postmenopausal.

Although the ovary may contain some follicles, the steroid synthesis and secretion of these units is negligible. Still, the postmenopausal women is not without circulating estrogen. Estrone (E_1) is produced from the androgen precursor androstenedione by a process of extraglandular and extrahepatic aromatization (Judd, 1976). The bulk of androstenedione is derived from adrenocortical secretion. The ovarian stromal compartment, however, does participate for a short time in the production of this androgen substrate, which is used in the extragonadal conversion to E_1.

The amount of estrogen available to the postmenopausal woman is affected by her age and body weight (Siiteri and MacDonald, 1973; Hemsell et al., 1976). With increasing weight to the point of obesity, there is a concomitant increased conversion of androstenedione to E_1. This conversion reflects the ability of filled fat cells to trap and aromatize androgens at efficient rates. There is also evidence that with increasing age, aromatization rates increase as well. Finally, not only is the peripheral production of estrogen increased by such factors as age and obesity, but higher concentrations of free estradiol (E_2) are available to target cells (endometrium, breast) because of a depression in sex hormone-binding globulin (SHBG) capacity in postmenopausal obese women (Nisker et al., 1980).

As a result, the postmenopausal level of estrogen can vary acutely not only within an individual woman, but from one woman to another. The two major influences modifying extent of available estrogen are an increase in (1) substrate availability, as in stress-induced increases in adrenal secretion or in conditions associated with reduced metabolic clearance of androstenedione (e.g., liver disease or congestive heart failure) or (2) the percent of androstenedione converted to estrogen, as in obesity or aging.

In the average postmenopausal women, estrogen production generated from these sources is sustained for several years at levels that support reasonably normal structure and function of such dependent tissues as breasts, urethra, vagina, and vulva.

With increasing age, however, the contribution of precursors for estrogen production becomes inadequate, first from the ovarian stroma and then, more importantly, from the adrenal cortex. Despite availability of aromatization, during this final stage of substrate reduction, estrogen levels are no longer sufficient to sustain secondary sex tissues. Atrophy of estrogen-dependent organs ensues.

In summary, the progressive elimination of estrogen with time leads to a sequential loss of such estrogen-dependent functions as anovulation, amenorrhea, and vaginal and vulvar tissue contraction, culminating in generalized atrophy of all estrogen-dependent tissue. This prolonged period, from age 40 to beyond the seventies, is called the female climacteric. The single point at which insufficient follicle maturity results in inadequate estrogen and cessation of menses is known as the menopause.

On the basis of a review of the physiology of the perimenopause, the physician can gain understanding of the diversity of clinical symptomatology seen in practice. For example:

- Young castrates may have more serious risks of long-term estrogen deprivation.
- Postmenopausal women may not display progressive genital atrophy.
- Thin women are more likely to suffer the liabilities of estrogen deprivation than obese women.
- Obese women are more susceptible to dysfunctional uterine bleeding (DUB), endometrial hyperplasia, and breast and uterine neoplasia.
- Hot flushes and elevated gonadotropin (FSH) may appear *before* true menopause and yet be followed by intervals of ovulatory menstrual cycles.
- Stress, whether acute or chronic, may induce menstrual flow in postmenopausal women.
- Endometrial hyperplasia and neoplasia may occur in untreated oophorectomized women.

Clinical Implications of Progressive Estrogen Withdrawal

It has been emphasized that as a woman progresses through perimenopause, the mechanism of production and the biological activity of estrogen change. Estrone production is not under the control of the usual servomechanism factors modifying pituitary-ovarian interactions that ordinarily maintain physiologically appropriate quantities of estradiol during the reproductive years. In the absence of feedback restraints, unusual modifiers, such as age, stress, and obesity, emerge as prime elements in the production of estrone. Not only does the biologic supply of estrogen decline, but superimposed variations in the rate of that decline hinge on prehormonal substrate production or on the efficiency of conversion of the prehormone into estrogen.

The clinical presentation is also complicated by the blurred distinction between the impact of aging and that of estrogen deficiency, a problem compounded further by the patient's vulnerability or resistance to the general physical deterioration inherent in the aging process. Finally, just as the physical reaction to the aging process is variable, so also is the psychological impact of the implications of the so-called change of life. In some women, accommodation to the pyschosocial and physical meaning of aging is simple and nontraumatic. Other women mistakenly assume their last menstrual period marks the onset of life characterized by increasing physical disability, conflicts with and estrangement from family, and dwindling sexual adequacy and interest.

Despite the inherent variability in the patient's presentation and personal perception of her situation, certain clinical categories are recognized: (1) disturbances of menstrual pattern, oligomenorrhea, and/or DUB; (2) vasomotor symptoms, involving heat, flush, and perspiration; (3) atrophy of specific systems, such as those that are directly estrogen dependent (i.e., secondary sexual characteristics) or estrogen modified (e.g., the skeletal system); and (4) psychological symptoms. These areas have been characterized according to functional impetus:

1. Those associated with *estrogen deprivation*, such as hot flashes, atrophic vaginitis, urethritis, and osteoporosis.
2. Those associated with relative *estrogen excess*, such as DUB, endometrial hyperplasia, and carcinoma.
3. Those associated with *estrogen replacement therapy*.

Symptoms and Signs of Reduced Endogenous Estrogen

Altered Menstrual Function

Oligomenorrhea followed by amenorrhea is usually the first clinical evidence of the female climacteric, even though fertility has declined since age 35 and many premenopausal women also note the transient presence of hot flushes. The diagnosis of permanent loss of menses requires sufficient follow-up time for retrospective confirmation. Usually 6–12 months or amenorrhea in a woman over 45 years of age is the commonly accepted rule of thumb for diagnosis of the menopause. Only rarely will vaginal bleeding reappear; when it does, organic pathology must be ruled out. Many patients will insist that pregnancy be ruled out and that confirmation of postmenopausal status (hypergonadotropinism) be obtained.

Vasomotor Symptoms

The vasomoter flush is viewed as the hallmark of the female climacteric (Molnar, 1975). The term *hot flush* is descriptive of a sudden onset of reddening of the skin over the head, neck, and chest, accompanied by a feeling of intense body heat, and concluded by sometimes profuse perspiration. The duration of a flush varies from a few seconds to several minutes. Its frequency may be rare to recurrent every 30 min. Finally, flushes appear to be more frequent and severe at night, when the women may be awakened from sleep, or during times of stress. Although the flush may occur in the premenopause, it lasts in most women for 1–2 years, but in some (as many as 25%) for longer than 5 years.

The physiology of the flush has been investigated (Sturdee *et al.*, 1978; Meldrum *et al*, 1980). It appears to coincide with a surge of LH, and not FSH, and is preceded by a subjective prodromal awareness that a flush is beginning. This aura is followed by measurable, increased heat over the entire body surface. Core temperature falls. In short, the flush is not a release of accumulated body heat but is, rather, a sudden inappropriate excitation of heat-release mechanisms. Its relationship to the LH surge and temperature change within the brain is not understood. The observation that flushes occur after hypophysectomy suggests that the mechanism is not triggered directly by release of LH. Estrogen therapy effectively decreases the frequency and severity of the subjective symptoms and objective signs of menopausal flushes.

With the fall in core temperature due to peripheral vasodilation and perspiration, one would expect the usual compensatory reactions to occur: subjective sensation of feeling "cold," addition of clothing, and shivering. That these are not seen adds credence to the central initiation and site of estrogen control of this phenomenon. The possibility that neurotransmitter controllers of GnRh (norepinephrine, dopamine, opioid peptides) that are also susceptible to estrogen modulation are involved in the physiology of the hot flush is under investigation.

Osteoporosis

Although the hot flush is the most common problem of the postmenopause, it presents no inherent health hazard. Osteoporosis, a change in bone structure characterized by a reduction in quantity rather than change in chemical composition results in mechanical fragility with subsequent compression and fracture. Bones get their strength from a structure of protein fibers combined with hard calcium phosphate crystals. A reduction of both bone protein and calcium results in osteoporosis. The osteoporotic disabilities sustained by castrate or postmenopausal white women include fractures of the vertebral body, humerus, upper femur, distal forearm, and ribs, back pain, and decreased height and mobility. Cross-sectional studies of white women in northern European countries reveal the following (Albright, 1940; Meema and Meema, 1965; Knowelden *et al.*, 1964; Meema *et al.*, 1976; Beals, 1972):

Spinal compression fracture: Symptomatic spinal osteoporosis causing pain, loss of height, postural deformities with consequent pulmonary, gastrointestinal, and bladder dysfunction is five times more common in white women than men. Approximately 25% of white women over 60 years of age display spinal compression fractures. The average nontreated postmenopausal white woman can expect to shrink $1\frac{1}{2}$ in.

Colles fracture: There is a tenfold increase in distal forearm fracture in white women (not men) as they progress from age 35 to 60 years.

Head of femur fracture: The incidence of hip fractures also increased with age in white women, rising from 0.3/1000 to 20/1000 from 45 to 80 years of age. The incidence is about ten times that of endometrial cancer of postmenopausal women. Eighty percent of all hip fractures are associated with osteoporosis. This fracture carries a heavy risk of morbidity and mortality. Between 15–20% of patients with hip fracture die due to the fracture or its complications (surgical, embolic, cardiopulmonary) within 3 months. In addition, the survivors are frequently severely disabled and are permanent invalids.

Loss of mineral content of bone happens in all aging individuals. However, white and oriental women start losing bone earlier and at a more rapid rate within 3–8 years after menopause or castration; not only a sexual but also a racial difference is displayed. The problem is particularly severe in thin, cigarette-smoking women castrated early in life or those with gonadal dysgenesis.

Unquestionably, exercise and diet have a beneficial effect on bone integrity. Equally impressive is the accumulating evidence not only that postmenopausal osteoporosis is related to estrogen loss but also that estrogen therapy has prophylactic value in retarding the process. The mechanism of this salutary action is thought to be an interference with the parathyroid hormone influence on bone resorption, requiring that calcium homeostasis be achieved in three ways (Nordin et al., 1980; Gallagher et al., 1980):

1. Increased intestinal absorption of dietary calcuim.
2. Reduced urinary loss due to increased renal tubular resorption of calcium.
3. Inhibition of bone resorption.

Estrogen also increases calcitonin levels. This potent inhibitor of bone resorption would also restrict calcium loss from bone during estrogen treatment.

Long-term follow-up of postmenopausal, castrated women reveals a significantly lower rate of bone mineral loss in those women receiving estrogen therapy (Meema et al., 1976). A pioneering double-blind prospective study showed that women receiving mestranol replacement also displayed marked reduction in the loss of mineral content of bone (Lindsay et al., 1976). In response to the pivotal question of whether estrogen therapy has an effect on bone turnover dynamics and reduces fracture risk, the answer is "yes" (Weiss et al., 1980; Hutchinson et al., 1979; Hammond et al., 1979; Johansson et al., 1975; Nachtigall et al., 1979). Users in every age group 50–75 can expect a 50%–60% reduction in the risk of fracture of the hip or arm. This decrease in fracture risk is identical with both 0.625 and 1.25 mg dosages of conjugated estrogens. However, this protection is maintained only while women continue taking estrogen.

It is now clear that if estrogen therapy is discontinued, bone loss resumes and the reduced mineral content of previously treated women rapidly reaches the low levels of nontreated controls (Lindsay et al., 1979). In other words, the same accelerated rate of bone loss occurs in the immediate years after stopping estrogen as seen in the first 3 postmenopausal years (Beals, 1972; Heaney *et al.*, 1978). Therefore, as a prophylaxis against osteoporosis, estrogen therapy must be started before the process has reached clinical significance, and it must be continued virtually indefinitely.

Although it has been estimated that at the present time, only 20% of castrate women and 30% of women experiencing natural menopause lose significant bone, it is not possible (aside from racial factors) to designate which individual is at risk for osteoporosis. Patients who already have osteoporosis should be screened for other conditions that lead to osteoporosis:

1. Serum calcium, phosphorous, and alkaline phosphatase—for primary hyperparathyroidism. Renal function tests—for secondary hyperparathyroidism with chronic renal failure.
2. Blood count and smear and sedimentation rate—for multiple myeloma.
3. Thyroid function tests—for thyrotoxicosis.

Unfortunately, estrogen therapy does not reverse the inroads of advanced osteoporosis, although further bone loss can be reduced.

The prevention of osteoporosis in the aging woman now constitiutes a major public health problem. By virtue of the large number of women living to their postmenopausal years, a significant reduction in the clinical manifestations of osteoporosis will have a very large impact on our health care system and our patients in terms of quality of life, mortality, and money saved.

An analysis of an aging woman's calcium needs indicates a greater requirement than previously appreciated (Heaney *et al.*, 1978). In order to remain in zero calcuim balance, women on estrogen replacement require approximately 1.0 g of calcium per day, while untreated women require 1.5 g/day. In other words, estrogen therapy alone may not avoid a negative calcium balance.

In a prospective study, the efficacy of supplementary calcium was studied on the rate of vertebral fractures with and without combinations of vitamin D, estrogen, and sodium fluoride (Riggs *et al.*, 1982). While calcium supplementation significantly reduced the number of fractures, it was even more effective when combined with estrogen (85% reduction), but the lowest fracture rate was achieved by combining calcium, estrogen, and sodium fluoride (90% reduction). Because fluoride is a potent stimulator of bone formation and estrogen inhibits bone resorption, it is not surprising that an additive effect results. The addition of Vitamin D to the therapeutic regimens did not significantly affect the fracture rate, but was associated with problems of hypercalcemia or hypercalciuria. However, a high rate (40%) of side affects is encountered with fluoride supplementation, including joint and tendon inflammation, anemia, and gastrointestinal disturbances. Therefore, the optimum regimen is a combination of estrogen and calcium.

Finally, the benefits of exercise are real. Physical activity, involving antigravity exercise, as little as 30 min a day for three days a week, will increase the bone mineral content in older women (Smith, 1982). The exercise need not be extreme. Walking and ordinary calisthenics will suffice.

Menopausal Syndrome

As noted, the major extragenital symptoms of the menopause are the hot flush and osteoporosis. There are additional problems encountered in the early postmenopause that are seen frequently, but their causal relation with estrogen is uncertain.

Called the menopausal syndrome, these problems include fatigue, nervousness, headaches, insomnia, depression, irritability, joint and muscle pain, dizziness, palpitations, and formication. Attempts to study the effects of estrogen on these problems have been hampered by the subjectivity of the complaints (high placebo reactions) and the "domino effect" to what reduction of hot flushes would do to frequency of insomnia or irritability. Using a double-blind cross-over prospective study format, Campbell and Whitehead (1977) have concluded that many symptomatic "improvements" ascribed to estrogen therapy result from relief of hot flushes—a domino effect. On the other hand, tonic effect—improvement in memory and reduction of anxiety—was also noted in these obsevations. The unreliability of claims regarding skin dryness or wrinkles was underscored. Fatigue, irritibility, headache, and depression are not thought to be estrogen-related phenomena.

Emotional stability during the perimenopausal period may be disrupted by poor sleep patterns. Estrogen therapy improves the quality of sleep time (Schiff *et al.*, 1979). Perhaps flushing may be insufficient to awaken a woman but sufficient to affect the quality of sleep and therefore diminish ability to handle the next day's problems and stresses.

Finally, there is a general clinical consensus that certain physical changes—redistribution of fat deposits, muscle tone loss, and loss of elastic tissue of the skin with wrinkling—are due to aging rather than estrogen-reduced or treatable conditions.

Cardiovascular Effects

The apparent sex differences between men and women with respect to coronary artery disease and stroke is a phenomenon characteristic only of affluent populations carrying a high burden of other risk factors: hyperlipidemia, diabetes, hypertension, and cigarette smoking. Inspection of the apparent sex difference in national mortality statistics shows that there is no abrupt change in the incidence of coronary mortality in women at the time of the menopause. The observed decrease in the ratio of male to female deaths is not due to an acceleration in female mortality but is due to a slowing in the rate in the male population with increasing age. Where the overall coronary mortality is low, sex differentials diminish. Nevertheless, the question of whether cessation of ovarian function is associated with an increased incidence of heart disease persists with vexing uncertainty.

Several factors account for this inconclusive status. Most important are the results of the Framingham study reporting data derived from biannual examination of 3000 women over a 24-year span (Gordon *et al.*, 1978). These data reveal a marked increase in incidence of heart disease at the menopause regardless of age. The increase is substantial and abrupt and continues only slowly thereafter. Estrogen therapy did not reduce morbidity or mortality in this population. These data, however, are derived from a time when estrogen therapy was uncommon, and it is likely that dosage was relatively high. In women who are castrated prior to the menopause, studies have indicated that premature loss of estrogen is associated with a higher incidence of coronary heart disease (Oliver, 1976). In addition, recent case–control studies have noted a general reduction in cardiovascular disease with estrogen replacement therapy (Ross *et al.*, 1981; Dannel, 1982).

An excellent study utilizing a very large retirement community has demonstrated significant protection against death by ischemic heart disease with estrogen therapy (Ross *et al.*, 1981). This protection was observed with both the 0.625 mg and 1.25 mg dosages of conjugated estrogens, and it was effective even in cigarette smokers, changing a high risk to a risk less than that of controls. This protective relationship coincides with a decrease in mortality rates from ischemic heart disease among American white women during the time of increased estrogen use in recent years. This decrease in women is greater than the decrease in men over the same period of time, despite a greater decrease in cigarette smoking in men compared to women (Ross *et al.*, 1980). The mechanisms of this protection may be related to the decrease in levels of serum levels of serum LDL cholesterol and the increase in HDL cholesterol seen with low doses of estrogen (Wallace *et al.*, 1979). Particularly in older age groups, an inverse relationship exits between HDL cholesterol and coronary mortality (Yaari *et al.*, 1981).

In the United States, the death rate from ischemic heart disease is over four times the combined death rates of breast cancer and endometrial cancer. The protective effect of replacement estrogen, if real, would be a significant benefit of therapy.

Finally, when estrogen replacement therapy is administered in appropriate dosage to postmenopausal women, there is no increase observed in stroke, thromboembolism, or myocardial infarction (Pfeffer *et al.*, 1978; Pfeffer *et al.*, 1979; Rosenberg *et al.*, 1976; Pfeffer and Van Den Noort, 1976). These data reinforce the dose–response relationship between estrogen and thrombosis, and it is not surprising that relatively physiologic doses of estrogen are free of thrombotic side effects.

Atrophic Changes

With extremely low estrogen production in the late postmenopausal stage or many years after castration, atrophy of all mucosal surfaces, notably that of the vagina, takes place, accompanied by vaginitis, pruritus, dyspareunia, and stenosis. Genitourinary atrophy leads to a variety of symptoms that affect the ease and quality of living. Urethritis with dysuria, urgency incontinence, and urinary frequency are further examples of mucosal thinning, in this instance, of the urethra. Vaginal relaxation with cystocele, rectocele, and uterine prolapse and vulvar dystrophies are not a consequence of estrogen deprivation.

Unless dermatologic conditions exist masquerading as menopausal atrophy, estrogen replacement is invariably successful in reversing these atrophic problems. Relief from these problems often results in significant improvements in general well-being. Dyspareunia seldom brings older women to our offices. A basic reluctance to discuss sexual behavior still permeates our society, even among older patients and physicians. Gentle questioning may lead to estrogen treatment of atrophy and prolonged sexual enjoyment.

Problems of Excess Estrogen

Not all climacteric women experience symptoms or signs of estrogen deprivation. Most are asymptomatic; some actually manifest estrogen excess via the presence of uterine bleeding—dysfunctional uterine bleeding.

Throughout the usual period of life identified with perimenopause (40–60), there is a significant incidence of dysfunctional uterine bleeding. Although the greatest concern provoked by this symptom is endometrial neoplasia, the usual finding is nonneoplastic tissue displaying estrogen effects unopposed by progesterone. This results from anovulation in the premenopausal woman and from extragonadal endogenous estrogen production or estrogen administration in the postmenopausal woman. There are four mechanisms that could result in increased endogenous levels:

1. Increased precursor androgen (functional and nonfunctional endocrine tumors, liver disease, stress).
2. Increased aromatization (obesity, hyperthyroidism, liver disease).
3. Increased direct secretion of estrogen (ovarian tumors).
4. Decreased levels of sex-binding globulin (SBG) leading to increased levels of free estrogen.

In all women, whether premenopausal or postmenopausal, whether on or off hormone therapy, specific organic causes (intrauterine tumor, carcinoma, complications of unexpected pregnancy, or bleeding from extrauterine sites) must be ruled out. In addition to careful history and physical examination, dysfunctional uterine bleeding beyond the age of 35 is evaluated by multiple-specimen endometrial aspiration biopsy. In the absence of organic disease, appropriate management is dependent upon the age of the patient and endometrial tissue findings. In the perimenopausal woman with dysfunctional uterine bleeding associated with proliferative or hyperplastic (uncomplicated by atypical or dysplastic constituents) endometrium, periodic oral progestin therapy is mandatory; 10 mg Provera is given daily the first 10 days of each month. Follow-up aspiration biopsy is required, and if progestin is ineffective and histologic regression is not observed, formal dilatation and curettage is an essential preliminary to alternate therapeutic surgical choices.

When monthly progestin therapy reverses hyperplastic changes and controls irregular bleeding, treatment should be continued until withdrawal bleeding ceases. This is a reliable sign (in effect, a bioassay) indicating the onset of estrogen deprivation and the need for the addition of estrogen. If vasomotor disturbances begin before the cessation of menstrual bleeding, the combined estrogen–progestin program may be initiated as needed to control the flushes.

Problems of Estrogen Therapy

General and Metabolic

In the foregoing paragraphs, it was apparent that substantial benefits may accrue if estrogen deprivation conditions are treated with estrogen replacement. But, as in all clinical situations, benefits of therapy must be balanced by evaluating the liabilities of adverse pharmacological effects of the hormone. Considerable information of this type has accumulated in relation to the use of oral contraceptives. The increased incidence of thromboembolic disease, hypertension, and altered carbohydrate metabolism during oral contraceptive usage is well documented and considered attributable to the estrogen component of the pill. Probably as a result of the lower potency of the lower dosage of estrogen, these metabolic effects are not seen in postmenopausal replacement therapy. Postmenopausal women on estrogen therapy are not at risk for myocardial infarction, idiopathic thromboembolism, or breast tumors (Rosenberg et al., 1976; Pfeffer and Van Den Noort, 1976). As in contraception, however, estrogen replacement therapy did not carry an increased incidence of gall bladder disease, and careful follow-up monitoring of the appearance of symptoms and signs of biliary tract disease is necessary.

Obviously, the general area of metabolic effects cannot be dismissed; high-risk patients need special attention when a decision in favor of estrogen treatment is undertaken. ''Metabolic'' contraindications to estrogen replacement therapy include acute liver disease or chronically impaired liver function, acute vascular thrombosis (with or without emboli), and neurophthalmic vascular disease. Relative contraindications (estrogens may have adverse effects on some patients) include seizure disorders, hypertension, familial hyperlipidemias, and migrainous headaches.

Breast Cancer

The possibility that estrogen use increases the risk of breast cancer must be intensively scrutinized. The epidemiologic data on the scope of human female breast cancer are astonishing: one of every 15 women will develop breast cancer in her lifetime; it is the leading type of cancer in women (25%) and the leading cause of cancer death in women (20%), about ten times the number of deaths from endometrial cancer. The mortality rate from breast cancer of 23:100,000 female population has not changed in 45 years.

Evidence for the involvement of endogenous estrogen in human breast cancer is persuasive (even if the mechanism is unknown):

1. The condition is 100 times more frequent in women than in men.
2. Breast cancer invariably occurs after puberty.
3. Gonadal dysgenesis and breast cancer are mutually exclusive.
4. Breast cancer frequently contains estrogen and progestin receptors, and these respond beneficially to endocrine ablative therapy.
5. Breast cancer risks decrease by 70% if castration occurs prior to the age of 35 years.
6. There is increased risk if first pregnancy is delayed beyond 30 years of age.

Although the relationship between endogenous estrogen and breast cancer does not require or imply a cause-and-effect carcinogenic action of the estrogen, it does suggest a permissive or supportive role of the hormone in the disease process. What of exogenous estrogen hormone and its effects on breast cancer incidence? There are several retrospective case–control studies (Hammond *et al.*, 1979; Craig *et al.*, 1974; Casagrande *et al.*, 1976; Sartwell *et al.*, 1977; Wynder *et al.*, 1978; Brinton *et al.*, 1979) that have shown no significant association between estrogen use and breast cancer risk. For a full critical analysis of all these studies, the reader is referred to Horwitz and Feinstein (1980) and Hertz (1979).

A practice-based (not statistically significant) follow-up study showed slightly increased rates in high-dose estrogen users after approximately 10–15 years (Hoover *et al.*, 1976). In two case–control studies, a slightly increased risk for developing breast cancer was noted if there was a high dose or prolonged exposure to estrogen (Ross *et al.*, 1980; Jick *et al.*, 1980). At particular risk were patients who had preexisting benign breast disease or who developed surgically proven benign breast disease during estrogen replacement therapy.

The incidence of breast cancer in general has not risen in the United States despite a massive increase in the use of estrogen during the 1960s and early 1970s. However, the latency period before breast cancer may be clinically manifested, prolonged, and greater than 10–20 years. Some reassurance has come from the 1977–1979 Connecticut study, showing no increased risk to estrogen users in a hospital-based study.

In a long and large prospective long-term follow-up study, there has been no evidence that estrogen therapy increases the risk of breast cancer. Even more impressive, as the study approached 15 years of follow-up, the incidence of breast cancer in women using an estrogen–progestin combination program was lower than the incidence in women using estrogen alone (Gambrell et al., 1980a,b). It is logical to expect that the same mechanism of estrogen receptor depletion should operate in both the endometrium and the breast and the protection against normal mitotic activity should exist in both target tissues.

Despite the acknowledged deficiencies and uncertainties in the accumulated epidemiologic data, prudent management of estrogen replacement therapy should include these elements with respect to the potential breast cancer problem:

1. Routine replacement dosage should not exceed 0.625 mg of conjugated estrogens (or its potency equivalent).
2. The development of surgically proven benign breast disease on the therapy warrants consideration of avoidance of further estrogen.
3. A history of previous benign breast disease may be a factor in avoiding estrogen therapy.
4. In view of the significant potential offered by periodic exposure to progestin, a combined program of estrogen–progestin should be utilized for all patients, including those women who have undergone hysterectomy.
5. However, it would be imprudent to assume that the "protective" effect of progestin, as seen in endometrium (see below), extends to the breast. In this respect, women at intrinsically high epidemiologic risk for breast cancer should not be given estrogen in the belief that progestin will eliminate their risk.

Endometrial Cancer

Within the last few years, the relationship between estrogen replacement therapy and the risk of recipients developing endometrial hyperplasia and carcinoma has been revealed (Smith et al., 1975; Ziel and Finkle, 1975; Mack et al., 1976), emphasized (Antunes et al., 1979), evaluated (Horwitz and Feinstein, 1978), and given clinical perspective (Smith et al., 1975; Cramer and Knapp, 1979). As a result of this scrutiny, extensive modification of diagnostic and treatment regimens were proposed, and preliminary evidence for their beneficial impact demonstrated (Gambrell *et al.*, 1979; Campbell *et al.*, 1978; Patterson *et al.*, 1980). Clinicians have witnessed similar rapid swings of the informational pendulum before. They have learned to "hold" decisions on management until time has passed and all the facts were in. Nevertheless, the swiftness of the evolution and outcome of the medical dialectic operating in the estrogen therapy–endometrial carcinoma issue was impressive. The catalysts compelling the solution were the prevalence of estrogen use, the pressures for informed consent, and the need to rationalize the polarizing elements of high utility vs. presumed high risks and, fortunately, the ease of diagnostic access to the endometrial cavity. The outcome?—low dose, cyclic, sequential estrogen and progestin therapy and periodic diagnostic biopsy.

Estrogen normally promotes mitotic growth of the endometrium. Abnormal progression of growth through cystic hyperplasia, adenomatous hyperplasia, atypia, and early carcinoma have been associated with unopposed estrogen activity. Some 10% of women with adenomatous hyperplasia progress to frank cancer, and adenomatous hyperplasia is found adjacent to adenocarcinoma of 40% of cases. Although a majority of cancers are found associated with inactive or atrophic endometrium appropriate for the age and hormonal condition of the patient, the morphologic progression (reflected in architectural disarray and cellular cytonuclear abnormalities) from hyperplasia to carcinoma *in situ* represents a disease continuum catalyzed by unopposed estrogen stimulation. The practicing gynecologist is concerned about such outcomes; hysterectomy is legitimate treatment for endometrial adenomatous hyperplasia in the perimenopausal woman.

Attention has focused on the relationships of this histologic progression to the use of estrogen replacement therapy. Retrospective studies have estimated that the risk of endometrial cancer in women on estrogen replacement therapy is increased by a factor of four to eight. In America, the actual incidence of endometrial cancer would therefore increase from one per 1000 postmenopausal women per year to four to eight per 1000 per year. The risk appears to increase with duration of exposure and dose of estrogen and decreases after estrogen is stopped. When adjusted for the recognized risk of obesity, hypertension, and diabetes, the increased risk with estrogen matched the risk associated with these conditions (estrogen risk was not achieved).

Estrogen replacement therapy based on the amount of sales of estrogen in the United States more than doubled from 1965 to 1974, and this had been reflected in a higher incidence of endometrial cancer in various parts of the country. The increase, however, is confined to early, highly localized disease (Hulka, 1980). While incidence is increasing, mortality from endometrial cancer is decreasing in study areas. Does estrogen increase early diagnosis of otherwise silent disease (selection bias)? Are pathologists "overdiagnosing" adenomatous hyperplasia (oberservation bias or misclassification)? Are women at intrinsically higher risk also being treated with estrogen (confounding factor bias)?

The data on this issue do not need to be definitive or conclusive. For the present time, it is clear that estrogen is associated with an increased risk of endometrial cancer, even though several biases may have falsely increased the magnitude of the association and risk is related to dosage, duration, and continuous as opposed to cyclic therapy, and may be reduced by the addition of progestin therapy.

Whereas estrogen promotes the growth of endometrium, progestins inhibit that growth. This countereffect is accomplished by progestin reduction in cytoplasic receptors for estrogens (by diminished replenishment) *and* by induction of a target cell enzyme (17-Ol-dehydrogenase) that converts estradiol to E_1. As a result, the number of estrogen receptor complexes that are translocated and retained in the endometrial nucleus is decreased, as is the overall intracellular availability of the powerful estradiol.

Reports of the clinical impact of additional progestin in sequence with estrogen replacement therapy include the reversal of hyperplasia with a diminished incidence of endometrial cancer (Thom *et al.*, 1979; British Gynaecological Cancer Group, 1981). Not all studies support this salutary effect: clearly, the dose and duration of progestin are important. Recent information indicates that a 10-day course of progestin achieves a maximal antiestrogenic effect (Whitehead *et al.*, 1981; Gambrell *et al.*, 1979; Gambrell *et al.*, 1980b).

Is there a "high-risk" group for endometrial cancer? Clearly, it is wise to avoid estrogen therapy in obese individuals, those with high endogenous estrogen levels, or those with other constitutional features that predispose them for the disease. If there is any doubt, a pretherapy biopsy to document endometrial activity should be performed. Treatment of all dysfunctional uterine bleeding of perimenopause with progestins only (after biopsy) is a necessary added precaution. When on estrogen replacement therapy, the cost effectiveness of periodic endometrial biopsy must be considered. Many patients and physicians opt for biopsy every 1–3 years to ensure endometrial stability, regardless of the high overall cost and relatively low case identification rates. In this high-risk patient, a combined estrogen–progestin program would be safer than leaving the patient to prolonged and uninterrupted endogenous estrogen exposure.

Postmenopausal Estrogen Replacement Therapy

In view of the above considerations, a clinical consensus has evolved: there is little question that women who suffer from hot flushes or atrophy of reproductive tissues can and should be relieved of their problems by use of estrogens. It is also now definite that the long-term disabilities of osteoporosis can be ameliorated by therapy with estrogen. The protective effects afforded by estrogen must be weighed against the increased incidence of cancer that may be associated with hormone use. Treatment with estrogen for women showing any stigma of hormone deprivation and hormonal prophylaxis against osteoporosis appear to be prudent clinical doctrine. The lowest dose of estrogen that reverses the deficiency should be used and monthly addition of a progestin is mandatory. In practice, therapy is not offered to those patients in whom estrogen is specifically contraindicated (estrogen-dependent tumors, impaired liver metabolism, and, in a matter of clinical judgement, patients with thromboembolic problems or conditions predisposing to thromboembolism). The recommendation that replacement therapy be give for the shortest period of time appears short-sighted in view of the impressive evidence that therapy has a profound impact on osteoporosis and that there are more beneficial than harmful effects.

Patients under the Age of 40 (Castrates and Patients with Gonadal Dygensesis)

In these women, the duration of estrogen deprivation is prolonged and the loss of estrogen acute. The cyclic use of estrogen is recommended for short-term reduction of vasomotor symptoms and for long-term prophylaxis against osteoporosis and target organ atrophy. In many young patients, 0.625 mg of conjugated estrogens is insufficient to allow menstrual bleeding because women of this age ordinarily are exposed to estrogen levels that stimulate endometrial growth and withdrawal bleeding, and, for psychological reasons, a dose of 1.25 mg conjugated estrogen or its equivalent is utilized until the menopause time of life. In those patients castrated because of endometriosis, recurrence of endometriosis has not been a problem with this regimen.

Perimenopausal Dysfunctional Uterine Bleeding

After exclusion of other gynecologic causes, dysfunctional bleeding is treated by progestin therapy and biopsy surveillance. Vasomotor reactions appearing in women despite the presence of menstrual bleeding (presumably the flushes are due to a relative decrease in estrogen) can be treated by the usual estrogen–progestin regimen.

The Early Postmenopause

Progestin therapy is administered periodically (every month) until withdrawal bleeding does not occur. If vasomotor reactions begin, however, estrogen therapy is begun along with sequential progestin. Because there is no clinically useful diagnostic test to isolate those patients at risk for osteoporosis, the long-term postmenopausal use of hormone therapy depends heavily on the patient's own informed assessment of the special problems that this prospect represents. Should therapy be accepted, the sequential estrogen–progestin replacement program is initiated. At the low doses of estrogen recommended for replacement, increased growth of uterine fibroids, endometriosis, or breast reactions are rarely a concern.

As a result of immediate responses in early climacteric symptoms, the patient enters the climacteric more confident of herself emotionally, sexually, and physically. In our view, this establishes or cements good patient–physician interchange and relations. The follow-up of the patient of effective estrogen replacement is more secure and certain. The practitioner offering estrogen replacement has a better and more reliable opportunity to act as a primary physician for these aging women. All monitoring of health systems will be improved as a result of this single involvement. Bowel, breast, cardiac, and various metabolic functions are scrutinized periodically as consistent with good health practice.

The Late Menopause

The onset of atrophic conditions can be effectively treated with local or oral therapy in low maintenance doses. It is beneficial to administer the replacement program to women who already have osteoporosis and who have not previously taken estrogen? If there is no apparent bases for the osteoporosis other than aging and ovarian failure, estrogen therapy and calcium supplementation are advisable. Further loss of bone mass may be prevented or slowed. In these older women, a higher dose of estrogen (1.25 mg) may be necessary. Assessment or impact of progress can be obtained by measuring bone density. Sensitive methods are now available, either photon absorptiometry measured at the lower radius, or limited CT scanning of the spine.

Which Drug Should Be Used?

There currently is no evidence that one form of estrogen is superior to another. The specific estrogen does not appear to be as important as the duration, dose, and continuity of exposure in the absence of progesterone. Which estrogen is administered is not as significant as the method with which it is used.

It is difficult to know the equivalent potency within the human body of the various estrogen preparations. Ethinyl estradiol in a dose of 5 μg is equivalent to 0.625 mg conjugated estrogens as measured by a variety of responses, including sex-binding globulin, urinary calcium/creatinine ratios, gonadotropins, thyroxine-binding globulin, and vaginal cytology. However, any estrogen properly monitored and administered is acceptable.

How to Treat?

Estrogens are administered on a cyclic basis, usually from the 1st through the 25th of each month, as a convenient aid to remembering the routine. For the last 10 days of estrogen administration, a daily dose of 10 mg of medroxy-progesterone acetate (Provera) is added. Some patients develop unwanted reactions to this dose of progestin, such as weight gain, edema, or depression. Other progestins may be given and in lower doses: norethindrone 2.5 mg, norgestrel 150 µg (Whitehead et al., 1981). The lower equivalent doses of medroxyprogesterone acetate have not been established by clinical studies, but it is probably 2.5–5 mg. A possible disadvantage of the 19-nortestosterone progestins is an adverse effect on HDL cholestrol (a decrease) reversing the beneficial effect of the estrogen (Hirvonen et al., 1981). For this reason, medroxyprogesterone acetate is our drug of choice.

In the absence of a uterus, some practitioners do not see the need to use cyclic administration or a progestin. However, in view of a possible impact on the breast, it seems best to adhere to a cyclic schedule including the terminal use of a progestin, for the reasons we have already noted. There are no clinical studies to guide us in dealing with a woman who is symptomatic during the days off medication. Our clinical judgement has led us to combine daily estrogen (every day through the month) with progestin daily for the first 14 days of each month. The use of progestin for up to 10 days each month will not provide complete protection when estrogen is given continuously (Hirvonen et al., 1981).

The dose of estrogen utilized is that which will provide sufficient estrogen to sustain physiologic functions, yet short of provoking a return of menstrual flow. An important principle of treatment is that relief of symptoms can usually be achieved by subbleeding doses of estrogen. For early climacteric where there is still considerable endogenous estrogen present, the usual effective dose is 0.625 mg conjugated estrogens per day. In late climacteric where endogenous estrogen may be very low at best, a higher dose, 1.25 mg, may be necessary. Even at these doses there are inconsistent effects of various target tissues: a pharmacologic effect on the liver and a subphysiologic effect on the vagina (Geola et al., 1980). These doses do not cause significant elevation of blood pressure except for a rare idiosyncratic reaction. A dose of 0.625 mg is adequate for prophylaxis against osteoporosis. For optimal protection against osteoporosis, the hormonal regimen is supplemented with 500 mg calcium daily (one Os-Cal 500 tablet). If withdrawal bleeding occurs at the 0.625 mg dose, the estrogen dose is reduced to 0.3 mg, to be increased in later years as the requirement also increases.

When estrogen is contraindicated, Depo Provera (150 mg ever 3 months) is effective in relieving vasomotor symptoms. Oral Provers (10-20 mg daily) may be utilized. The effect on calcium excretion is less than that seen with estrogen although progestins do exert some inhibition of bone loss (Lindsay *et al.*, 1978). In addition, progestin will not improve vaginal atrophy. When estrogen administration is not possible, calcium should be supplemented at the rate of 1.0 gm per day (two Os-Cal 500 tablets).

Clonidine was initially found to reduce the frequency of flushes, but when studied in a double-blind cross-over study, it was ineffective (Laufer *et al.*, 1982). Propranolol is also ineffective (Cooper *et al.*, 1978).

Estrogens are absorbed efficiently from the vagina and high blood levels are achieved with the usual recommended doses. Until dose–response studies are available indicating the physiologic range for vaginal, parenteral, and other types of administration (e.g., transdermal), the oral program must be followed in the interests of patient safety.

A striking and consistent finding in most studies dealing with menopause and estrogen replacement is a marked placebo response in a variety of symptoms including flushing. A significant clinical problem encountered in our referral practice is the following: occasionally a woman will undergo an apparent beneficial response to estrogen, only to have the response wear off in several months. This leads to a sequence of periodic visits to the physician and ever-increasing doses of estrogen therapy. When a patient reaches a point of requiring large doses of estrogen (2.5 mg congugated estrogens or more), a careful enquiry must be undertaken for a basic psychoneurotic problem.

Hormonal treatment for decreased libido should be discouraged in that psychosocial reasons are usually the blame. However, we have found that occasionally the addition of androgen (methyltestosterone, up to 5 mg daily) in addition to the estrogen, may provide an increased sense of well-being, along with an increase in libido. The patient should be cautioned that hirsutism may develop.

We find no need to monitor dosage by any means other than symptoms and bleeding; assessing vaginal cytology is not useful.

When to Biopsy?

Aspiration endometrial biopsy is recommended prior to instituting therapy in the perimenopausal period. This aggressive approach will identify those cases of hyperplasia already present. If hyperplasia is encountered, therapy is initiated and rebiopsy is performed 4 months later. This approach will isolate those patients with severe atypical changes resistant to progestin, allowing a clear-cut decision in favor of surgical extirpation. The appearance of unscheduled, irregular, breakthrough bleeding demands biopsy at any time or at any age during postmenopausal estrogen therapy. In the absence of abnormal bleeding, sampling of the endometrium every 2–3 years is recommended by many authorities, but individualization is certainly in order, and a certain amount of trust in the protection effected by the progestin is justified.

The cost effectiveness of routine perimenopausal biopsies can be argued. It has been estimated that 30,000 biopsies are necessary to detect an invasive lesion in an asymptomatic woman. A reasonable economic moderation would be to limit pretreatment biopsies to patients at higher risk for endometrial changes, those patients with conditions associated with chronic estrogen exposure: obesity, dysfunctional uterine bleeding, anovulatory infertility, hirsutism, high alcohol intake, hepatic disease, metabolic problems such as diabetes mellitus and hypothyroidism, and family history of endometrial cancer and breast cancer.

References

Aksel, S., Schomberg, D. W., Tyrey, L., and Hammond, C. B., 1976, Vasomotor symptoms, serum estrogens and gonadotropin levels in surgical menopause, *Am. J. Obstet. Gynecol.* **126**:165.

Albright, F., 1940, Postmenopausal osteoporosis, *Trans. Assoc. Am. Physicians* **55**:298.

Antunes, C. M. F., Strolley, P. D., Rosenshein, N. B., *et al.*, 1979, Endometrial cancer and estrogen use. Report of a large case-control study, *N. Engl. J. Med.* **300**:9.

Beals, R. K., 1972, Survival following hip fracture: Long term follow up of 607 patients, *J. Chron. Dis.* **25**:235.

Brinton, L. A., Williams, R. R., Hoover, R. N., Stegens, N. L., Feinleib, M., and Fraumeni, F. J., Jr., 1979, Breast cancer risk factors among screening program participants, *J. Natl. Cancer Inst.* **62**:37.

British Gynaecological Cancer Group, 1981, Oestrogen replacement and endometrial cancer, *Lancet* **1**:1359.

Campbell, S., Whitehead, M., 1977, Estrogen therapy and the menopausal syndrome, *Clin. Obstet. Grnecol.* **4**:31.

Campbell, S., McQueen, J., Menardi, J., and Whitehead, M. I., 1978, The modifying effect of progestogen and the response fo the postmenopausal endometrium to exogenous estrogen, *Postgrad. Med. J.* **54**:59.

Casagrande, J., Gerkins, V., Henderson B. E., Mack, T., and Pike, M. C., 1976, Estrogens and breast cancer in women with antural menopause (Brief communication), *J. Natl. Cancer Inst.* **56**:839.

Cooper, J., Williams, S. T., and Patterson, J. S., 1978, Effectiveness of propanolol in hot flashes, *Br. J. Obstet. Gynaecol.* **85**:472.

Craig, T. J., Comstock, G. W., and Geiser, P. B., 1974, Epidemiologic comparison of breast cancer patients with early and late onset of malignancy and general population control, *J. Natl. Cancer Inst.* **53**:1577.

Cramer, D. W., and Knapp, N. C., 1979, Review of epidemiologic studies of endometrial cancer and exogenous estrogen, *Obstet. Gynecol.* **54**:521.

Gallagher, J. C., Riggs, B. L., and DeLuca, H. F., 1980, Effect of estrogen on calcium absorption and serum vitamin D metabolites in postmenopausal osteoporosis, *J. Clin. Endocrinol. Metab.* **51**:1359.

Gambrell, R. D., Jr., Massey, F. M., Castaneda, T. A., and Boddie, A. W., 1980a, Estrogen therapy and breast cancer in postmenopausal women, *J. Am. Geriatr. Soc.* **28**:251.

Gambrell, R. D., Jr., Massey, F. M., Castaneda, T. A., Ugenas, A. J., Ricci, C. H., and Wright, J., 1980b, Use of the progestogen challenge test to reduce the risk of endometrial cancer, *Obstet. Gynecol. Surv.* **55**:503.

Gambrell, R. D., Jr., Massey, F. M., Castaneda, T. A., Ugenas, A. J., and Ricci, C. H., 1979, Reduced incidence of endometrial cancer among postmenopausal women treated with progesterones, *J. Am. Geriatr. Soc.* **27**:389.

Geola, F. L., Frumar, A. M., Tartaryn, I. V., Lu, K. H., Hersham, J. M., Eggena, P., Samhli, M. P., and Judd, H. L., 1980, Biologic effects of various doess of conjugated estrogens in postmenopausal women, *J. Clin. Endocrinol. Metab.* **51**:620.

Gordon, T., Kannel, W. B., Hjortland, M. C., and McNamara, P. M., 1978, Menopause and coronary heart disease, *Ann. Int. Med.* **89**:157.

Hammond, C. B., Jelovsek, F. R., Lee, K. L., Creaseman, W. T., and Parker, R. T., 1979, Effects of long-term estrogen replacement therapy. I. Metabolic effects, *Am. J. Obstet. Gynecol.* **133**:525.

Hammond, C. B., Jelovsek, F. R., Lee, K. L., Creaseman, W. T., and Parker, R. T., 1979, Effects of long-term estrogen replacement therapy. II. Neoplasia *Am. J. Obstet. Gynecol.* **133**:537.

Heaney, R. P., Recker, R. R., and Saville, P. D., 1978, Menopausal changes in calcium balance performance, *Lab. Clin. Med.* **92**:953.

Hemsell, D. L., Grodin, J. M., Brenner, P. F., Siiteri, P. K., and MacDonald, P. C., 1976, Plasma precursors of estrogen. II. Correlation of the extent of conversion of plasma and rostenedione to estrogen with age, *J. Clin. Endocrinol. Metab.* **38**:476.

Hertz, R., 1979, The steriod cancer hypothesis, *J. Steroid Biochem.* **11**:435.

Hirvonen, E., Malkonen, M., and Manninen, V., 1981, Effects of different progestogens on lipoproteins during postmenopausal replacement therapy, *N. Engl. J. Med.* **304**:560.

Hoover, R., Gray, L. A., Sr., Cole, P., and MacMahon, B., 1976, Menopausal estrogens and breast cancer, *N. Engl. J. Med.* **295**:401.

Horwitz, R. I., and Feinstein, A. R., 1978, Alternative analytic methods for case control studies of estrogens and endometrial cancer, *N. Engl. J. Med.* **299**:1089.

Horwitz, R. I., and Feinstein, A. R., 1980, The clinical epidemiology of breast and uterine cancer. In *The Menopause and Postmenopause* (N. Pasello, U. R. Paoli, and J. L. Ambrus, eds.), MTD Press, Ltd. Lancaster.

Horsman, A., Nordin, B. E. C., and Crilly, R. G., 1979, Effect on bone of withdrawal or oestrogen therapy, *Lancet* **2**:33.

Hulka, B. S., 1980, Effect of exogenous estrogen on postmenopausal women: the epidemiological evidence, *Obstet. Gynecol. Surv.* **35**:389.

Hutchinson, T. A., Polansky, S. M., and Feinstein, A. R., 1979, Postmenopausal estrogens protect against fractures of hip and distal radius. A case control study, *Lancet* **2**:705.

Jick, H., Walker, A. M., Watkins, R. N., D'Ewart, D. C., Hunter, J. R., Danford, A., Madsen, S., Dinan, B. J., and Rothman, K. J., 1980, Replacement estrogens and breast cancer, *Am. J. Epidemiol.* **112**:586.

Johansson, B. W., Kaij, L., Kullander, S., Lenner, H. C., Svanberg, L., and Asted, B., 1975, On some late effects of bilateral oophorectomy in the age range 15-30 years, *Acta Obstet. Gynecol. Scand.* **54**:449.

Judd, H. L., 1976, Hormonal dynamics associated with the menopause, *Clin. Obstet. Gynecol.* **19**:775.

Kannel, W. B., 1982, Meaning of the downward trend in cardiovascular mortality, *JAMA* **247**:877.

Kelsey, J. L., and LaVolsi, V., 1981, Estrogen replacement therapy and breast cancer incidence, *J. Natl. Cancer Inst.* **67**:327.

Knowelden, J., Buhra J., and Dunbar, O., 1964, Incidence of fractures in persons over 35 years of age, *Br. J. Prev. Soc. Med.* **18**:130.

Laufer, L. R., Erlik, Y., Meldrum, D. R., and Judd, H. L., 1982, Effect of clonidine on hot flashes in postmenopausal women, *Obstet. Gynecol.* **60:**583.

Lindsay, R., Aitken, J. M., Anderson, J. B., Hart, D. M., MacDonald, E. B., and Clarke, A. C., 1976, Long-term prevention of postmenopausal osteoporosis by oestrogen, *Lancet* **1:**1038.

Lindsay, R., Hart, D. M., Purdie, D., Ferduson, M. M., Clark, A. S., and Kraszeweki, A., 1978, Comparative effects of oestrogen and a progestogen on bone loss in postmenopausal women, *Clin. Sci. Mol. Metab.* **54:**193.

Lindsay, R., MacLean, A., Kroszewski, A., Clark, A. C., and Garwood, J., 1979, Bone oestrogen therapy, *Lancet* **2:**33.

Mack, T. M., Pike, M. C., Henderson, B. E., Pfeffer, R. I., Gerkins, V. R., Arthur, M., and Brown, S. E., 1976, Estrogens and endometrial cancer in a retirement community, *N. Engl. J. Med.* **294:**1262.

Meema, S., and Meema, H. E., 1965, Loss of compact bone due to menopause, *Obstet. Gynecol.* **26:**33.

Meema, S., Bunker, M. L., and Meema, H. E., 1976, Preventive effect of estrogen on postmenopausal bone loss, *Arch. Int. Med.* **135:**1436.

Meldrum, D. R., Tataryn, I. V., Frumer, E., Erlik, Y., Lu, K. H., and Judd, H. L., 1980, Gonadotropins, estrogens, and adrenal steroids during the menopausal hot flush, *J. Clin. Endocrinol. Metab.* **50:**685.

Molnar, G. W. L., 1975, Body temperature during menopausal hot flashes, *J. Appl. Physiol.* **38:**499.

Nachtigall, L. E., Nachtigall, R. H., Nachtigall, R. D., and Beckman, E. M., 1979, Estrogen replacement therapy I: A 10-year prospective study in the relationship of osteoporosis, *Obstet. Gynecol.* **53:**277.

Nisker, J. A., Hammond, G. L., Davidson, B. J., Frumar, A. M., Takaki, N. K., Judd, H. L., and Siiteri, P. K., 1980, Serum sex hormone-binding globulin capacity and the percentage of free estradiol in postmenopausal women with and without endometrial carcinoma, *Am. J. Obstet. Gynecol.* **138:**637.

Nordin, B. E. C., Horsman, A., Crilly, R. G., Marshall, R. G., and Simpson, M., 1980, Treatment of spinal osteoporosis in postmenopausal women, *Br. Med. J.* **280:**453.

Oliver, M. F., 1976, The menopause and coronary heart disease, in: *The Management of the Menopause and Postmenopausal Years* (S. Campbell, ed.), University Press Park, Baltimore, pp. 175–184.

Patterson, M. E. L., Wade-Evans, S. T., Surdee, D. W., Thom, M. H., and Studd, J. W. W., 1980, Endometrial disease after treatment with oestrogens and progestogens in the climacteric, *Br. Med. J.* **1:**822.

Pfeffer, R. I., and Van Den Noort, S., 1976, Estrogen use and stroke risk in postmenopausal women, *Am. J. Epidemiol.* **130:**455.

Pfeffer, R. I., Whipple, G. H., Kurosaki, T. T., and Chapman, J. M., 1978, Coronary risk and estrogen use in postmenopausal women, *Am. J. Epidemiol.* **107:**479.

Pfeffer, R. I., Kurosaki, T. T., and Charlton, S. K., 1979, Estrogen use and blood pressure in later life, *Am. J. Epidemiol.* **110:**469.

Riggs, B. L., Seeman, E., Hodgson, S. F., Taves, D. R., and O'Fallon, W. M., 1982, Effect of the fluoridecalcium regimen on vertebral fracture occurence in postmenopausal osteoporosis, *N. Engl. J. Med.* **306:**446.

Rosenberg, L., Armstrong, B., and Jick, H., 1976, Myocardial infarction and estrogen therapy in postmenopausal womem, *N. Engl. J. Med.* **294:**1256.

Ross, R. K., Mack, T. M., Paganini-Hill, A., Arthur, M., and Henderson, B. D., 1981, Menopausal oestrogen therapy and protection from death from ischaemic heart disease, *Lancet* **1:**858.

Ross, R. K., Mack, T. M., Paganini-Hill, A., Gerkins, V., Mack, T. M., Pfeffer, R., Arthur, M., Henderson, B. E., 1980, A case-control study of menopausal estrogen therapy and breast cancer, *JAMA* **243:**1635.

Sartwell, P. E., Arthes, F. G., and Tonascia, J. A., 1977, Exogenous hormones, reproductive history, and breast cancer, *J. Natl. Cancer Inst.* **59:**1589.

Schiff, I., Regestein, Q., Tulchinsky, D., and Ryan, K. J., 1979, Effects of estrogens on sleep and psychological state of hypogonadal women, *JAMA* **242:**2405.

Sherman, B. M., West, J. H., and Korenman, S. G., 1976, The menopausal transition: Analysis of LH, FSH, estradiol, and progesterone concentrations during menstrual cycles of older women, *J. Clin. Endocrinol. Metab.* **42:**629.

Siiteri, P. K., and MacDonald, P. C., 1973, Role of extraglandular estrogen, in: *Handbook of Physiology* Section 7, *Endocrinology: Human Endocrinology*, (S. R. Geiger, E. B. Astwood, and R. O. Greep, eds.), American Physiological Society, Baltimore,

Smith, D. C., Prentice, R., Thompson, J. D., and Herrmann, W. L., 1975, Association of exogenous estrogens and endometrial carcinoma, *N. Engl. J. Med.* **293:**1164.

Smith, E. L., 1982, Exercise for prevention of osteoporosis: A review, *Physician and Sports Med.* **10:**72.

Sturde, D. W., Wilson, K. A., Pipil, E., and Crocker, A. D., 1978, Physiologic aspects of the menopausal hot flush, *Br. Med. J.* **2:**79.

Sturdee, D. W., Wade-Evans, T., Paterson, M. E. L., Thom, M., and Studd, J. W. W., 1978, Relations between bleeding pattern, endometrial histology, and oestrogen treatment in menopausal women, *Br. Med. J.* **1:**1575.

Thom, M. H., White, P. J., Williams, R. M., Sturdee, D. W., Paterson, M. E., Wade-Evans, T., and Stud, J. W., 1979, Prevention and treatment of endometrial disease in climacteric women receiving estrogen, *Lancet* **2:**455.

Vermeulen, A., 1976, The hormonal activity of the post-menopausal ovary, *J. Clin. Endocrinol. Metab.* **42:**247.

Wallace, R. B., Hoover, J., Barrett-Conner, E., Rifkind, B. M., Hunninghake, D. B., Mackenthun, A., Heiss, G., 1979, Altered plasma lipid and lipoprotein levels associated with oral contraceptive and oestrogen use, *Lancet* **2:**112.

Weiss, N., Ure, L., Ballard, J. H., Williams A. R., and Baling, J. R., 1980, Estimated incidence of fractures of lower forearm and hip in postmenopausal women, *N. Engl. J. Med.* **303:**1195.

Whitehead, M. I., Townsend, P. T., Pryse-Davies, J., Ryder, T. A., and King, R. J. B., 1981, Effects of estrogens and progestins on the biochemistry and morphology of the postmenopausal endometrium, *N. Engl. J. Med.* **305:**1599.

Wynder, E. L., MacCormack, F. A., and Stellman, S. D., 1978, The epidemiology of breast cancer in 785 United States caucasian women, *Cancer* **41:**2341.

Yaari, S., Even-Zohar, S., Goldbourt, U., and Neufeld, H. N., 1981, Association of serum high density lipo-protein and total cholesterol with total, cardiovascular, and cancer mortality in a 7-year prospective study of 10,000 men, *Lancet* **1:**1011.

Ziel, A. K., and Finkle, W. D., 1975 Increased risk of endometrial carcinoma among users of conjugated estrogens, *N. Engl. J. Med.* **293:**1167.

17

ESTROGEN THERAPY AND THE RISK OF BREAST, OVARIAN, AND ENDOMETRIAL CANCER

NANCY C. LEE, PHYLLIS A. WINGO,
HERBERT B. PETERSON, GEORGE L. RUBIN, and
RICHARD W. SATTIN

Introduction

Millions of women have used estrogens for relief of menopausal symptoms. Theoretical considerations, clinical observations, and epidemiologic information raise concern as to possible associations between use of exogenous estrogens and cancers of the female reproductive tract (Thomas, 1978). Because of the large number of women who have been using or will use estrogens for menopausal symptoms, any risk of or protection from these cancers could have major health impact on women in their postmenopausal years.

Convincing epidemiologic and clinical evidence indicates that estrogen therapy is a risk factor for the development of endometrial cancer (Hulka, 1980). Available studies do not indicate that estrogen therapy is a major risk factor for breast cancer, although there is some evidence that the risk may be increased in women who have received higher-dose or long-term estrogen therapy (Thomas, 1984). Little epidemiologic evidence exists concerning the association of estrogen therapy with ovarian cancer.

In this chapter, we summarize the available epidemiologic information

NANCY C. LEE, PHYLLIS A. WINGO, HERBERT B. PETERSON, GEORGE L. RUBIN, and RICHARD W. SATTIN • Division of Reproductive Health, Center for Health Promotion and Education, Centers for Disease Control, Atlanta, Georgia 30333.

concerning the relationship between estrogen therapy and cancer of the breast, ovary, and endometrium. In addition, we present preliminary information from the Cancer and Steroid Hormone Study, a multicenter, case–control study of breast, ovarian, and endometrial cancer. In general, our analyses support the results from other studies. However, we still do not know whether postmenopausal estrogen therapy is a risk factor for breast and ovarian cancer. Our evidence and the evidence of others suggest that it may be; clearly, these questions need to be carefully examined through studies especially designed for this purpose.

The Cancer and Steroid Hormone Study

The Cancer and Steroid Hormone Study is a large, case–control study coordinated by the Division of Reproductive Health at the Centers for Disease Control, with support from the National Institute of Child Health and Human Development and the National Cancer Institute (Centers for Disease Control Cancer and Steroid Hormone Study, 1983a, b, c). Although the study was designed primarily to examine the relationship between the use of oral contraceptives (OCs) and breast, endometrial, and ovarian cancer, information was also collected on noncontraceptive estrogen use. Study participants were enrolled between December 1980 and April 1983 from the geographic areas of eight regional tumor registries of the Surveillance, Epidemiology and End Results Centers of the National Cancer Institute. These registries were located in the urban areas of Atlanta, Detriot, Seattle, and San Francisco, and the states of Utah, Iowa, Connecticut and New Mexico.

The study methods have been described in detail elsewhere (Centers for Disease Control Cancer and Steroid Hormone Study, 1983a, b, c). Briefly, a case was defined as a woman aged 20–54 years with histologically confirmed breast, endometrial, or ovarian cancer newly diagnosed during the study period who resided in one of the eight areas at the time of diagnosis. The control group consisted of women selected by randomly telephoning households in the same geographic areas as the cases to provide a group of women 20–54 years of age reflecting the general population of these areas (Waksberg, 1978). These women were frequency-matched by 5-year age group to the expected age distribution of the women with breast cancer.

All cases and controls were administered an hour-long questionnaire concerning their reproductive, medical and family histories. A calendar marking important life events was also used to aid the women's recall of their use of OCs and other female hormones. To help them remember specific brands and doses of the OCs and exogenous estrogens they used, the women were shown a book of photographs of the major brands marketed in the United States.

Breast Cancer

This year 37,000 women will die of breast cancer in the United States: 75% of these deaths will occur among women over the age of 54 (Silverberg and Lubera, 1983). If estrogen therapy increases a woman's risk of breast cancer, this could have important public health impact. Despite a number of epidemiologic studies, the relationship between use of estrogens and subsequent breast cancer risk is not clear.

One important retrospective cohort study examined the relationship of estrogen therapy to breast cancer risk by comparing the observed number of breast cancer cases in the study cohort to the expected number in the general population, derived from national survey data (Hoover *et al.*, 1976). The study cohort was a group of women in one physician's practice who had received conjugated estrogen therapy during the period from 1939 to 1972. During this time, 49 women developed breast cancer, whereas 39 would have been expected from the national data, resulting in a risk of 1.3 of developing breast cancer in the cohort of women, compared to the general population of women. This finding was not statistically significant. However, the risk was 2.0, and statistically significant, in women who had been followed in the study for at least 15 years.

At least 12 case–control studies of breast cancer and estrogen therapy have been published since 1974 (Boston Collaborative Drug Surveillance Program, 1974; Craig *et al.*, 1974; Casagrande *et al.*, 1976; Sartwell *et al.*, 1977; Ravnihar *et al.*, 1979; Jick *et al.*, 1980; Ross *et al.*, 1980; Brinton *et al.*, 1981; Hoover *et al.*, 1981; Kelsey *et al.*, 1981; Hulka *et al.*, 1982; Kaufman *et al.*, 1984). The studies vary in methodology and in size, but in almost all, the relative risk of breast cancer in women who had ever used estrogens was approximately 1 compared with the risk for women who had never used estrogens, suggesting little overall association between estrogen use and breast cancer.

Table I presents information from published studies concerning the risk of breast cancer, considering the duration or dosage of estrogen therapy. No trend in risk by duration of use is apparent; however, Hoover found a steady trend of increasing risk of breast cancer with increasing number of estrogen prescriptions filled (Hoover *et al.*, 1981). Likewise, some studies show increasing risks with increasing daily dose or increasing cumulative dose of estrogen (Ross *et al.*, 1980; Brinton *et al.*, 1981; Hoover *et al.*, 1981). These studies suggest that higher-dose estrogen therapy may increase a woman's risk of breast cancer by up to twofold. However, other studies have not found evidence for a dose-response relationship (Kelsey *et al.*, 1981; Hulka *et al.*, 1982; Kaufman *et al.*, 1984).

Additional evidence that high dosages of estrogens can cause breast cancer comes from a report that two transvestite men who received massive doses of

Table I

Published Risk Estimates of Breast Cancer in Relation to Various Doses of Estrogens[a]

Measurement of dose	Study	Dose category	Relative risk[b]
Duration of use	Ravnihar *et al.* (1979)	<2 years	0.7
		≥2 years	0.7
	Ross *et al.* (1980)	≥7 years	1.8
	Brinton *et al.* (1981)	<5 years	1.2
		5–9 years	1.3
		≥10 years	1.2
	Hulka *et al.* (1982)[c]	≤3 years	2.1
		4–9 years	1.5
		≥10 years	1.7
	Kaufman *et al.* (1984)[c]	<1 year	0.9
		1–4 years	0.9
		5–9 years	0.7
		≥10 years	1.3
Number of prescriptions	Hoover *et al.* (1981)	1	1.0
		2–4	1.3
		5–9	1.5
		≥10	1.7
Daily dose	Hoover *et al.* (1981)	<1.25 mg	1.1
		≥1.25 mg	1.5
	Brinton *et al.* (1981)	0.3 mg	0.8
		0.6 mg	1.3
		1.25 mg	1.2
		2.5 mg	1.7
	Hoover *et al.* (1976)	0.3 mg	1.2
		0.625 mg	1.1
		>0.625 mg	1.9
	Hulka *et al.* (1982)[c]	≤0.625 mg	1.9
		>0.625 mg	1.0
	Kaufman *et al.* (1984)[c]	<1.25 mg	1.2
		≥1.25 mg	0.7
Cumulative dose	Ross *et al.* (1980)	1–1499 mg/life	0.8
		≥1500 mg/life	1.9
	Kelsey *et al.* (1981)	1–49 mg-months	0.8
		≥50 mg-months	0.6

[a] Modified from Thomas (1984).
[b] Risk among users relative to never-users.
[c] Study limited to women with a natural menopause.

estrogens to induce breast development subsequently developed breast cancer (Symmers, 1968).

The estrogens used by women in these studies were predominantly con-jugated estrogens; the findings thus reflect largely the influence of conjugated estrogens on the risk of breast cancer.

Since the Cancer and Steroid Hormone Study was designed primarily to examine OC use, we did not enroll cases or controls over the age of 54, because older women were unlikely to have ever used OCs. For the analysis of breast cancer and postmenopausal estrogens, we included only women who had had a surgical or a natural menopause. By definition, women with a surgical menopause had had a hysterectomy causing cessation of their menses; they may or may not have had intact ovaries. Additionally, if a woman had used estrogens before her last menstrual period, she was excluded from the analysis. After these exclusions we had data for analysis from 1352 women with breast cancer and 1617 controls. The results are derived from a logistic regression analysis that controlled for current age, number of ovaries, type of menopause, age at last menstrual period, age at first full-term pregnancy, history of benign breast disease, and family history of breast cancer.

As expected, compared with controls, the women with breast cancer were more likely to be nulliparous (14% vs. 9%), to have a history of benign breast disease (21% vs. 13%), and to have a family history of breast cancer (27% vs. 20%). About two thirds of both cases and controls had had a surgical menopause.

The highest relative risks of breast cancer were found in those women with the longest duration of estrogen use, but none of the risk estimates was significantly different from 1 (Table II). Note that the number of women using estrogens for 15 or more years was small, so that the risk estimates for these duration categories are unstable. Because the enrollment cutoff in the study was at age 54, most of the postmenopausal women who ever used estrogens used them after a surgical rather than a natural menopause. The results, therefore, largely reflect the risks among women having a surgical menopause. We examined the risk of breast cancer among only the women with a surgical menopause and

Table II
Risk of Breast Cancer by Duration of Postmenopausal Estrogen Use[a]

Duration of Estrogen use[b]	Cases	Controls	Relative risk[c] (95% confidence interval)
Never	959	1135	1 (referent)
< 1 year	74	87	1.1 (0.8–1.5)
1–4 years	159	193	1.0 (0.8–1.3)
5–9 years	98	113	1.1 (0.8–1.6)
10–14 years	33	57	0.8 (0.5–1.3)
15–19 years	17	20	1.4 (0.7–2.8)
≥ 20 years	7	6	1.6 (0.5–5.0)

[a] From Centers for Disease Control Cancer and Steroid Hormone Study.
[b] Excludes five cases and six controls with unknown duration of use.
[c] Logistic regression estimate, adjusted for age, number of ovaries, type of menopause, age at last menstrual period, age at first full-term pregnancy, history of benign breast disease, and family history of breast cancer.

found similar results; that is, the risk was still highest among the women with a surgical menopause who had used estrogens for 15 or more years. There were too few women with a natural menopause who had a long duration of estrogen use to determine the risks for this group.

We considered not only the duration of estrogen received but also the time that had elapsed since a woman first received postmenopausal estrogen therapy, called the latent period. Here, the highest risks were seen for women who had first used estrogens 15 or more years ago (Table III).

The risk of breast cancer associated with postmenopausal estrogen use was not altered by a woman's parity, her history of benign breast disease, or her family history of breast cancer. Neither was it altered by the presence or absence of ovaries.

This preliminary analysis has not resolved the controversy surrounding the possibility that postmenopausal estrogen therapy may increase a woman's risk of breast cancer. The study's primary limitation is that women over the age of 54 were not enrolled; hence, large numbers of postmenopausal women who may or may not have taken estrogens were not eligible for the study. Moreover, most women in the study who took estrogens for a long period of time had a surgical menopause, and estrogen's effects on these women may not be similar to its effects on women who had a natural menopause. Even with these limitations, this is still the largest study of the question, providing data on a substantial number of women with a surgical menopause who have used estrogens for at least 15 years. We have not found a large protective or large harmful effect. We have found a suggestion of a one and one-half to two-fold increased risk in those women who have received long-term estrogen therapy. This association, however, may be due to chance or an undetected bias.

Whether estrogen therapy increases a woman's risk of breast cancer needs

Table III

Risk of Breast Cancer by Time Since First Postmenopausal Estrogen Use[a]

Time since first estrogen use[b]	Cases	Controls	Relative risk[c] (95% confidence interval)
Never	959	1135	1 (referent)
< 1 year	22	31	0.9 (0.5–1.6)
1–4 years	114	138	1.0 (0.8–1.4)
5–9 years	142	161	1.1 (0.9–1.5)
10–14 years	67	99	0.9 (0.7–1.3)
15–19 years	27	32	1.3 (0.7–2.2)
≥ 20 years	16	15	1.5 (0.7–3.2)

[a] From Centers for Disease Control Cancer and Steroid Hormone Study.
[b] Excludes five cases and six controls with an unknown time since first use.
[c] Logistic regression estimate, adjusted for age, number of ovaries, type of menopause, age at last menstrual period, age at first full-term pregnancy, history of benign breast disease, and family history of breast cancer.

to be studied further through studies specifically designed to answer the question, with special consideration of estrogen's effects on women at high risk for breast cancer and on women with long duration of estrogen use.

Ovarian Cancer

Ovarian cancer is much less common than endometrial cancer or breast cancer; however, because of the higher case-fatality rate, ovarian cancer accounts for a disproportionately large number of cancer deaths each year. This year about 11,500 women will die of ovarian cancer.

Of the three cancers we have discussed, least is known about the relationship between estrogen therapy and ovarian cancer. In the late 1970s, before any epidemiologic evidence was available, Stadel proposed that use of estrogens might lower a woman's risk of ovarian cancer because estrogens lowered pituitary gonadotropins (Stadel, 1975). Since then, at least seven epidemiologic studies have been published that examine the relationship between estrogen use and ovarian cancer (Hoover *et al.*, 1977; Annegers *et al.*, 1979; McGowan *et al.*, 1979; Hildreth *et al.*, 1981; Weiss *et al.*, 1982; LaVecchia *et al.*, 1982; Cramer *et al.*, 1983). Six of the seven were case–control studies.

The one cohort study by Hoover (Hoover *et al.*, 1977) used the same cohort of women previously described in the breast cancer section. They followed up 908 women who had been treated with conjugated estrogens during the period from 1939 to 1969. During that time eight cases of ovarian cancer developed, whereas 3.4 cases would be expected, yielding a relative risk of 2.4 for this group of women compared to the general population of women. The researchers found no increased risk with increasing duration of use but did find an increased risk with increasing strength of estrogen tablet used. Of note, in the small group of women who had used diethylstilbestrol (DES) for 1 year or more in addition to conjugated estrogens, three women developed cancer whereas 0.1 case was expected. This gave a relative risk of 30, which was highly statistically significant.

Table IV summarizes results from the six case–control studies of ovarian cancer and estrogen therapy. Note that the number of ovarian cancer cases in each study was small. No statistically significant overall increased risk was found in any of the studies, although Cramer's study suggested that the risk may be higher in women with a long duration of use and in those women who had had a natural menopause (Cramer *et al.*, 1983). Annegers' study was the only one that found a protective effect, but this effect disappeared if women had used estrogens for more than six months (Annegers *et al.*, 1979). Regarding type of estrogen used, Weiss did not find an increased risk with DES use (Weiss *et al.*, 1982); Cramer found similar risks associated with conjugated and nonconjugated estrogens (Cramer *et al.*, 1983). Three of the studies examined the

Table IV

Risk Estimates from Case–Control Studies of Estrogen Therapy and Ovarian Cancer

Study	Age range	No. of cases	Ever use: relative risk[a]	Duration: relative risk[a]
Annegers *et al.* (1979)	≥ 15	116	0.5[b]	≥ 6 months, 1.0
McGowan *et al.* (1979)	Average, 52	197	No association	Not stated
Hildreth *et al.* (1981)	45–74	62	0.9	No trend
Weiss *et al.* (1982)	50–74	207	1.3	No trend
LaVecchia *et al.* (1982)	40–69	135	1.0	No trend
Cramer *et al.* (1983)	Natural menopause	92	2.0	Not stated
	40–80	173	1.6	≤ 1 yr, 1.5
				2–5 yr, 1.0
				> 5 yr, 2.8

[a] Risk relative to never-users.
[b] $P < 0.05$.

association with estrogen use by histologic type of ovarian cancer; all found that estrogen therapy was a stronger risk factor for endometrioid ovarian cancer than for ovarian cancer of other types (Weiss *et al.*, 1982; LaVecchia *et al.*, 1982; Cramer *et al.*, 1983).

Overall, the results from these seven studies are inconclusive. However, the weight of evidence does not support the hypothesis that estrogen therapy protects a woman from ovarian cancer.

To examine the relationship between ovarian cancer and postmenopausal estrogen therapy using data from the Cancer and Steroid Hormone Study, we limited the study group to women who had either a natural or surgical menopause. However, because women who do not have ovaries are not at risk for developing ovarian cancer, women with no ovaries were also eliminated from the control group. Hence, all women in this analysis had at least one ovary. After these exclusions, we had data on 160 women with ovarian cancer and 1223 controls. A panel of three expert pathologists reviewed histology slides for over 90% of the ovarian cancer cases. We used logistic regression techniques to adjust the risk estimates for age, parity, OC use, and type of menopause.

Parity has an important association with ovarian cancer risk, so we were not surprised that women with ovarian cancer were more likely to be nulliparous or of low parity than were the controls. Controls (48%) were more likely to have had a surgical menopause than were cases (33%). These women did not have a surgical climacteric, since we included in the ovarian cancer analysis only women with at least one ovary.

Women who had ever used postmenopausal estrogens had a risk of 1.3 of developing ovarian cancer as compared with women who had never used post-

Table V

Risk of Ovarian Cancer by Duration of Postmenopausal Estrogen Use[a]

Duration of estrogen use[b]	Cases	Controls	Relative risk[c] (95% confidence interval)
Never	134	1007	1 (referent)
Ever	26	203	1.3 (0.8–2.1)
<2 years	9	83	1.1 (0.5–2.3)
2–5 years	8	63	1.3 (0.6–2.8)
≥6 years	9	57	1.7 (0.8–3.6)

[a] From Centers for Disease Control Cancer and Steroid Hormone Study.
[b] Excludes four controls with unknown duration of use.
[c] Logistic regression estimate, adjusted for age, parity, type of menopause, and history of oral contraceptive use.

menopausal estrogens, although this difference was not statistically significant (Table V). There was a trend of increasing risk with increasing duration of estrogen use, with a relative risk of 1.7 for women who had used estrogens for more than 6 years as compared with women who had never used estrogens.

When we examined the risk of ovarian cancer by time since first use of estrogens — the latent period — we found that compared with women who had never used estrogens, women who had first used estrogens more than 10 years ago had a risk of ovarian cancer of 2.0 (95% confidence interval: 1.0–4.1), while women who had first used estrogens more recently had a risk of 1.0 (0.6–1.8). The effect of increased latency was apparent regardless of whether a woman had used estrogens for a short or a long period of time.

Because OC use is known to protect a woman from developing ovarian cancer, we wondered if a past history of OC use might somehow modify the association of postmenopausal estrogen therapy and ovarian cancer. We computed the risk estimates for women who had and had not used postmenopausal estrogens by their past history of OC use (Table VI). Compared with women who had never used estrogens or OCs, those who had ever used both estrogens and OCs had an overall relative risk of 0.7 of developing ovarian cancer, whereas women who had used estrogens but had not used OCs had an overall risk of 1.7. Neither of these were statistically significant. When we examined the risks by duration of estrogen use we found that among the women who had never used OCs, those who had used estrogens for 6 or more years had a risk of 3.1 as compared with women who had never used estrogens. This was a statistically significant increased risk. Of note, in this group there were eight cases and 24 controls.

Almost all of the women in the study who had postmenopausal estrogen therapy had used conjugated estrogens. We had no case and only one control who had ever used DES. Hence, we could not analyze our data by type of estrogen used.

Because three previous studies suggested that the endometrioid type of

Table VI

Risk of Ovarian Cancer by Postmenopausal Estrogen Use and by Use of Oral Contraceptives[a]

Duration of estrogen use[b]	Use of oral contraceptives			
	Never		Ever	
	Cases/controls	Relative risk[c]	Cases/controls	Relative risk[c]
Never	87/502	1 (referent)	47/505	0.7
Ever	18/86	1.7	8/117	0.6
<2 years	6/34	1.4	3/49	0.5
2–5 years	4/28	1.1	4/35	1.0
≥6 years	8/24	3.1[d]	1/33	0.2

[a] From Centers for Disease Control Cancer and Steroid Hormone Study.
[b] Excludes 4 controls with unknown duration of estrogen use.
[c] Logistic regression relative risk, adjusted for age, parity, and type of menopause.
[d] $P < 0.05$.

ovarian cancer had a higher risk associated with estrogen therapy than did the other types of ovarian cancer (Weiss *et al.*, 1982; LaVecchia *et al.*, 1982; Cramer *et al.*, 1983), we examined our data by histologic type of ovarian cancer. Although the numbers of women studied were small, we found no difference in risk for the different histologic types.

In this analysis, we are limited by the small number of cases of ovarian cancer and by the enrollment cutoff at age 54. Our findings are consistent with those of Cramer (Cramer *et al.*, 1983), which suggest that there may be an increased risk of ovarian cancer with long duration of estrogen therapy. However, our findings are also consistent with no effect of estrogen therapy on the risk of ovarian cancer.

These eight epidemiologic studies indicate that estrogen therapy neither protects against ovarian cancer nor has a large harmful effect. If an increased risk does exist, OC use may minimize the risk. But no study to date has been able to satisfactorily examine the effects of long duration and long latency of estrogen therapy on the risk of ovarian cancer. Resolution of these questions will require a study with a much larger number of ovarian cancer cases in the postmenopausal age group.

Endometrial Cancer

Much evidence exists concerning the relationship between estrogen therapy and endometrial cancer. Fortunately, mortality rates are low from this cancer; still, 39,000 new cases are diagnosed in the United States each year.

Epidemiologic evidence from at least 14 case–control studies published since the 1970s has consistently shown a positive relationship between estrogen therapy and adenocarcinoma of the endometrium, by far the most common type of endometrial cancer (Table VII) (Smith *et al.*, 1975; Ziel and Finkel, 1975; Mack *et al.*, 1976; Gray *et al.*, 1977; McDonald *et al.*, 1977; Horwitz and Feinstein, 1978; Hoogerland *et al.*, 1978; Antunes *et al.*, 1979; Hulka *et al.*, 1980; Shapiro *et al.*, 1980; Jelovsek *et al.*, 1980; Spengler *et al.*, 1981; Stavraky *et al.*, 1981; Kelsey *et al.*, 1982). Each one of these studies has found that estrogen therapy, especially long-term therapy, is a risk factor for developing endometrial cancer. The relative risks for women who have ever used estrogens compared to never-users have ranged from 2 to 12, depending on such factors as the control group used and the durations considered. For long-term estrogen therapy, the risks have ranged from 4 to 15. Most of these studies have examined risks associated with use of conjugated estrogens, but there is no reported evidence of differences in risk associated with different types of estrogens.

Convincing evidence indicates that adding progestin therapy to estrogen therapy prevents the development of hyperplastic changes in the endometrium, considered the precursor to adenocarcinoma (Paterson *et al.*, 1980). Although the hypothesis is biologically plausible, sufficient epidemiologic evidence has not yet been published to determine whether combined estrogen–progestin therapy actually reduces the risk of endometrial cancer.

Table VII

Risk Estimates from Case–Control Studies of Estrogen Therapy
and Endometrial Cancer

Study	Relative risks[a]	
	Ever	Long-term
Smith *et al.* (1975)	4.5	—
Ziel and Finkel (1975)	7.6	13.9
Mack *et al.* (1976)	5.6	8.8
Gray *et al.* (1977)	3.1	11.6
McDonald *et al.* (1977)	2.0	7.9
Horwitz and Feinstein (1978)	12.0	—
Hoogerland *et al.* (1978)	2.2	6.7
Antunes *et al.* (1979)	6.0	15.0
Hulka *et al.* (1980)	—	3.6
Shapiro *et al.* (1980)	3.9	6.0
Jelovsek *et al.* (1980)	2.4	4.8
Spengler *et al.* (1981)	3.2	8.6
Stavraky *et al.* (1981)	4.2	14.4
Kelsey *et al.* (1982)	—	8.2

[a] Risk relative to never-users.

Thus the evidence for a causal association between estrogen therapy and endometrial cancer is strong: the relative risks are high and the results have been confirmed by many studies, using different methodologies. In addition, dose response has been well documented. Importantly, the association is biologically plausible since it is known that unopposed estrogens stimulate hyperplastic changes in the endometrium (Whitehead et al., 1979).

In this chapter, we are presenting preliminary data from yet another case–control study of endometrial cancer and estrogen therapy. As in all the other studies, we also have found an increased risk with use of estrogens.

This analysis was limited to the 136 women with endometrial cancer and the 558 control women enrolled in the Cancer and Steroid Hormone Study who had undergone a natural menopause, since women who had had a hysterectomy were not at risk for developing endometrial cancer. We excluded four cases and 20 controls who had a past history of use of sequential OCs, since these are also reported to be associated with an increase risk of endometrial cancer. In the logistic regression model, we controlled for the potentially confounding variables of age, parity, obesity, and OC use.

Women with endometrial cancer were more likely to be nulliparous (32% vs. 11%), obese (29% vs. 13%), and hypertensive (36% vs. 22%) than were control women. Also, cases were less likely to have ever used OCs (30% vs. 42%), which is not unexpected since OC use has been shown to protect women from endometrial cancer.

The number of cases and controls who had ever used estrogens was small, but note that women who had ever used menopausal estrogens had a relative risk of 3.2 in comparison with women who had never used estrogens (Table VIII). A trend of increasing risk was associated with increasing duration of estrogen use. For the few women who had used estrogens for 6 years or more, the risk of endometrial cancer was seven times greater than for women who had

Table VIII

Risk of Endometrial Cancer by Duration of Postmenopausal Estrogen Use[a]

Duration of estrogen use[b]	Cases	Controls	Relative risk[c] (95% confidence interval)
Never	104	493	1 (referent)
Ever	26	41	3.2 (1.8–5.8)
<2 years	8	19	2.1 (0.9–5.3)
2–5 years	9	15	2.8 (1.1–7.1)
≥ 6 years	9	7	7.2 (2.5–21.0)

[a] From Centers for Disease Control Cancer and Steroid Hormone Study.
[b] Excludes six cases and 34 controls with unknown duration of use.
[c] Logistic regression estimate, adjusted for age, parity, obesity, and history of oral contraceptive use.

Table IX

Risk of Endometrial Cancer by Use of Postmenopausal Estrogens and by Past History
of Oral Contraceptive Use[a]

Duration of estrogen use[c]	Use of oral contraceptives[b]			
	Never		Ever	
	Cases/controls	Relative risk[d]	Cases/controls	Relative risk[d]
Never	74/298	1 (referent)	30/195	0.8
Ever	20/20	3.8[e]	6/21	1.9
<2 years	5/10	1.7	3/9	2.5
2–5 years	7/7	3.7[e]	2/8	1.3
≥6 years	8/3	11.5[e]	1/4	2.1

[a] From Centers for Disease Control Cancer and Steroid Hormone Study.
[b] Excludes one case and two controls with unknown OC use.
[c] Excludes six cases and 34 controls with unknown duration of estrogen use.
[d] Logistic regression relative risk, adjusted for age, parity, and obesity.
[e] $P < 0.05$.

never used estrogens. We also found a trend of increasing risk of endometrial cancer with increasing time since first use of estrogens.

Because epidemiologic evidence suggests that use of combination OCs prevents a woman from subsequently developing endometrial cancer, we examined the data to see if women who used OCs in addition to estrogens had a different risk of endometrial cancer than women who had used estrogens but had no history of OC use (Table IX). Compared with women who had never used estrogens or OCs, those who had used estrogens and OCs had an overall risk of 1.9 whereas women who had used estrogens but had no history of OC use had an overall risk of 3.8. We had few cases and controls in each duration category; however, the highest risk estimates were seen among the women who had longer duration of estrogen use, but no history of past OC use.

Insufficient numbers of past users existed to examine the risks after discontinuation of estrogen therapy. Likewise, few women in the study had used concurrent progestin therapy, so we were unable to determine whether this altered the risk estimates.

The results from this study support the results obtained in other case–control studies of estrogen therapy and endometrial cancer, both in the magnitude of the relative risks and the trend toward increasing risk with increasing duration of therapy. To date, no published study has had results suggesting that women with a history of both OC use and estrogen therapy have a different risk of endometrial cancer than do women who have used estrogens but never used OCs. We emphasize that this particular finding is not conclusive, but rather suggestive. Some previously published studies of estrogen therapy and endometrial cancer have collected information on use of OCs; reanalysis of their data may clarify this question.

Conclusion

In summary, the preliminary results from the Cancer and Steroid Hormone Study are generally similar to the findings of previous studies concerning the association of estrogen therapy to breast, ovarian, and endometrial cancer. Our findings suggest that long-term estrogen users may have an increased risk of breast and ovarian cancer, although the risk elevations may well be due to chance or some unrecognized bias. Also, our results support the well-established findings that estrogen therapy is a risk factor for endometrial cancer. We have reported for the first time results that suggest that past OC use may modify the association between estrogen therapy and endometrial and ovarian cancers.

Studies are needed to definitively characterize the risks of cancer associated with estrogen therapy. For endometrial cancer, convincing information is needed to demonstrate that combined estrogen–progestin therapy does in fact reduce the increased risk of endometrial cancer associated with unopposed estrogen therapy. If it does, what are the long-term sequelae of combined therapy on other systems, particularly the cardiovascular system? Also, does past OC use really modify the association between endometrial cancer and estrogen therapy?

More puzzling questions surround breast and ovarian cancer: Does estrogen therapy increase a woman's risk of developing breast or ovarian cancer? How might combined estrogen and progestin therapy affect the risk of developing these two cancers? Obtaining answers to these questions will require specially designed studies with sufficient numbers of study subjects to detect differences of the magnitude demonstrated today.

If long-term estrogen therapy, with or without concomitant use of a progestin, is promoted to prevent osteoporosis and its serious sequelae, then we need to have a clear understanding of the association of long-term therapy with cancer. More information about the effects of estrogen therapy on different organ systems is needed so that the benefits and risks of estrogen therapy can be placed in perspective.

References

Annegers, J. F., Strom, H., Decker, D. G., Dockerty, M. B., and O'Fallon, W. M., 1979, Ovarian cancer: Incidence and case–control study, *Cancer* **43**:723–729.

Antunes, C. M. F., Stolley, P. D., Rosenshein, N. B., Davies, J. L., Tonascia, J. A., Brown, C., Burnett, L., Rutledge, A., Pokempner, M., and Garcia, R., 1979, Endometrial cancer and estrogen use: Report of a large case–control study, *N. Engl. J. Med.* **300**:9–13.

Boston Collaborative Drug Surveillance Program, 1974, Surgically confirmed gallbladder disease, venous thromboembolism, and breast tumors in relation to postmenopausal estrogen therapy, *N. Engl. J. Med.* **290:**15–18.

Brinton, L. A., Hoover, R. N., Szklo, M., and Fraumeni, J. F., 1981, Menopausal estrogen use and risk of breast cancer, *Cancer* **47:**2517–2522.

Casagrande, J., Gerkins, V., Henderson, B. E., Mack, T., and Pike, M. C., 1976, Brief communication: Exogenous estrogens and breast cancer in women with natural menopause, *J. Natl. Cancer Inst.* **56:**839–841.

Centers for Disease Control Cancer and Steroid Hormone Study, 1983a, Long-term oral contraceptive use and the risk of breast cancer, *JAMA* **249:**1591–1595.

Centers for Disease Control Cancer and Steroid Hormone Study, 1983b, Oral contraceptive use and the risk of ovarian cancer, *JAMA* **249:**1596–1599.

Centers for Disease Control Cancer and Steroid Hormone Study, 1983c, Oral contraceptive use and the risk of endometrial cancer, *JAMA* **249:**1600–1604.

Craig, T. J., Comstock, G. W., and Geiser, P. B., 1974, Epidemiologic comparison of breast cancer patients with early and late onset of malignancy and general population controls, *J. Natl. Cancer Inst.* **53:**1577–1581.

Cramer, D. W., Hutchinson, G. B., Welch, W. R., Scully, R. E., and Ryan, K. J., 1983, Determinants of ovarian cancer risk: I. Reproductive experiences and family history, *J. Natl. Cancer Inst.* **71:**711–716.

Gray, L. A., Christopherson, W. M., and Hoover, R. N., 1977, Estrogens and endometrial carcinoma, *Obstet. Gynecol.* **49:**385–389.

Hildreth, N. G., Kelsey, J. L., LiVolsi, V. A., Fischer, D. B., Holford, T. R., Mostow, E. D., Schwartz, P. E., and White, C., 1981, An epidemiologic study of epithelial carcinoma of the ovary, *Am. J. Epidemiol.* **114:**398–405.

Hoogerland, D. L., Buchler, D. A., Crowley, J. J., and Carr, W. F., 1978, Estrogen use — risk of endometrial carcinoma, *Gynecol. Oncol.* **6:**451–458.

Hoover, R., Gray, L. A., Sr, Cole, P., and MacMahon, B., 1976, Menopausal estrogens and breast cancer, *N. Engl. J. Med.* **295:**401–405.

Hoover, R., Gray, L. A., Sr, and Fraumeni, J. F., Jr, 1977, Stilboestrol (Diethylstilbestrol) and the risk of ovarian cancer, *Lancet* **2:**533–534.

Hoover, R., Glass, A., and Finkle, W. D., et al., 1981, Conjugated estrogens and breast cancer risk in women, *J. Natl. Cancer Inst.* **67:**815–820.

Horwitz, R. I., and Feinstein, A. R., 1978, Alternative analytic methods for the case–control studies of estrogens and endometrial cancer, *N. Engl. J. Med.* **299:**1089–1094.

Hulka, B. S., 1980, Effect of exogenous estrogen on postmenopausal women: The epidemiologic evidence, *Obstet. Gynecol. Surv.* **35:**389–399.

Hulka, B. S., Grimson, R. C., Greenberg, B. G., Kaufman, D. G., Fowler, W. C., Hogue, C. J. R., Berger, G. S., and Pulliam, C. C., 1980, "Alternative" controls in a case-control study of endometrial cancer and exogenous estrogen, *Am. J. Epidemiol.* **112:**376–387.

Hulka, B. S., Chambless, L. E., Deubner, D. C., and Wilkinson, W. E., 1982, Breast cancer and estrogen replacement therapy, *Am. J. Obstet. Gynecol.* **143:**638–644.

Jelovsek, F. R., Hammond, C. B., Woodard, B. H., Draffin, R., Lee, K. L., Creasman, W. T., and Parker, R. T., 1980, Risk of exogenous estrogen therapy and endometrial cancer, *Am. J. Obstet. Gynecol.* **137:**85–91.

Jick, H., Walker, A. M., Watkins, R. N., D'Ewart, D. C., Hunter, J. R., Danford, A., Madsen, S., Dinan, B. J., and Rothman, K. J., 1980, Replacement estrogens and breast cancer, *Am. J. Epidemiol.* **112:**586–594.

Kaufman, D. W., Miller, D. R., Rosenberg, L., Helmrich, S. P., Stolley, P., Schottenfeld, D., and Shapiro, S., 1984, Noncontraceptive estrogen use and the risk of breast cancer, *JAMA* **252:**63–67.

Kelsey, J. L., Fischer, D. B., Holford, T. R., LiVolsi, V., Mostow, E. D., Goldenberg, I. S., and White, C., 1981, Exogenous estrogens and other factors in the epidemiology of breast cancer, *J. Natl. Cancer Inst.* **67**:327-333.

Kelsey, J. L., LiVolsi, V. A., Holford, T. R., Fischer, D. B., Mostow, E. D., Schwartz, P.E., O'Connor, T., and White, C., 1982, A case-control study of cancer of the endometrium, *Am. J. Epidemiol.* **116**:333-342.

LaVecchia, C., Liberati, A., and Franceschi, S., 1982, Noncontraceptive estrogen use and the occurrence of ovarian cancer, *J. Natl. Cancer Inst.* **69**:1207.

Mack, T. M., Pike, M. C., Henderson, B. E., Pfeffer, R. I., Gerkins, V. R., Arthur, M., and Brown, S. E., 1976, Estrogens and endometrial cancer in a retirement community, *N. Engl. J. Med.* **294**:1262-1267.

McDonald, T. W., Annegers, J. F., O'Fallan, W. M., Dockerty, M. B., Malkasian, G. D., and Kurland, L. T., 1977, Exogenous estrogen and endometrial carcinoma: Case-control and incidence study, *Am. J. Obstet. Gynecol.* **127**:572-580.

McGowan, L., Parent, L., Lednar, W., and Norris, H. J., 1979, The woman at risk for developing ovarian cancer, *Gynecol. Oncol.* **7**:325-344.

Paterson, M. E. L., Wade-Evans, T., Sturdee, D. W., Thom, M. H., and Studd, J. W. W., 1980, Endometrial disease after treatment with oestrogens and progestogens in the climacteric, *Br. Med. J.* **280**:822-824.

Ravnihar, B., Seigel, D. G., and Lindtner, J., 1979, An epidemiologic study of breast cancer and benign breast neoplasias in relation to the oral contraceptive and estrogen use, *Europ. J. Cancer* **15**:395-405.

Ross, R. K., Paganini-Hill, A., Gerkins, V. R., Mack, T. M., Pfeffer, R., Arthur, M., and Henderson, B. E., 1980, A case-control study of menopausal estrogen therapy and breast cancer, *JAMA* **243**:1635-1639.

Sartwell, P. E., Arthes, F. G., and Tonascia, J. A., 1977, Exogenous hormones, reproductive history and breast cancer, *J. Natl. Cancer Inst.* **59**:1589-1592.

Shapiro, S., Kaufman, D. W., Slone, D., Rosenberg, L., Miettinen, O. S., Stolley, P. D., Rosenshein, N. B., Watring, W. G., Leavitt, T., and Knapp, R. C., 1980, Recent and past use of conjugated estrogens in relation to adenocarcinoma of the endometrium, *N. Engl. J. Med.* **303**:485-489.

Silverberg, E. and Lubera, J. A., 1983, A review of American Cancer Society estimates of cancer cases and deaths, *CA* **33**:2-25.

Smith, D. C., Prentice, R., Thompson, D. J., and Herrmann, W. L., 1975, Association of exogenous estrogen and endometrial carcinoma, *N. Engl. J. Med.* **293**:1164-1167.

Spengler, R. F., Clarke, E. A., Woolever, C. A., Newman, A. M., and Osborn, R. W., 1981, Exogenous estrogens and endometrial cancer: A case-control study and assessment of potential biases, *Am. J. Epidemiol.* **114**:497-506.

Stadel, B. V., 1975, The etiology and prevention of ovarian cancer, *Am. J. Obstet. Gynecol.* **123**:772-773.

Stavraky, K. M., Collins, J. A., Donner, A., and Wells, G. A., 1981, A comparison of estrogen use by women with endometrial cancer, gynecologic disorders and other illnesses, *Am. J. Obstet. Gynecol.* **141**:547-555.

Symmers, W. S., 1968, Carcinoma of the breast in transsexual individuals after surgical and hormonal interference with the primary and secondary sex characteristics, *Br. Med. J.* **2**:82.

Thomas, D. B., 1978, Role of exogenous female hormone in altering the risk of benign and malignant neoplasms in humans, *Cancer Res.* **38**:3991-4000.

Thomas, D. B., 1984, Do hormones cause breast cancer? *Cancer* **53**:595-604.

Waksberg, J., 1978, Sampling methods for random digit dialing, *J. Am. Stat. Assoc.* **73**:40-46.

Weiss, N. S., Lyon, J. L., Krishnamurthy, S., Diftert, S. E., Liff, J. M., and Daling, J. R., 1982, Noncontraceptive estrogen use and the occurrence of ovarian cancer, *J. Natl. Cancer Inst.* **68:**95–98.

Whitehead, M. I., King, R. J. B., McQueen, J., and Cambell, S., 1979, Endometrial histology and biochemistry in climacteric women during oestrogen and oestrogen/progestogen therapy, *J. R. Soc. Med.* **72:**322–327

Ziel, H. K., and Finkel, W. D., 1975, Increased risk of endometrial carcinoma among users of conjugated estrogens, *N. Engl. J. Med.* **293:**1167–1170.

VI

CONCLUSION

18

OVERALL SUMMARY AND NEW DIRECTIONS FOR RESEARCH

LUIGI MASTROIANNI, Jr.

Numerous facets of aging as it relates to reproduction and to the climacteric were considered and discussed. It is evident that with the exception of reproduction, full bodily function can be maintained over time. This is especially true if proper attention is given to the management of the menopause.

The review of work in rodent models provided by Finch and Gosden in Chapter 1 focused attention on important endocrine approaches to aging research. Work in animal models even suggests that reproductive mechanisms can be preserved for a longer time than would occur naturally. The startling fact that old animals, oophorectomized early, resume cyclic functioning when they receive young ovaries leads to the suggestion that medical modification of ovarian activity in the early reproductive years could in some way preserve reproductive function into old age.

The rodent model allows relatively more rapid assessment of age effects than would be possible in a primate model. The subhuman primate, a spontaneous ovulator, and, in the case of the Old World Monkey, a menstruator, also offers exciting opportunities. Age-related issues might well be explored in parallel with other ongoing research in primates with a modest additional investment of resources.

Changes in cellular mechanisms associated with aging may in fact be little different for reproductive tissues than for other bodily functions. Reproduction does offer the possibility of providing markers, however. Menopause is but one culminating event that provides exciting opportunities for imaginative research.

LUIGI MASTROIANNI, Jr. • Department of Obstetrics and Gynecology, Hospital of the University of Pennsylvania, Philadelphia, Pennsylvania 19104.

Methodologic issues were reviewed by Minaker and Rowe in Chapter 2. They emphasized the importance of recognizing the effect of exogenous influences—alcohol, drugs, medications—and anthropomorphic covariables—bodily fat changes and the like. Bremner *et al.* (Chapter 3) focused attention on age-related changes in the male reproductive system in both rat and man. Both testicular and CNS changes occur. These are obviously interrelated, although a decline in testosterone levels seems to be independent of CNS changes.

Wise (Chapter 6) summarized studies that clearly support the notion that the ovary is not the only organ reflecting changes. There are hypothalamic changes that compromise function, such as loss of rhythm or modifications in oscillations. Basic mechanisms of any change may occur at the receptor level as was suggested by the elegant work of Roth (Chapter 7).

Nieschlag and Michel (Chapter 4) presented new information on reproductive function in grandfathers. Their carefully completed studies focused attention on the importance of longitudinal evaluation in human investigation. Of particular note was the marked deterioration in health of elderly subjects over a short 4-year interval. In general, changes in the ejaculate were related to disease and not age, per se. Those who remained healthy remained potent with androgen production. There were, however, some changes in seminal fructose suggesting failure of function at the seminal vesicle level. Motility changes also occurred, but these were difficult to evaluate because of the known relationship between percentage motility and frequency of ejaculation. There is no question that motile spermatozoa are produced well on into the late years.

Surprisingly, a careful analysis of age-related production of genetic defects by the male carried out by Hook (Chapter 8) failed to establish a clear-cut causal relationship. The suggested use of the hamster egg penetration test to evaluate haploid chromosomes offers possibilities. Further research into such systems is needed to develop practical methods to evaluate male haploid chromosomes. A more efficient system to analyze haploid chromosomes, one which would not require the cumbersome use of viable mammalian oocytes, would be most useful clinically.

Lipsett (Chapter 5) pointed out that although ovarian changes are relatively abrupt, adrenal changes are not. Cortisol production remains exquisitely regulated throughout normal life. He cited recent information on RNA synthesis in the follicle that decreases with age. This raises fascinating research possibilities in human systems. The tissues are available incidental to indicated surgical procedures and should be examined.

The morphological changes in the ovary and follicle during aging were reviewed by Nicosia (Chapter 11). There is clearly a progressive decrease in number of follicles, both primordial and growing, with age. This process of atresia is poorly understood. It is of tremendous fundamental importance and research in this area should be accorded highest priority.

In the ovulating ovary, the corpus luteum is a prime example of a structure with a predictable lifespan, 14 plus or minus 2 days. What regulates it? What turns it off? Is that mechanism akin to an aging process?

Regulation of ovarian cells, specifically the epithelium, is also worthy of continued attention. Some new approaches to this tissue have been developed in Nicosia's laboratory.

Nicosia also focused attention on the ovarian stroma. He reminded us that the stroma proliferates maximally 10 years after the menopause. The ovary does not cease to function when it stops ovulating. Proliferating cells are present in the ovarian hilus. These have contact with nerves that may stimulate transformation to steroid-producing cells. This interrelationship should be explored further. The endocrine function of the human postmenopausal ovary is not to be discounted.

Siiteri (Chapter 12) brought us back to the adrenal. The adrenal takes on an increasingly important role in the postmenopausal woman. No doubt it is the adrenal that allows some postmenopausal women to remain asymptomatic. When the adrenal is not functioning properly, when it is suppressed, for example, osteoporosis can be accelerated markedly.

Siiteri also stressed the importance of bodily fat and its function in conversion of androgen to estrogen. He emphasized the role of sex hormone-binding globulin in estrogen availability. A little bit of adiposity could offer advantages.

Menken and Larsen (Chapter 9) focused attention on the cumulative hazards of aging in women. Based on a careful survey of all the data, they presented an optimistic outlook for the reproductive potential of women over 35. With regard to first-birth probabilities, 1941 was similar to 1980. They stressed the importance of intercurrent disease, especially salpingitis, in terms of reproductive capacity.

Age-related changes in gestation and in pregnancy outcome were considered by Resnik (Chapter 10). Of particular note is a remarkably improved fetal outcome in recent years. Intercurrent conditions such as hypertension are, of course, tremendously important.

Judd (Chapter 13) focused attention on the menopause and the hot flush, a common debilitating symptom that can last for years unless treated. The etiology of the hot flush and the endocrine relationships surrounding this important and occasionally frustrating and debilitating symptom was considered.

Fordney (Chapter 14) evaluated the declining rate of sexual activity with age in women. At age 75, 20% of women still seem to be continuing to have satisfactory coitus. The problem is that the older woman has an increased perception of herself as asexual. The impact of the environment is most forceful in terms of sexuality in older women. There is no question that there are physical changes, but, interestingly, there is no change in the ability to achieve orgasm or even multiple orgasm, although there are qualitative differences that are perhaps significant.

It seems that women who are sexually active are on the whole 10 kg heavier than those who are not. Thus, in terms of estrogen storage and conversion, as well as in terms of sexual activity, some adiposity is probably important. This focuses attention on the importance of diet. We really need to look at the impact of nutrition on some of the age-related processes.

Fordney pointed out that androgens are known to increase sex drive. This raises some interesting and important therapeutic possibilities that should be explored in carefully controlled studies.

The survey by McKinlay and Mckinlay (Chapter 15) left no doubt that individuals who have had an artificial menopause have high rates of disease and of hospitalization. On the other hand, women approaching the menopause who are generally healthy tend to stay that way.

As for treatment, Kase (Chapter 16) made a good case for substitution therapy, both estrogen and progesterone, in all postmenopausal symptomatic women. He took the position that those who are asymptomatic are probably not candidates for substitution therapy. Longitudinal studies are needed on such women, however.

Lee *et al.* (Chapter 17) addressed issues of safety. Their presentation focused attention on the importance of epidemiologic studies.

In his opening remarks, Dr. Franklin Williams, Director of the National Institute of Aging, focused on some major issues in aging research. He emphasized the importance of separating out age-related factors from the intercurrent disease that inevitably occurs in time in the lives of all individuals. The latter inexorably results in death. After all, death is the fate of all living organisms. A better understanding of the aging process will lead to programs to make that period of life euphemistically referred to as "the declining years" comfortable and productive.

INDEX